WAFER FABRICATION: FACTORY PERFORMANCE AND ANALYSIS

THE KLUWER INTERNATIONAL SERIES IN ENGINEERING AND COMPUTER SCIENCE

MICROELECTRONICS MANUFACTURING

Series Editor
Arjun N. Saxena
Rensselaer Polytechnic Institute

Other books in the series:

YIELD AND VARIABILITY OPTIMIZATION OF INTEGRATED CIRCUITS, *J. C. Zhang and M. A. Styblinski*

The Kluwer Microelectronics Manufacturing Series will aim to provide a cohesive set of books on all relevant aspects of microelectronics manufacturing for both practitioners (engineers, managers and operators) and students. Topics to be addressed include, but are not limited to, the following:

Process Design for Manufacturability
Process Design for Reliability
Semiconductor Manufacturing Equipment
FABS for Semiconductor Manufacturing
Ultraclean Manufacturing
Test Structures/Circuit Elements for Manufacturing
Design Rules
SPC/Yield Enhancement
Simulations
Economics of Semiconductor Manufacturing
Management of Semiconductor Manufacturing
Semiconductor Manufacturing in the Current and Future Global Economy

WAFER FABRICATION:
FACTORY PERFORMANCE AND ANALYSIS

Linda F. Atherton
Robert W. Atherton

In-Motion Technology

Kluwer Academic Publishers
Boston/Dordrecht/London

Distributors for North America:
Kluwer Academic Publishers
101 Philip Drive
Assinippi Park
Norwell, Massachusetts 02061 USA

Distributors for all other countries:
Kluwer Academic Publishers Group
Distribution Centre
Post Office Box 322
3300 AH Dordrecht, THE NETHERLANDS

Library of Congress Cataloging-in-Publication Data

Atherton, Linda F.
Wafer fabrication : factory performance and analysis / Linda F. Atherton, Robert W. Atherton.
p. cm. -- (The Kluwer international series in engineering and computer science : SECS 339. Microelectronics manufacturing)
Includes bibliographical references and index.
ISBN 0-7923-9619-7 (alk. paper)
1. Semiconductor wafers--Design and construction. 2. Factory management. I. Atherton, Robert W. II. Title. III. Series : Kluwer international series in engineering and computer science : SECS 339. IV. Series: Kluwer international series in engineering and computer science. Microelectronics manufacturing.
TK7871,85.A844 1995
621.3815'2--dc20

95-40381
CIP

Printed on acid-free paper.

We dedicate this book to our parents,

Charles and Mary Tarrant

William and Pauline Atherton

and in memory of a friend,

J. F. Sebastian

Contents

Chapter 3. Processes for Wafer Fabrication

Chapter 4. Fab Graphs and Static Analyses

Chapter 5. Factory Performance Basics

Chapter 6. Advanced Performance Concepts

Chapter 7. Factory Modeling

Chapter 8. Cluster Tools

Chapter 9. Performance Economics

About the Authors

Linda F. Atherton is a principal with the firm In-Motion Technology, Los Altos, CA. Providing assistance on wafer fab operations on a world-wide basis, Linda created the short courses on which the book is based. Previously, Ms. Atherton was a project manager at the Electric Power Research Institute in Palo Alto, CA, where she supervised some seventy research projects. Linda holds several patents for energy processes and biotechnology. Earlier she worked in operations research at Lockheed Research Laboratory, Palo Alto. Linda Atherton holds B.A. and M.Ch.E. degrees from Rice University and the degree of Engineer from Stanford University, all in chemical engineering

Robert W. Atherton is a principal with the firm In-Motion Technology, Los Altos, CA. He is an expert on the automation of electronics factories. Beginning with control theory applications at Systems Control, Inc., Palo Alto, he became involved with the semiconductor industry in 1977. Mr. Atherton developed wafer fabrication equipment, including plasma etchers, at Applied Materials, Inc. in Santa Clara, CA. At the Fairchild Research Laboratory, Palo Alto, he developed automation systems for wafer fabrication factories. Mr. Atherton holds two patents for electronics factory automation, and he received the Donald P. Eckman Award for contributions to automatic control. Robert Atherton holds B.A. and M.Ch.E. degrees from Rice University and Engineer and Ph.D. degrees from Stanford University, all in chemical engineering.

Preface

This book is concerned with wafer fabrication and the factories that manufacture microprocessors and other integrated circuits. With the invention of the transistor in 1947, the world as we knew it changed. The transistor led to the microprocessor, and the microprocessor, the *guts* of the modern computer, has created an epoch of virtually unlimited information processing. The electronics and computer revolution has brought about, for better or worse, a new way of life. This revolution could not have occurred without wafer fabrication, and its associated processing technologies.

A microprocessor is fabricated via a lengthy, highly-complex sequence of chemical processes. The success of modern chip manufacturing is a miracle of technology and a tribute to the hundreds of engineers who have contributed to its development. This book will delineate the magnitude of the accomplishment, and present methods to analyze and predict the performance of the factories that make the chips.

The set of topics covered juxtaposes several disciplines of engineering. A primary subject is the chemical engineering aspects of the electronics industry, an industry typically thought to be strictly an electrical engineer's playground. The book also delves into issues of manufacturing, operations performance, economics, and the dynamics of material movement, topics often considered the domain of industrial engineering and operations research.

Hopefully, we have provided in this work a comprehensive treatment of both the technology and the factories of wafer fabrication. Novel features of these factories include long process flows and a dominance of processing over operational issues.

Many of the topics discussed are seemingly eclectic, but all are instrumental in describing wafer fabrication. Performance analysis usually encompasses factory capacity, bottleneck identification, cycle time, and throughputs. In addition, chip factories require treatment of atypical subjects such as: the design and use of cluster tools, the impact of multi-level interconnect technologies, and wafer fab economics. To our knowledge, the coverage provided by this treatise is unique.

The first three chapters of the book provide an overview of the electronics industry and wafer processing. Chapters Four through Nine contain original material. Chapter Four introduces the fab graph, the key analytical tool for performance analysis. The fab graph reflects the mapping of process flows onto equipment. Chapters Five and Six show the way toward detailed performance analysis of wafer fabrication factories, as well as providing some quick rules of thumb gleaned from many years in the business. Chapter Seven describes how to go about modeling wafer fabs for accurate factory-specific performance analyses. Chapter Eight discusses the design, configuration, and use of cluster tools. And finally, Chapter Nine relates operations dynamics to factory economics.

The primary audience for this book includes the engineers and managers working in the electronics industry. They need an overall understanding of the wafer fab, as it is commonly called. The details of fabrication processes and the related aspects of electronic devices can be overwhelming without a systematic overall view of the technology.

The second major audience for this book is provided by engineers and other interested professionals outside of electronics. The electronics industry is expanding at a rapid pace, and it will continue to absorb professionals from other areas. For such individuals, this work provides a perspective on an exciting possible new career.

A third major audience is engineering students looking for a picture of the real world of engineering and technology. Their teachers will also find the book of interest, for it contains many unsolved problems.

The book's origin began as a series of short courses for professionals in the electronics industry. We want to thank our many students for the enlightenment provided by their intense interactions. Ken Tennity at Kluwer heard of the course and persuaded us to commit to a book, despite our fears of the size of the project. Our concerns were justified, but the effort has been worthwhile, and we thank Ken for his eloquence and persistence. After Ken's untimely death, John Bodt continued as our editor at Kluwer, and we wish to thank him for his support.

This book reflects many professional interactions over the course of a career. We would like to thank Rice University and its department of chemical engineering for a thorough and superb education. Stanford University provided our introduction to *high technology*, and we have enjoyed continuing interaction with our respective research advisors Douglas J. Wilde and George M. Homsy. Walter Benzing of Applied Materials introduced us to the semiconductor industry. Rice engineers Matthew S. Buynoski and Robert B. Herring have been invaluable colleagues in our journey through the mazes of Silicon Valley. We want to acknowledge many stimulating discussions on the semiconductor industry with Douglas Peltzer. Mark A. Pool, our friend and associate at In-Motion Technology, has assisted us in ways too numerous to list.

Finally, we would like to thank Don Briner and Don Yeaman of Briner/Yeaman Engineering, Inc. for providing the original 24-step simplified CMOS flow.

We now come to that part of the preface where the authors take responsibility for the inevitable errors of commission and omission. Despite our best efforts, a volume of this size will contain some of both, and we invite our readers to point them out to us.

Linda F. Atherton
Robert W. Atherton

Manufacturing is the formidable competitive weapon.
-Skinner

Chapter 1
Introduction to the Electronics Industry

1.0 Introduction

This book is concerned with the manufacture of the engine of transformation of the information age: *the microprocessor*. Said manufacturing, otherwise known as *wafer fabrication*, involves a sequence of physico-chemical processes that build or fabricate three-dimensional microstructures on the surface of wafers of single-crystal silicon. If the fabrication process is performed successfully, the three-dimensional structure, located on a portion of the wafer called a chip, will perform as an electronic device such as a microprocessor.

The key role of wafer fabrication, per se, and the associated chemical processing, has been largely overlooked in the microprocessor revolution. On the most fundamental level, advanced microprocessors containing millions of transistors simply could not exist without the fabrication technology associated with wafer processing. On a more pragmatic, economic level, the efficiencies provided by wafer fabrication have resulted in the geometric cost reduction of computers.

Wafer fabrication factories or *wafer fabs* (or sometimes simply 'the fab') are factories with unusual features in terms of process difficulty and manufacturing complexity. Currently, there are approximately one thousand wafer fabs world-wide. It is estimated that by the end of the century there will be 2000 wafer fabs. Many of the complex features of wafer fabs are shared by other factories making modern electronic products, such disk drives and flat panel displays. The expanding demand for such products means that these complex factories will become increasingly more common.

The unusual features and complexity of wafer fabs creates challenges for operations management, performance analysis, and mathematical modeling. The purpose of this book is to enable engineers to design and operate wafer fabs more effectively.

1.1 Wafer Fabrication

The fundamental objective of wafer fabrication is the manufacture of electrical circuits. In modern microelectronic manufacturing, integrated circuits containing interconnected electrical devices such as transistors, capacitors, resistors, and diodes are produced. Interconnections are provided by layers of patterned conductors. Starting from the very simple integrated circuits of the 1960's, the state of the manufacturing art now allows the routine fabrication of devices containing hundreds of thousands of transistors. Such a complex device is properly called an integrated system. More advanced fabrication lines produce microprocessors and memories (DRAM's) containing transistor counts in the millions.

Semiconductor wafer fabrication is perhaps the most complex of modern manufacturing processes. A major process flow in a semiconductor-device factory may contain over 500 separate steps or operations. Examples of these steps include patterning (photolithography), thin-film etching, thin-film deposition, and oxidation.

Considered separately, each of these processes is very demanding. Impurities in chemicals and materials are monitored to parts per billion. Geometric features are controlled to feature sizes less than one millionth of a meter (one micron).

But the real complexity of wafer fabrication comes from the requisite knowing, controlling, and utilizing of the interactions of the process steps, both planned and unplanned. Since, the electrical properties of the fabricated structure are key to the success of the product device, IC manufacturing **must** be concerned with the overall integrated process flow, not myopic unit operations.

Process control of wafer manufacturing is a challenging and, as yet, unconquered problem. As discussed later under yield, these complex structures do not automatically perform correctly as microprocessors. Wafers have been manufactured with no working devices. In fact, entire lots of wafers have, on occasion, been found dead. Such a production disaster is called a *yield crash*. One of the 'dirty little secrets' of wafer fabrication is how often yield crashes still occur.

In describing wafer fabrication, the following hierarchy of concepts is useful: process flow, process step, processing equipment. Process flow or process technology, or sometimes simply 'technology', refers to the totality of the sequence of process steps and recipes required to make a working device. This term represents an awesome amount of knowledge, and actually encompasses the lower members of the hierarchy. The flow consists of many process steps, each of which is a physico-chemical method for transforming the surface of the wafer; and each process step is performed in a specialized type of equipment.

A wafer completing the manufacturing sequence may contain approximately one hundred microprocessors or other advanced integrated circuits (Fig. 1.1). The portion of the silicon wafer containing one integrated circuit, whatever its complexity, is called a *die* or a *chip*. A lot of twenty-five wafers may contain 2500 integrated-circuit chips. Consequently, a wafer fab is sometimes called a *silicon-chip factory*.

1.2 Wafer Fabrication Factories

The first milestone for wafer fabrication is the development of a small set of functioning devices such as microprocessors. The second milestone is a working factory producing thousands of functioning microprocessors per day in a cost-effective manner. A famous aphorism is that manufacturing is 'doing the thing right', where the 'thing' is the mass production of microprocessors. What then is involved in a successful wafer fabrication factory?

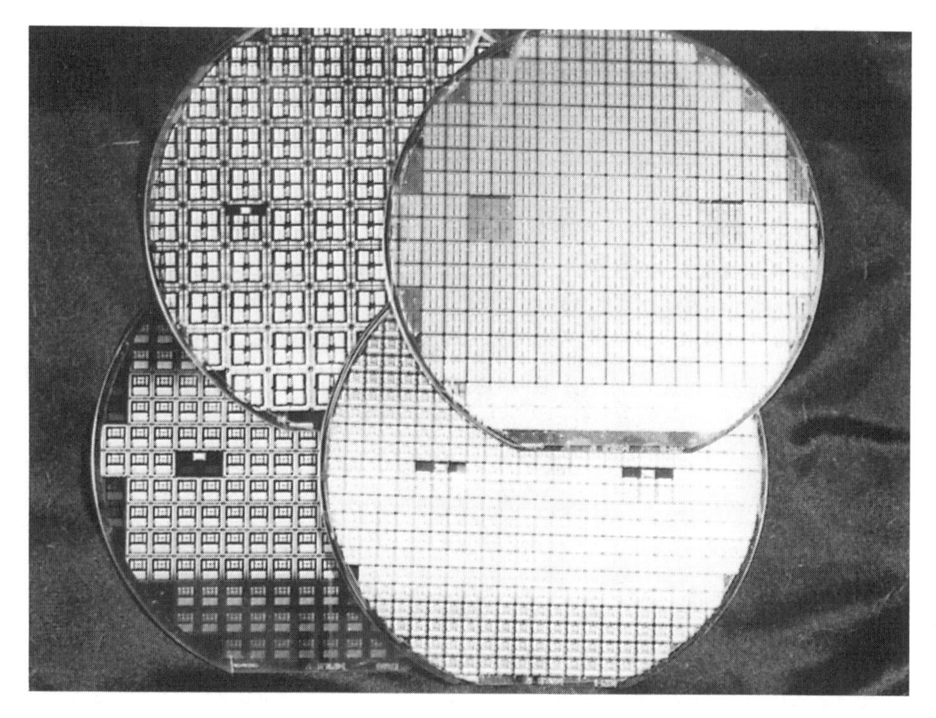

Figure 1.1: Patterned Wafers, the End Product of Fabrication

Physically, a wafer fab is a large intricate building containing specialized machinery. The intricacy comes from the requirements for clean-room areas where particulate contamination is tightly controlled. In addition, the building provides extensive piping for the delivery of reagent chemicals as well as ducting and exhaust for the removal of by-products.

The specialized machinery is wafer-processing equipment for performing each of the specialized physico-chemical processes required to build the microstructure. Approximately fifty different types of equipment may be required for advanced fabrication processes. The range of equipment types is quite large. For example, wafer cleaning requires liquid processing at atmospheric pressure; wiring for circuits involves metals deposition in high-vacuum chambers. Perhaps, the most exotic equipment is the plasma etchers which use glowing electric discharges to remove unwanted material from patterned layers.

The number of equipment types required in the fab is determined by the complexity of the fabrication processes; the number of machines of each type is determined by the fab's required throughput.

Typical semiconductor factories may have from one to ten major fabrication process flows and may produce 20,000 or more wafers per month. Highly complex fabs can operate with more than fifteen flows and produce 30,000 wafers per month.

The product of a wafer fab is a finished wafer containing on its surface from approximately one hundred to a few thousand devices (chips). Contemporary processing equipment allows fabrication with silicon wafers that are nominally eight inches (200 mm) in diameter. The next generation of fabrication equipment will allow processing wafers with twelve-inch diameters (300 mm). The area of a silicon chip, containing a large complex device such as a microprocessor, is approximately 250 square millimeters.

Functioning silicon chips are housed in packages that are seen prominently on circuit boards when a personal computer is opened. Because of the required packaging, the wafer fab is often called the 'backend' factory. To go from a finished wafer to a packaged electronic component requires three more factories (or factory operations): wafer sort, assembly, and final test (Fig. 1.2).

In wafer sort, each chip or die on a wafer is tested to see if it performs as the designed electronic device. The wafer is then sawed into chips, and the good chips are sent to assembly to be mounted into packages. In final test, the completed component is tested for thorough functioning. Assembly and final test are often called 'frontend' processing. Wafer sort may go with either factory.

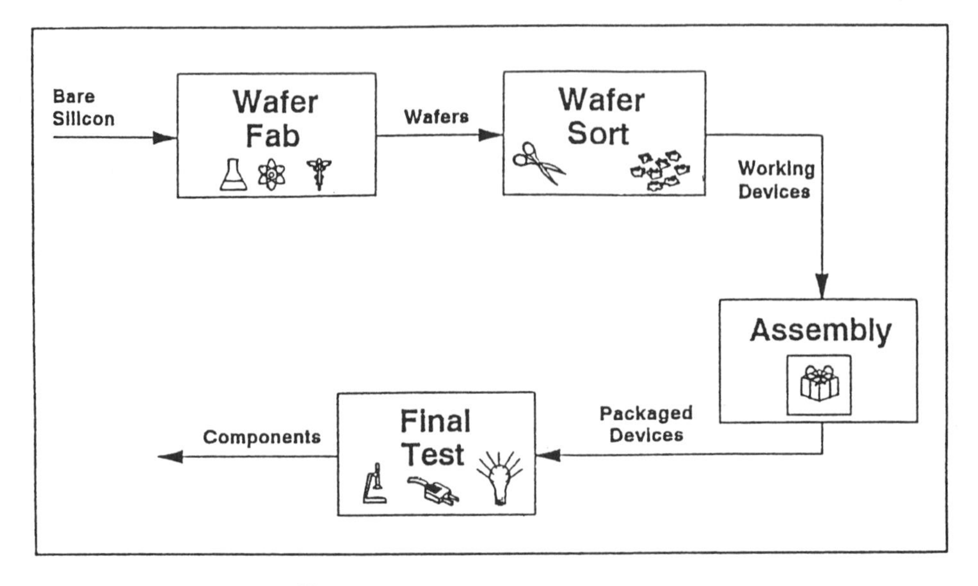

Figure 1.2: Electronics Factories

Wafer fabs may exist at separate sites or in separate companies, or they may be integrated in factory complexes with assembly and test factories. Electronics companies that own and operate all four factories are called integrated electronic companies. Since the wafer fab is a chemical plant that makes electronic parts, there has been a recent trend toward the formation of two other types of companies: pure electronic companies and wafer foundries.

Pure electronic companies, also called *fabless* semiconductor companies, design and sell electronic circuits. Typically, their only manufacturing is final test and product-packaging. Wafer fabrication and component assembly are handled by contract manufacturers. A contract wafer fab is otherwise know as a *foundry*.

Wafer foundries, sometimes called 'silicon foundries' will manufacture device-covered wafers to customer specifications. The main business of the wafer foundry is chemical processing, and its final products are finished wafers. Since the customer is generally the fabless electronic companies that design and sell high-end microprocessors, the asking price of these wafers is expensive. But the high price allows additional degrees of specialization and emphasis not always seen in integrated electronics companies.

1.3 Wafer Fabrication Equipment

A wafer fab contains a few hundred pieces of specialized equipment for thin-film processing. This equipment is the product of another specialized industry sector, the semiconductor equipment industry. Wafer foundries and integrated electronic companies outfit their factories by purchasing equipment from these companies. Naturally, there is a strong interaction and dependency between the wafer fabs and their suppliers. At particular points in the evolution of wafer fabs, the lack of performance of specific equipment types, such as plasma etchers, has limited fab performance and, to some extent, slowed device evolution.

A brief survey of equipment types is given in Table 1.1. The first column lists the equipment type. Next is given the thin-film application used to build the microstructures. Finally, the fundamental physico-chemical processes involved in producing the thin-film effect are listed.

Photoresist processing equipment provides coaters for the application of polymeric layers to wafers. Photolithography equipment allows the transfer of patterns from masks to the polymeric layer via exposure to ultra-violet light. Plasma etchers expose the wafers to complex atmospheres of chemicals in order to remove selectively portions of a patterned layer.

Ion implanters are miniature particle-beam accelerators; they provide a beam of charged particles onto exposed areas of the wafer in order to control impurity distributions in such applications as forming transistors. Specialized oxidation and deposition equipment provide layers of insulators or metallic conductors as well as protective layers.

TABLE 1.1: WAFER TRANSFORMATION PROCESSES		
Equipment Type	Thin-Film Application	Physico-Chemical Effect
Deposition	Deposit thin-film layer on wafer surface	Chemical reaction Evaporation Glow discharge
Etching	Remove thin-film layer from wafer surface	Chemical reactions at interface of gas or liquid phase
Photoresist	Apply photoresistive organic-polymer film	Liquid coating on solid surface
Photolithography	Inscribe pattern on photosensitive layer	Precision optics Photochemistry
Resist development	Remove negative or positive image from patterned coating	Liquid etching of organic layer
Wafer clean	Remove unwanted layer or particles	Etching Scrubbing
Plasma etching	Etch underlying layer according to pattern resist	Glow discharge Plasma chemistry
Resist stripping	Remove photoresist layer of hardened organic film	Liquid etching of organic layer
Ion implantation	Implant dopants (ions) into exposed layer	Plasma physics Ion-beam optics
Oxidation	Grow silicon dioxide layer	Chemical reactions Fluid dynamics Diffusion
Diffusion	Introduce dopants into exposed layer	Diffusion Fluid dynamics
Chemical-vapor deposition (CVD)	Deposit layers of dielectrics, metals, semiconductors	Chemical reactions Transport phenomena
Sputtering	Deposit layers of dielectrics, metals	Glow discharge Chemical reactions
Chemical-mechanical polishing (CMP)	Smooth or remove a patterned layer	Abrasion Chemical reaction

Each equipment type represents a major technological accomplishment and provides technical areas for specialization and sub-specialization. In addition to thin-film processing capability, wafer fab equipment provides automated wafer handling, computer control, and some degree of process control.

Not surprisingly, each piece of capital equipment can cost one to four million dollars. A *minimal* set of equipment to manufacture microprocessors can easily cost over a hundred million dollars.

The level of automation in fabrication equipment has encouraged a new trend toward process-step integration. Individual process steps that were traditionally performed in discrete equipment are being grouped, or clustered, into a single machine. (Chapter 8 has details on this type of equipment.)

1.4 Circuit Design to Wafer Fab: The Interface

The wafer fabrication factory has been described as a chemical processing plant which manufactures million-transistor integrated circuits. Furthermore, silicon foundries can manufacture complex circuits to order. An interesting, and often misunderstood link is how knowledge of the device is communicated from the circuit designers to the factory and the manufacturing process. The short answer is a set of some twenty or so mask images.

The integrated system is defined in terms of the geometry of patterns defined on each of some twenty layers. The image of a single layer of a device is called a *reticle*, and an array of reticles is a *mask image*. During fabrication, photolithography processing transfers the mask image to a layer of polymeric material.

Circuits for integrated systems like microprocessors are now designed, simulated, and evaluated using software for computer-aided engineering. Mask and reticle images for each layer are the end result of this circuit-design effort, and they are generated by the software.

Of course, a mask set only has meaning in terms of the underlying process technology. It is therefore not uncommon for design iteration and debug to be required in order to produce working devices in silicon. In integrated electronic companies, these steps are performed by coordination between design and manufacturing.

One term for the software used to go from circuit concept to mask sets is *silicon compiler*. Companies providing software related to silicon compilers are yet another industry, and provide a second key building block for the fabless electronic companies. By purchasing silicon compilers and computers, such companies can be formed for capital costs equal to only a few pieces of wafer fab equipment. These fabless electronic companies can then provide mask sets to silicon foundries with the appropriate process technology, and the foundries can produce device wafers in the specified quantity. Prices for finished wafers can be $15,000 or more.

Using their spare capacity, many integrated electronics companies also provide silicon foundry services. Conversely, integrated electronics companies may also purchase foundry services in order to gain additional capacity or operational flexibility.

1.5 Manufacturing Economics

The miracle of wafer fabrication is the cost per transistor resulting from circuit manufacturing. The useful unit of cost for a single transistor in a microprocessor is the micro-dollar. Manufacturing costs for such a transistor are now in the tens of micro-dollars, and such costs are dropping monthly.

Unfortunately, there is also a downside to the economics of electronics manufacturing: fabrication costs are increasing, and have been for some time. The capital and operating costs of the wafer fabs producing miracle chips are rising at a significant rate. Historically, increased wafer costs were overcome by placing more devices, with denser circuits, on larger wafers. Thus, one of the key cost drivers in electronics has been to reduce the feature size of the transistors and wires in the microstructures.

But, as indicated in the discussion of wafer processing equipment, fabrication machinery is becoming increasingly more expensive. The escalating capital cost of the factory is being driven, in large part, by the increasing complexity of the process technology. The trends are for circuit features and device geometries to shrink and for device complexity to increase. As device geometries shrink, each individual process step becomes more demanding and more expensive. Furthermore, increasing device complexity requires longer process flows to satisfy requirements for more circuit interconnections. Thus, both these trends lead to more expensive wafer fabs.

A wafer fab can easily cost a few hundred million dollars. New wafer fabs for advanced technologies designed for a throughput of 20,000 wafers (eight inch) per month can cost one billion dollars or more. But as frightening as the cost is, such capital investments can result in manufacturing costs for a single chip ranging from under one dollar to over forty dollars. Realization of these manufacturing efficiencies, however, does not come without engineering effort. In particular, the design and operation of a billion-dollar factory with a high degree of operational difficulty requires considerable insight.

One of the key concerns in electronics manufacturing is whether rapidly-increasing wafer fabrication costs will now overcome the advantages of denser circuits on larger wafers. As an example, if transistor costs stabilize, then the cost of devices will scale with the number of transistors. New competitive rules will dominate the electronics business. Die shrinks alone will not solve manufacturing cost problems.

This discussion of manufacturing economics has focused on wafer cost. However, a second manufacturing parameter, die yield, is required to convert wafer cost to die cost. *Die yield* is the number of good die per wafer. Comprehensive treatment of yield encompasses a wide variety of complex phenomena. For this discussion, it is sufficient to observe that a wafer fab with a low yield is not economical.

We can now summarize the conundrum facing the electronics industry: operating billion-dollar chemical plants to produce high device yields at an acceptable wafer cost. Excellence in wafer-fab operations is now critical for the entire industry. Insights gained from performance analysis of wafer fabrication can determine the success of individual companies as well as the course of the entire industry.

1.6 Fab Performance Analysis

Fig. 1.3 provides a roadmap for many of the issues discussed to this point. An integrated system such as a microprocessor has been designed. It is to be implemented in silicon as a specific microstructure. A process flow specifies the steps required to build the microstructure. A minimum set of equipment types is required to implement this process flow. This equipment set is placed in a clean-room factory environment.

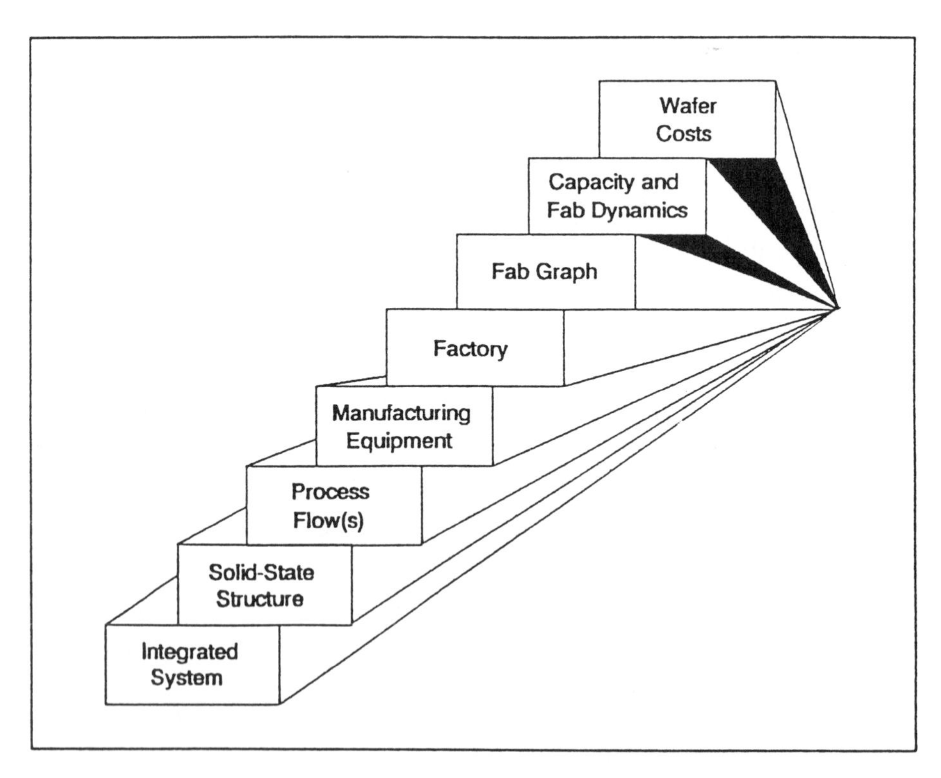

Figure 1.3: Intellectual Journey of Wafer Fabrication

Thus, the design, the flow(s), the factory are in place; we are now concerned with the performance of the factory as it works through the flow to implement the design. Performance will be measured in terms of capacity, throughput, cycle time, wafer costs, and yield.

The fundamental consideration in fab performance analysis is the silicon process flow. Equipment types must be able to perform all the required steps in the flow. The result of the manufacturing operation, i.e., working through the flow, must be the production of wafers containing a sufficient yield of working devices. Further, the wafers completing the process sequence must be produced at a reasonable cost.

Performance analysis is important in every phase of a fab's life cycle, including design, operation, and evolution. To complicate matters further, fab design, fab operation, and fab evolution, as well as fab diagnosis, present different challenges for performance analysis.

Not surprisingly, the initial design of the factory has a large impact on operating cost. Performance analysis at this phase of the plant's life provides a means of projecting the wafer cost that will result from alternative designs. Systematic performance analysis encourages consideration of radical alternatives which may provide significant strategic advantages.

Once the wafer fab is built and operating, performance analysis is critical for obtaining effective use of the massive investment in buildings and equipment. An operating fab is a tremendous generator of data and measurements. These measurements include material movement, equipment performance, and processing results. In current practice this data is collected on networks of computers which provide systems for computer-integrated manufacturing (CIM).

The purpose of fab diagnosis is to determine the real operating state of the factory. Actual factory performance can be considerably different from design projections, performance standards, or the platonic-ideals in managers minds. Fab diagnosis involves culling information from factory data in order to support improved factory operation. Information from fab diagnosis includes current dynamic capacity and its causes, bottlenecks, and the effective availability and utilization of equipment sets. While fab diagnosis for device-yield improvement is standard practice, diagnosis of operations is not straightforward and often haphazardly performed, despite the fact that it can provide a basis for improved factory operation.

If fab diagnosis is properly done, operational problems are identified, and a set of proposals for improved operation can be generated. The impact of these proposals may also be evaluated using performance analysis, and the best proposal selected for factory implementation. Once implemented, the impacts are tracked using the CIM system. The changed, hopefully improved, factory behavior then provides new opportunities for fab diagnosis (Fig. 1.4).

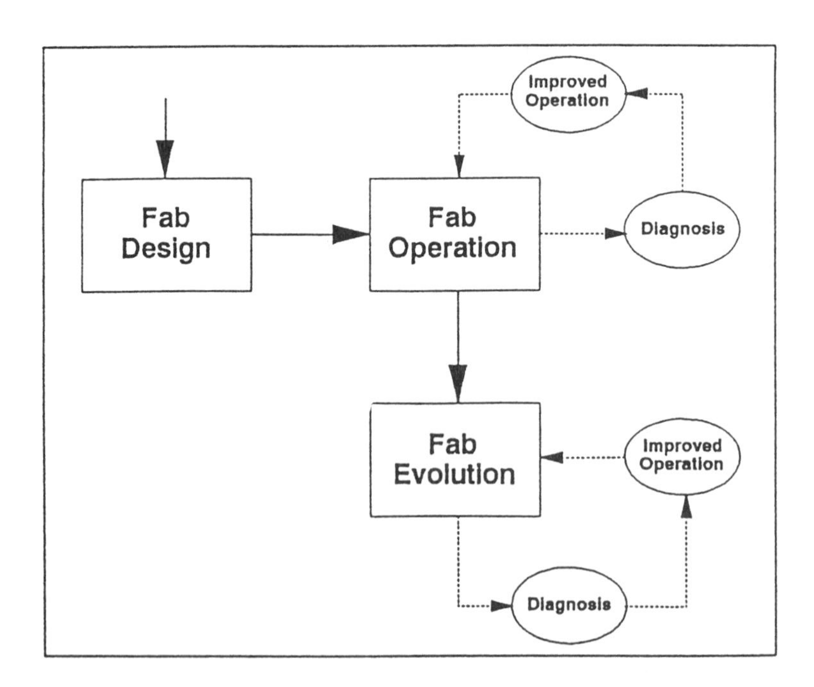

Figure 1.4: Cycles of Diagnosis and Improvement

Fab evolution results from changes in process flows and equipment over the lifetime of the factory. Process flows and process technology can dominate operational dynamics. In many fabs, the bottleneck equipment set is the result of a series of changes in equipment use, changes which are dictated by processing considerations for device-yield improvement. Thus, the resulting bottleneck and its associated factory dynamics are not the result of planning by industrial engineers, but evolves because of process engineering considerations. Indeed, the degree of process technology dominance over operational dynamics is one of the unique features of wafer fabs.

Fab evolution also occurs naturally over longer time-scales during the multi-year lifetime of the factory. The original process flow, the basis for factory design, changes. New, additional process technologies are added to the factory. New equipment types are added. Each change to the design, the flow, or the equipment provides significant opportunities for performance analysis and the use of cycles of fab diagnosis for operations enhancement.

One additional need for performance analysis results from the trend in newer fab designs to install automated material handling. For example, equipment, instruments, and robots are being tied together in areas called *bays* to create workcells. A second example of this type of combined automation/integration is the new integrated processing equipment that contain multiple process chambers and material movement.

Integrated processing equipment is also known as modular equipment, cluster tools, or production-integrated processing equipment (PIPE). PIPEs are capable of performing multiple, independent processing steps without leaving a controlled environment. By performing sequences of processing steps in a single machine, substantial yield improvements may be achieved. The automated processing reduces contamination and unnecessary human handling. The complexity of integrated equipment, however, introduces new problems in equipment performance analysis.

As manufacturing modules are connected together into complex systems, the performance of the overall entity becomes more difficult to predict. Simplistic calculations can no longer represent the variety of dynamic behavior exhibited. In order to analyze, design, and control these new complex manufacturing systems, detailed performance analysis is often required.

1.7 Methods for Performance Analysis

As the highly complex systems that they are, wafer fabs generate huge quantities of performance measurements. Without appropriate analytical tools, however, these data tend to be archived, rather than used. Which is unfortunate because the data are one of three essential ingredients for achieving effective fab performance, as well as effective performance of automated bays and cluster tools. The other two key ingredients: an appropriate set of tools and a systematic methods of analysis.

1.7.1 Fab Graphs

The fab graph is the fundamental tool of performance analysis for wafer fabs. As earlier noted, the quintessential feature of wafer fabs is the complex silicon process flow. However, a process flow can not be performed without equipment. Consideration of how the process flow is spread across the set of equipment leads naturally to the fab graph. The *fab graph* is a graphical mapping of a process flow onto an equipment set.

The fab graph is formed by first drawing a circle for each equipment type; then, an arrow or arc entering a circle represents a single process step performed at that equipment. The ordered set of paths illustrates the entire process flow. An example of a fab graph is shown in Fig. 1.5. The structure of the fab graph is discussed below.

In addition to providing a pictorial representation of the process flow, the fab graph is a legitimate mathematical entity.

As such, it provides a basis for modeling and performance analysis. In addition, the fab graph can be encoded in a data structure and used for computing and simulation.

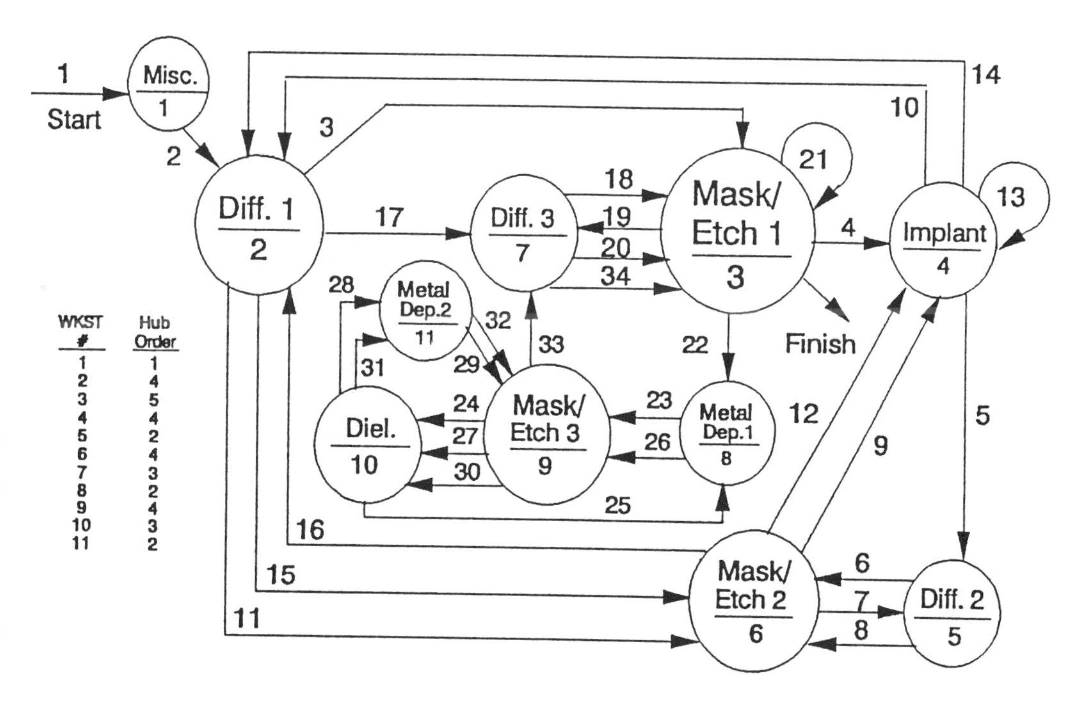

Figure 1.5: Example of a Fab Graph

1.7.2 Fab Structure

The underlying structure of the fab and its process flows provides fundamental information for performance analysis. Understanding this structure is especially important for fabs with multiple process flows, since flows 'share' or 'compete for' the common set of equipment. Much of the structure of a multi-flow fab can be understood in terms of the fab graph. Fig. 1.6 shows a fab with three simplified flows; each flow is a path through the factory.

A second feature of fab structure is that many equipment types, such as photolithography, may perform more than one process step, even within a flow. Such multi-use work centers are called *hubs*, and the set of process steps performed there is the *hub-set*. The twenty masking operations of a CMOS flow, for example, constitute a hub-set.

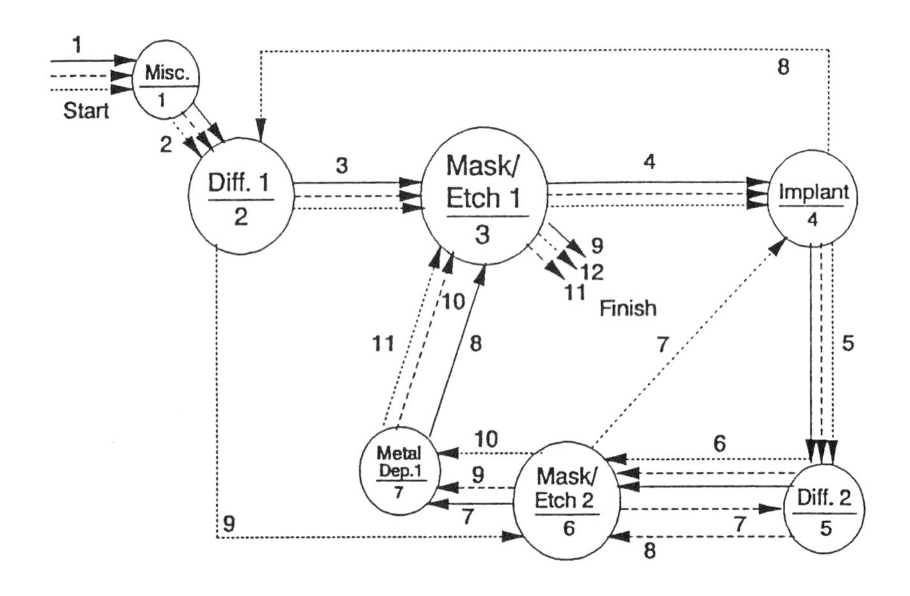

Figure 1.6: Ultra-Simplified 3-Flow Fab Graph

1.7.3 Simulation with a Dynamic Model

The next level of detail in analyzing performance is to move wafers, grouped as lots, through the process flow from work center to work center. To do so, however, fundamental events such as start-a-lot, move-a-lot, and load-a-machine must be understood and described in terms of if-then rules and other relationships. Such rules provide a dynamic model of the interaction of lots with machines.

To simulate the fab, the rules may be implemented simply using pencil, paper, and calculator, or, more efficiently, using a computer. The computer simulation provides fab performance measures such as dynamic capacity, cycle time, and equipment utilization.

1.7.4 Synthetic Experience

Using a dynamic model, a wide variety of operations scenarios may be evaluated. In making these evaluations, centuries of operations will be simulated. Needless to say, the modeling results represent a rich source of information. Organized correctly, this information provides valuable *synthetic experience*, insight, and rules-of-thumb on fab operation. The synthetic experience may be further organized into a knowledge base using signature analysis and sensitivity analysis.

1.8 Archetypes of Wafer Fabs

Wafer fabs can exhibit a variety of complex dynamic behavior. To aid in understanding the complex behavior, this book contains extensive case studies on dynamic performance. The case studies are based, in part, on data from real manufacturing plants. In addition, there are a variety of performance analyses based upon detailed models of a set of simplified, but realistic factories.

The case studies in the following chapters analyze the root causes of such performance measures as factory throughput, dynamic capacity, and bottlenecks. The impact of equipment dynamics and process flows on these measures is explored in some detail. Aspects of equipment dynamics include setup, batching, dispatching, loading rules, and availability. The process issues include multiple process flows, process creep, process migration, and overall fab evolution. In particular, the presence of engineering lots means that even the simplest fab must deal with multiple processes, which adds dimensions to the problem of fab complexity. Since wafer costs are a strong function of the operational dynamics of the factory, all aspects of the fab's performance must be considered for sound economic decisions.

The simplified wafer factories, CMOS-1 and CMOS-2, are important as paradigms of wafer fabrication. CMOS-1 deals only with twenty-four manufacturing process steps using nine types of equipment. However, the equipment set that was chosen and the sequence of process steps allow the toy fab to exhibit all the complex dynamics of a full microprocessor factory.

Further Reading and Notes

1. More background on the microelectronics revolution can be found in the book by Gilder [1.2], and in Chapter 1 of Glaser and Subak-Sharpe [1.3]. See also Ferguson and Morris [1.1].

2. Silicon compilers and the design of integrated systems are treated by Mead and Conway [1.4] and by Weste and Eshraghian [1.9]. General background on microelectronics is provided by Milnes [1.7]. The bibliographies of these books provide an entry to the extensive literature of solid-state electronics.

3. Wafer fabrication factories, yield, process technologies and other issues are discussed in Chapter 2. The unit operations of wafer processing, process steps, are treated in Chapter 3.

4. There is some discussion of wafer fabrication machinery in Chapter 3. Discussions and pictures of equipment can be found in the articles and advertisements of the trade publications **Semiconductor International** and **Solid State Technology.**

5. Facilities aspects of wafer factories, including clean room design and delivery of pure chemicals, are discussed in Middleman and Hochberg [1.6].

6. A survey of some contemporary fabs is given in Leachman [1.4].

7. Factory modeling is the subject of Chapters 4 and 7.

8. Performance analysis is treated in Chapters 5 and 6.

9. Cluster tools are discussed in Chapter 8.

10. Wafer fab economics is discussed in Chapter 9.

References

[1.1] Ferguson, Charles H., and Charles R. Morris, **Computer Wars: How the West Can Win in a Post-IBM World**, Random House, New York, 1993.

[1.2] Gilder, George, **Microcosm: The Quantum Revolution in Economics and Technology**, Simon and Schuster, New York, 1989.

[1.3] Glaser, A. B., and G. E. Subak-Sharpe, **Integrated Circuit Engineering: Design, Fabrication, and Applications**, Addison-Wesley, Menlo Park, CA, 1977.

[1.4] Leachman, R. C., (ed.), **The Competitive Semiconductor Manufacturing Survey: Second Report on Results of the Main Phase**, Report CSM-08, Engineering Systems Research Center, UC Berkeley, CA, September, 1994.

[1.5] Mead, Carver, and Lynn Conway, **Introduction to VLSI Systems**, Addison Wesley, Menlo Park, CA, 1980.

[1.6] Middleman, Stanley, and A. K. Hochberg, **Process Engineering Analysis in Semiconductor Device Fabrication**, McGraw-Hill, New York 1993.

[1.7] Milnes, A.G., **Semiconductor Devices and Integrated Electronics**, Van Nostrand Rheinhold, New York, 1980.

[1.8] Skinner, Wickham, **Manufacturing: The Formidable Competitive Weapon**, Wiley, New York, 1985.

[1.9] Weste, N.H.E., and K. Eshraghian, **Principles of CMOS VLSI Design: A Systems Perspective**, (2nd. ed.), Addison Wesley, Menlo Park, CA., 1993.

Exercises

1. To appreciate the magnitude of the efficiency of circuit manufacturing, compare transistor costs to the cost per circuit resulting from a household wiring job. If a contractor charges $2500 for twelve major circuits, calculate the cost per custom circuit. For an arbitrarily complex computer, estimate its manufacturing cost if each electronic component is priced like household wiring.

2. Consider circular silicon wafers with diameters of 200 mm and 300 mm. A die requires a manufacturing area that is 16 mm by 16 mm. For each diameter wafer, estimate the number of die sites per wafer. In making the estimate, consider the placement of die and losses near the edge.

3. A CMOS manufacturing flow requires 525 operations on 71 types of equipment. Let the hub number be the number of manufacturing operations performed per type of equipment. What is the average hub number? If 28 masking operations are performed, what is the range of hub numbers? Estimate the median hub number.

4. The CMOS technology above requires 180 hours of total processing to finish a wafer. The local silicon foundry runs a 24-hour operation, and has quoted a delivery time for your new microprocessor design of seven weeks. Calculate the ratio of manufacturing cycle time to process time. If four test cycles are required to perfect the design, how much time will be required to complete the product? Is the local foundry an appropriate choice?

5. List the factories required to go from bare silicon to a finished microprocessor. As an electronics entrepreneur, decide which factories you will own and operate and which you will utilize via contracting.

6. On a contract basis, a foundry provides a wafer containing 100 working microprocessors for $15,000. The microprocessor contains 1.5 million transistors. What is the average cost per transistor?

7. A sole wafer fabrication machine (semiconductor capital equipment) cost $2.1 million. It is to be amortized over five years. Over a year it will process 19 wafers per hour, and produce wafers for 6000 hours. What is the amortization charge per wafer?

8. If the cost for a manufacturing step ranges from $1 to $7 per wafer, estimate the manufacturing cost for a fabrication sequence involving 350 steps. If a silicon foundry wants to make a 60% gross margin, what will it charge for these wafers?

9. A wafer with a silicon foundry price of $10,000 has 200 die locations. Die yield fraction ranges from 0.1 to 1.0. Plot cost of good die versus die yield for this range.

Things are the way they are, because they got that way.
-Boulding

Chapter 2
Factory Phenomena

2.0 Introduction: Peeling the Onion

Chapter 1 provides an overview of the complexities involved in wafer fabrication. Dealing with the numerous, complex aspects is perhaps best tackled as one would peel an onion. As each layer of understanding is mastered, then peeled away, the fab appears differently. It is these different, illuminating perspectives that enable the engineer to deal with the world's most complex factory.

The wafer fab is viewed in different ways by different individuals (Fig. 2.1). Factory managers take a global view and see it as a process flow or technology, as a factory producing wafers, or as a set of distinct equipment. One level deeper, process integration engineers think of the fab in terms of the device microstructures being made.

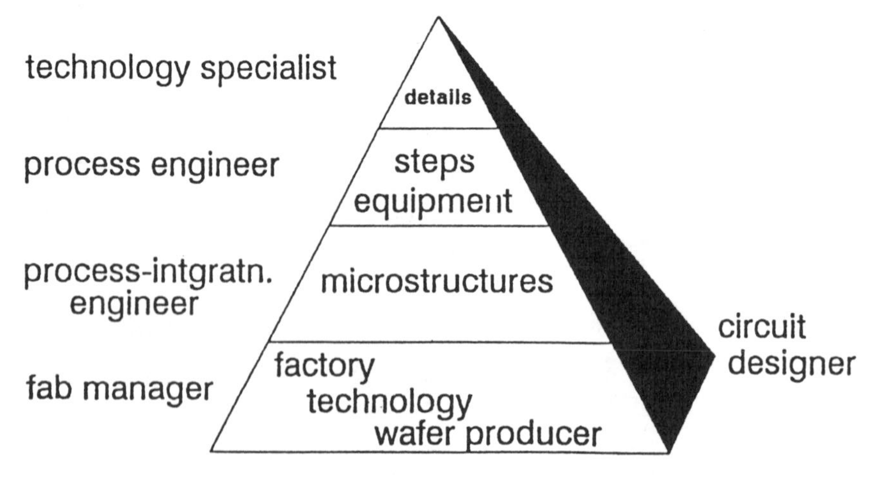

Figure 2.1(a): Views of the Wafer Fab

Deeper still are the process engineers who are concerned only with one process step and its associated equipment. Finally, there are the technology specialists who are concerned with specific details like photoresist spinning and etch plasma physics.

At the opposite extreme are the circuit designers who view the fab as a box that is useful for making circuits (Fig. 2.1(b)).

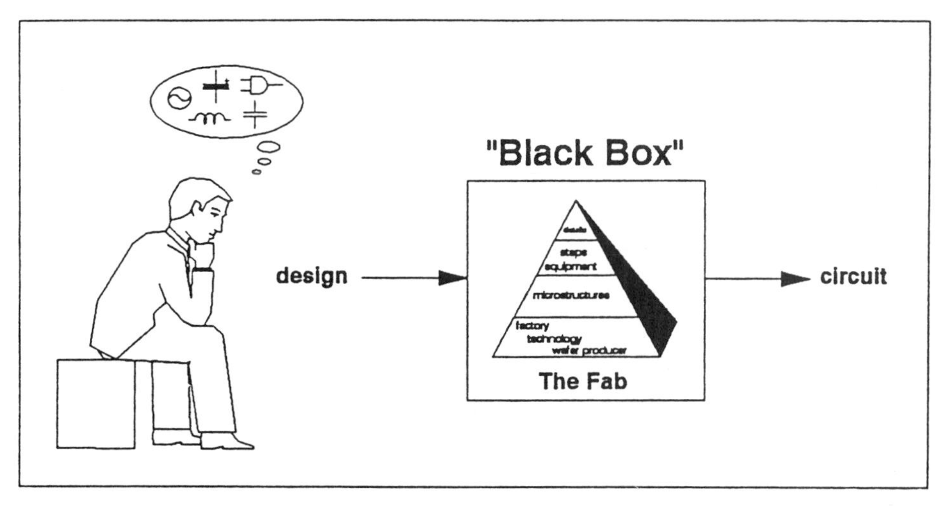

Figure 2.1(b): Circuit Designer's View of Fab

Of course, all these views are correct and useful on some level. Not surprisingly, strong arguments can be made about which view is more fundamental; i.e., what is at the core of the onion. Clearly, the circuit designers are correct that the purpose of the factory is to manufacture circuits. But the bridge from circuit design to wafer processing can be tortuous. In practice, circuit dimensions and topology are communicated to the manufacturing process via a set of mask images. Each of perhaps twenty masks determines the geometry of a layer. Correct fabrication of the layers results in microstructures, e.g., electronic devices.

Process engineers, on the other hand, argue that the core of the fab is the set of process technologies that allow fabrication of the structures. Certainly, individual processes like plasma etching are sufficiently rich intellectually that an engineer can devote an entire career to its pursuit. With great pride in their specialties, process engineers are often further denoted as epi engineers, photo engineers, diffusion engineers and so forth. Process engineers certainly exercise a veto over the factory, since any single malfunctioning step can cause failed circuits.

Another group of proud specialists are the yield enhancement engineers. They analyze fabricated microstructures to see if they are being built correctly. They ensure the quality of the millions of transistors in the microprocessor. Working with the process engineers and process-integration engineers, these engineers help to fine-tune the fabrication processes to increase device yield, circuit performance, and reliability.

Process integration engineers often claim the pinnacle of the fab. While all individual operations may be performed correctly, the interaction of steps may prevent successful fabrication of working microprocessors. Thus, these integration experts must understand transistors, interconnect mazes, and chemical processes; and they often take responsibility for the die yield (fraction of working chips). To these master technologists the view of the single-step expert is myopic.

Perhaps the most global view is that of the factory operations manager. In addition to die yield, the fab manager is concerned with wafer throughput, factory capacity, equipment bottlenecks, and cycle time. This engineer must deal with the strong interactions between the process technologies and the operations dynamics. This chapter addresses many issues of concern to the fab manager. However, the overview is valuable to anyone involved with wafer fabrication.

2.1 Fundamental Entities in Wafer Fabrication

In wafer fabrication, as in any manufacturing operation, there are three fundamental entities involved: the material being processed, the sequence of processing steps leading to a finished product, and the equipment or machines performing those steps. Each of these fundamental entities is integral to the phenomena occurring in the factory, and will be discussed in some detail below.

2.1.1 Process Flows

A distinctive feature of wafer fabs is the complex processing flows and their strong impact on operating dynamics. The sequence of process operations is determined by the need to build working structures. A technology sequence lists the major structures in order of fabrication. For example, the technology sequence for a BiCMOS flow [2.1] that could be used to build microprocessors is listed in Table 2.1.

TABLE 2.1: BiCMOS Flow I
Structure as Major Steps

	Step	Equipment
1	Starting wafer and N+ Burried Layer	Module 1
2	Epitaxial Layer Deposition	Module 2
3	Twin-well Formation	Module 3
4	Active Region and Channel Stop	Module 4
5	Deep N+ Collector and Base	Module 5
6	Polysilicon Emitter	Module 6
7	Gate and LDD Formation	Module 7
8	Sidewall Oxide and Final Junction	Module 8
9	Silicide and Local Interconnect	Module 9
10	First Level Metal	Metals Module
11	Second Level Metal	Metals Module
12	Third Level Metal	Metals Module
13	Fourth Level Metal	Metals Module
14	Final Passivation and Contacts	Module 10

The structures that are the simplest to explain are the metallization levels. Such structures consist of interconnected layers of wires with insulators between the wires and between the levels.

In Table 2.2 the dielectric structures for each level are explicitly listed; interconnect represent approximately half the flow.

TABLE 2.2: BiCMOS Flow II
Structure as Major Steps

	Step	Equipment
1	Starting wafer and N+ Buried Layer	Module 1
2	Epitaxial Layer Deposition	Module 2
3	Twin-well Formation	Module 3
4	Active Region and Channel Stop	Module 4
5	Deep N+ Collector and Base	Module 5
6	Polysilicon Emitter	Module 6
7	Gate and LDD Formation	Module 7
8	Sidewall Oxide and Final Junction	Module 8
9	Silicide and Local Interconnect	Module 9
10	First Dielectric	Diel Module
11	First Level Metal	Metals Module
12	Second Dielectric	Diel Module
13	Second Level Metal	Metals Module
14	Third Dielectric	Diel Module
15	Third Level Metal	Metals Module
16	Fourth Dielectric	Diel Module
17	Fourth Level Metal	Metals Module
18	Final Passivation and Contacts	Module 10

While simple in terms of device physics, interconnect structures require a significant and increasing portion of fabrication resources. Furthermore, the electrical performance of these structures is a limiting technology in the development of more complex integrated systems. Consequently, interconnect structures will be the source of many examples in discussing manufacturing issues.

Since equipment is fundamental to the factory, a list of machine requirements is given for both BiCMOS technology flows. However, the modules listed represent hypothetical equipment. But even a hypothetical list is a necessary first step for defining a factory. A process flow only has meaning in terms of the hardware that can implement the process operations. The factory is ultimately defined in terms of its set of equipment.

To move from a technology flow, in terms of structures, to a process flow, in terms of real equipment, requires that much more detail be given. Specifically, we must describe how to build each structure in terms of unit process operations. Supplying this detail is called exploding the technology sequence. Process sequences which will build specific structures are given in Tables 2.3, 2.4, and 2.5. Note, however, that these examples in no way cover all cases.

The processes involved in depositing and patterning a layer of metals or dielectrics [2.2, 2.3] are given in Table 2.3. These steps are considered standard in current practice. Each listed step, however, could well consist of multiple process steps. The two clean steps, denoted 'A' and 'B', are likely to be different processes performed in different equipment 'Y' and 'Z'.

TABLE 2.3: Pattern a Layer
Fundamental Sequence I

	Step	Equipment
1	Clean (C)	Clean (X)
2	Deposit Layer	Deposit
3	Measure Film	Metrology (X)
4	Pattern	Pattern
5	Etch	Etch
6	Resist Removal	Stripper
7	Clean (B)	Clean (Z)
8	Measure Feature	Metrology (Z)

Process Sequence II (Table 2.4) lists the steps involved in introducing a controlled patterned impurity into a layer.

TABLE 2.4: Dope a Layer
Fundamental Sequence II

	Step	Equipment
1	Clean (A)	Clean (Y)
2	Pattern	Pattern
3	Dope	Implanter
4	Measure Dopant	Metrology
5	Resist Removal	Stripper
6	Clean (B)	Clean (Z)

Process Sequence III (Table 2.5) describes a more complex structure where a patterned, doped layer is formed.

TABLE 2.5: Dope a Patterned Layer
Fundamental Sequence III

	Step	Equipment
1	Clean (C)	Clean (X)
2	Deposit Layer	Deposit
3	Measure Film	Metrology (X)
4	Pattern	Pattern
5	Etch	Etch
6	Resist Removal	Stripper
7	Clean (B)	Clean (Z)
8	Measure Feature	Metrology (Z)
9	Clean (A)	Clean (Y)
10	Pattern	Pattern
11	Dope	Implanter
12	Measure Dopant	Metrology (Y)
13	Resist Removal	Stripper
14	Clean (B)	Clean (Z)

At this level of description, the process sequences that are building structures require respectively eight, six, and fourteen steps. If the average structure requires twelve process steps, then the full process flow for eighteen structures will require 216 steps. However, there are additional levels of processing detail associated with patterning layers, specifically photolithography.

With the exception of epitaxial deposition each of the structures listed in Table 2.2 will require at least one photolithography step. The number of photolithography steps, or masking levels, is one measure of the difficulty of a process flow. The technology flow described here will require over twenty masking levels. Each of these masking steps is one operation in a patterning sequence.

Table 2.6 lists the steps and equipment involved in patterning. Patterning invariably involves photolithography and inspection. However, the final step in the table, resist hardening, is not used at every layer.

TABLE 2.6: Pattern Sequence

	Step	Equipment
1	Photolithography	Coater/Stepper
2	After-Develop Inspect	Metrology
3	Overlay Check	Checker
4	Measure Feature	Metrology
5	Resist Hardening	Processor

The **key** patterning step, photolithography, represents not one but a sequence of operations, such as those listed in Table 2.7. The first four steps in the table form a photosensitive polymeric layer on the wafer. In the fifth step, ultra-violet radiation is passed through a mask image to pattern the wafer.

Since a fuller description of patterning will require adding perhaps five steps per masking level, a complete description of all masking operations will add some one hundred additional steps, bringing the total sequence to over 300.

TABLE 2.7: Photolithography Sequence

	Step	Equipment
1	Wafer Cleaning	Clean
2	Baking & Priming	Primer
3	Coating	Coater
4	Softbake	Bake
5	Exposure	Stepper
6	Development	Developer

It is becoming clear that adding continuing levels of processing detail to the outline by structures can accumulate large numbers of process operations.

Combining unit process operations into a process flow provides a sequence of manufacturing steps necessary to fabricate a complex system of integrated circuits. Since this sequence may involve five hundred steps, it is important to study the manufacturing process in a systematic manner. To do so requires the next fundamental entity in wafer fabrication: the material that is moved through the processing sequence.

2.1.2 Lots, Wafers, and Devices

In wafer fabrication, the starting material is the *wafer*, a near-circular slice of single-crystal silicon. Currently, most fabs process nominal 6" and 8" wafers; 12" wafers are projected as the next standard size.

In order to understand material flow in current wafer fabs, two other concepts are necessary: cassettes and lots. *Cassettes* are a physical carrier that can contain up to twenty-five wafers. Cassettes became popular during the days of 2" wafers as holders for liquid-phase processing.

A *lot* is a group of wafers that traverse the factory together in cassettes. Hence, silicon wafers enter the factory in lots, and the lots follow specific process flows, moving from work area to work area. Typical lot sizes are twenty-four or twenty-five wafers, and require one cassette. Smaller lot sizes are known. A few fabs run fifty (or forty-eight wafer lots) using two cassettes. Maintaining lot integrity during processing is very important for process control and yield analysis.

Each 8" wafer (200 mm, to be more precise) may contain approximately one hundred microprocessors or other advanced integrated circuits. The portion of the silicon wafer containing one integrated circuit, whatever its complexity, is called a *die* or a *chip*. Thus, a lot of twenty-five wafers may possess 2500 integrated-circuit chips. A packaged die sold in an electronics store is often called a *component*. *Device* can refer to a single transistor, a die, or a component.

A key measure of the efficiency of a wafer fabrication operation is cost per wafer. If more die can fit on a wafer, the cost per die drops dramatically. Similarly, processing larger wafers is an effective cost cutter, if the additional number of potential chips increases faster than the cost per wafer. Of course, the equipment to process these larger wafers must be available.

For instance, the next expected standard wafer size will be a nominal 12" (300 mm). However, 12" fabs will not be built until a full set of fabrication machinery is available for processing the large substrates.

The impact of the larger wafer size on lots, cassettes, equipment, and fabs as a whole is expected to be significant. Lot sizes may become small, tending toward one wafer. Cassettes may disappear and be replaced by carriers in automated handling systems. The trend toward single-wafer processing may very well accelerate.

There are other factors driving the industry towards single-wafer processing besides just larger wafers. Most notable, of course, is anticipated improvements in yield.

Die yield is the fraction of die on a wafer that function to specification as electronic systems. Many factors may impact die yield; some are controllable, many are uncontrollable.

Yield loss may result from unplanned plant upsets such impurities in a gas tank or sudden, unexpected equipment failures. Or, it may be the result of poor fab operating practices like carrying excessive inventory which leaves wafers sitting in long queues, waiting. In fact, two parameters which are virtually inseparable are yield and cycle time. Long cycle time leads to lower yield, and the old aphorism is true: "only bad things happen while wafers wait."

2.1.3 Equipment: Fabrication Machinery

The third and final fundamental entity in wafer fabrication is the machinery or equipment. In wafer fabs, the machinery capable of fabricating microstructures with feature sizes on the order of 0.1 micron is highly specialized. The design of the equipment is strongly influenced by the nature and size of the wafer substrates that it must process. A further complication for the equipment is the set of distinct chemical processes that are required to build the microstructure.

Over fifty different types of equipment may be required for advanced fabrication processes. The range of types is quite large, differing primarily in terms of wafer-handling capabilities and processes of transformation. Typically, more than one piece of equipment of each type is required to meet throughput requirements. (A set of like equipment is called a *workstation* or *work center*.)

Another distinction of wafer fabrication, not often found in other types of manufacturing, is that the equipment may intrinsically process wafers either individually or in multi-wafer batches. In fact, one of the more recent innovations introduced into wafer fabrication is integrated equipment, or cluster tools. These machines not only perform single-wafer processing, but they provide multiple-process chambers for more than one step.

(See Chapter 8 for more on Cluster Tools.)

It is difficult to discuss wafer processing equipment independently of the process steps and process sequences required to build the microstructures. Each type of equipment must satisfy stringent processing requirements. Meeting these requirements demands design of a sophisticated chemical reactor that is capable of controlling heat transfer, fluid flow, and chemical reactions in the neighborhood of the wafer surface. Hence, a brief overview of key process equipment follows; and Chapter 3 gives more detail on individual processing steps.

Photoresist processing equipment provides coaters for the application of polymeric layers. *Photolithography* equipment allows the transfer of patterns from masks to the polymeric layer via exposure to ultra-violet light. In current practice, these processes are performed on individual wafers. However, the equipment takes cassettes as input and unloads and reloads them.

Plasma etchers expose wafers to complex atmospheres of chemicals in order to remove selectively portions of a patterned layer. Plasma etch equipment may use single-wafer processing or multi-wafer batches.

Ion implanters are miniature particle-beam accelerators; they provide a beam of charged particles onto exposed areas of the wafer in order to control impurity distributions in such applications as forming p-n junctions or transistors. Wafers are loaded via cassettes, and the proceeding is performed either by single wafers or in batches.

Specialized *oxidation* and *deposition* equipment provide layers of insulators or metallic conductors as well as protective layers. Some of these processes, as well as *diffusion*, can be performed in large-capacity furnaces. Furnaces can hold batches as large as 200 wafers. Such batches are formed by unloading cassettes onto a carrier.

In recent years, a new and revolutionary type of equipment has been introduced in the semiconductor industry. Individual process steps that were traditionally performed in discrete machines are being grouped, or clustered, into a single machine. *Cluster tools* are capable of performing a sequence of processing steps on a wafer without leaving a controlled environment. Hence, they may provide many processing and operational advantages over conventional equipment. They may also provide an unusual opportunity for radical redesign of wafer fabs.

In a wafer fab starting 5000 wafers/week, the processing equipment is likely to consist of 125-175 machines, of which there are likely to be 50-75 different kinds. The quantity and type of machines will be a function of: the product mix being manufactured, the complexity of the various products, the inventory level the fab is carrying, and, to some extent, the batching and loading rules of the machines. (Fab sizing is covered in Chapter 4.)

2.2 Factory Dynamics

Wafer fabs are dynamic; conditions in the factory and requirements for the factory are constantly changing. Equipment fails and is repaired. There is a demand for increased output. The fab does not have sufficient equipment to satisfy this demand, but an effort is made anyway. Changes occur as a result of process evolution: steps become longer, new steps are added. Entirely new process flows may be introduced into the factory.

The complex dynamics in wafer fabrication result from the interactions of lots with machines as the lots traverse the long process flows. Capacity, throughput, cycle time, inventory, etc., are all time-varying, and thus dynamic variables.

As is now known, complex dynamic systems can exhibit chaotic behavior. Wafer fabrication, being up there on the complexity scale, is no exception. Fig. 2.2 (discussed in detail in Chapter 6) displays a classic example of chaos; the figure shows the non-periodic oscillations of inventory at the bottleneck workstation.

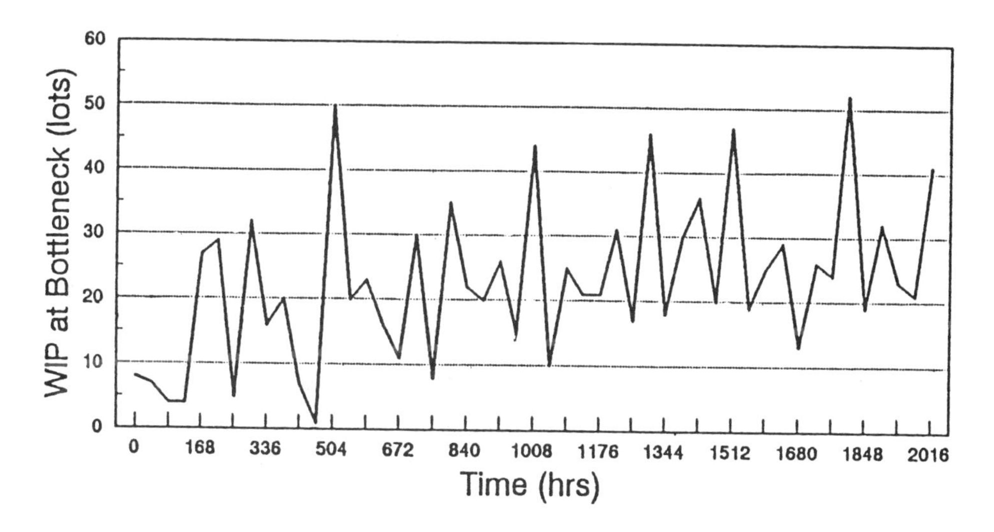

Figure 2.2: "Chugging" Behavior of WIP

Operational dynamics makes for an interesting struggle in analyzing wafer fab performance. Some key dynamic variables are described below.

2.2.1 Dynamic Capacity

Factory capacity is not constant over time; it changes from shift to shift, and from hour to hour, and hence is the single, biggest contributor to factory dynamics. Some of the capacity changes are brought about with intentional operational changes, such as lengthening of process flows, changes in product mix, additional engineering lots, etc. Some of the capacity changes, however, result from unexpected and sometimes uncontrollable occurrences, such as equipment failure, operator error, facilities problems, etc. Whatever the cause, the dynamics that occur in the fab generally cause capacity to be less than design, sometimes by 30% or more.

Factories are designed and built to meet a given throughput of products; production schedules are based upon design capacity. But if actual capacity is only 30% of design, it is obvious that production cannot be met. In fact, if lot starts are not reflective of the dynamic capacity, inventory will build, as in another physical system shown in Fig. 2.3.

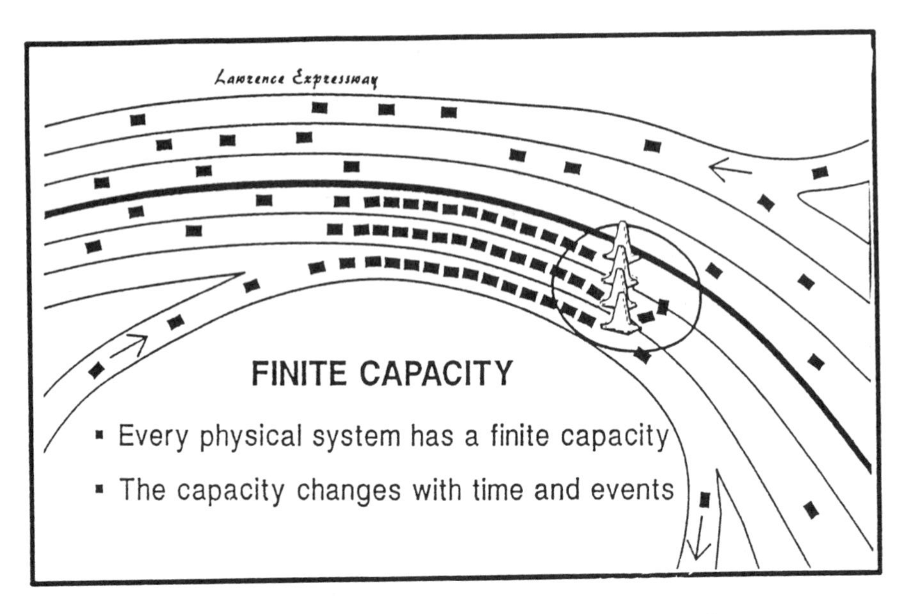

Fig. 2.3: "Inventory" Building in Another Physical System

Once inventory builds, many fabs have great difficulty clearing it in a rational manner. These fabs then either continue to operate with excessive inventory or they get into a 'binge-purge' mode of fab operation. (See Chapter 5.)

Perhaps, the best way to determine if demand matches the fab capacity is to observe the inventory build (or loss) rate over some time horizon. If inventory is building, the start rate is too high for the fab's capacity. If inventory is being consumed, there is excess capacity. Either condition is not good for long-term stable operation. (Of course, these observations only apply to a factory with WIP; an empty fab demonstrates special behavior.)

Actions taken to control or modify dynamic capacity include:

- adding equipment
- decreasing/increasing lot starts
- decreasing/increasing inventory
- revising maintenance procedures
- improving machine availabilities
- etc.

Of these actions, the first two, adding equipment and/or altering lot starts, are the most direct, and will have the biggest impact. Of course, adding equipment will only work if it is added at the bottleneck or controlling workstation(s).

2.2.2 Bottlenecks

The bottleneck of a factory is the workstation (or workstations) that throttle or choke the flow of lots through the fab. In most instances, the bottleneck will have the minimum capacity relative to processing needs. However, certain circumstances may cloud the relationship between capacity and bottlenecks, such as an upstream workstation with a long mean-time-to-repair profile. Other situations may include: insufficient granularity of equipment, poorly-matched equipment batch sizes, mishandling of inventory, large swings in number of engineering or hot lots, etc.

Granularity, a term peculiar to wafer fabrication, refers to the discrete levels of adding or subtracting capacity. The capacity of a workstation is strongly a function of the number of machines. Hence, the *impact* of a machine's failure depends on the number of machines present. The failure of a machine at a workstation having two or less units can be catastrophic. Workstations with five or more machines typically have sufficient granularity to deal with reasonable perturbations.

Bottlenecks in wafer fabrication generally occur for one of two reasons: either the factory was under-designed to begin with, or the demands have changed over time.

Initial fab underdesigns occur because most are still based on spreadsheet calculations. These calculations cannot possibly deal with the dynamic phenomena taking place in a real factory. Spreadsheet designs provide a capacity which turns out to be the best the fab is ever likely to achieve. Day-to-day operations will generally see capacities 20-30% lower, primarily due to such things as queuing losses, unplanned engineering lots, rework losses, and load-factor and batching losses.

Operating an under-designed fab at the planned lot-start rate will cause the first under-capacity workstation to bottleneck, choking the flow of lots to all downstream workstations. De-bottlenecking an under-designed fab presents a special set of headaches and must be done either one workstation at a time or with a logical plan.

The bottlenecks that result because of changing factory demands are generally easier to identify and eliminate than original misdesigns. The common sources of these bottlenecks are: process migration or creep, changes in product mix, and/or the addition of new processes. Fab graphs, preferably in conjunction with a factory model, should help identify which workstations have been impacted by demand changes. Studying production data for suspect workstations should quickly zero in on new bottlenecks. (Chapter 4 covers fab graphs in detail.)

In some special instances, the age of the factory may also contribute to bottlenecking. As equipment ages, mean-time-to failure may shorten and mean-time-to-repair may lengthen, making the designed-for availability is no longer valid. If equipment usage is monitored over time, it is generally possible to identify when changes in availability cause a bottleneck shift to another workstation.

2.2.3 Throughput

The next dynamic variable, throughput, is the **flux** of material through a system or component over a given time interval. For a wafer fab, throughput may be measured in terms of lots, wafers, or die (chips), and the time interval may be of basically any length (hours, quarters, etc.). In addition to the throughput of the entire fab, there are also throughputs associated with processes, process steps, and workstations.

The throughput of a specific process will be a function of the product mix in the factory and each flow's priority. Process-step throughput will be a function of these two, as well as equipment availability at the workstation used by the step. Finally, workstation throughput will be a function of equipment availability and operating control rules.

If the fab is not glutted with work-in-progress (WIP), then short-term throughput will be tightly tied to inventory level and inventory staging. Throughput will also be highly sensitive to equipment availability and operating control rules.

2.2.4 Cycle Time

Whereas throughput is the flux of material through a system, cycle time is the material's **residence time** within the system. In a wafer fab, the cycle time is the time that elapses between when a lot of bare silicon wafers enters the fab until the lot exits, having completed its entire process sequence. The cycle time, therefore, includes all process times, setup times, transport times, and wait times. Of these four, the one that is most unpredictable and the one which cannot be scheduled is wait. Waits occur based on the state of the factory at any given moment, and may vary from process step to process step, and from lot to lot.

Processing time is the total number of working hours required to process a lot, assuming all equipment is available as needed. Thus, processing time can be viewed as the *theoretical minimum cycle time*, and the difference between the cycle time and the processing time is wait time or time spent in queues.

Processing times in wafer fabrication typically range from 90 to 150 hours, upwards to 300 hours. In terms of calendar time, we may consider the processing time to be approximately one week. Quoted cycle times range from three to ten weeks, giving a wait-time ranging from two to nine weeks.

It is a common misconception to assume that throughput and cycle time either mean the same thing, or else that they at least perform in the same direction as a result of operational policies. An often cited fallacy: if cycle time is markedly shortened, more cycles will be run a year, and thus more wafers will be produced. (See Section 5.1.2 for a fallacy-busting example of this misconception.)

2.2.5 Inventory

Inventory, also known as work-in-process, work-in-progress, and WIP, consists of the lots and wafers that have started fabrication but have not yet completed the final step. Such lots will be processing in machines, waiting at inventory points (racks) throughout the factory, or will be resident on transport systems.

Historically, fabs have carried large to huge inventory because it acts as a levelizer, or a dampener of sensitivity for the entire fab. In addition to making fab operation more stable, large WIP generally leads to higher throughputs and higher equipment utilization, two outcomes that are sure to please any financial manager.

But not all is sunshine and roses. In exchange for stable fab operation, high throughputs, and high equipment utilization, these large inventory also results in longer cycle times, lower yields, and, more often than not, bottlenecks. In fact, in extreme cases where excessive inventory swamps the fab, even the anticipated advantages may disappear. It is unwise to allow inventory to 'just happen'.

2.2.6 Schedule Stability

The factory is a physical system, and this simple observation has several consequences for scheduling. Clearly, the factory has a physical limit or capacity. If a production plan exceeds the capacity of the factory, then following the production plan will not increase output, but will only build inventory. In the terminology of wafer fabrication: increasing wafer starts beyond the capacity of the fab will not increase wafers out, but will build WIP.

Schedule stability is the ability to meet production plans and ship scheduled lots on time. Not meeting schedules means disappointing customers, which is something most wafer fabs will go to extremes to prevent. The most common method to insure reliable and stable scheduled production is to carry a sizeable WIP. In fact, up to a certain point, large inventories are synonymous with high throughputs and good schedule stability. The price, however, may be excessively long cycle times, and a coincident drop in yield.

Which brings us full circle. Do we operate the fab for high throughputs and stable schedules, knowing we may suffer yield losses? Or do we run lean, looking for faster cycle times, higher yields, and be willing to accept less stability? Chapters 5 and 6 will give us tools to help us decide such issues. But first, we need a better understanding of the critical issues of yield and process control.

2.3 Yield and Process Control

Process control is generally defined in terms of ensuring stable, controlled operation of a process step, like metal deposition or plasma etching. In an obvious generalization, we can speak of process control for a sequence of process steps and building a microstructure. The sequence is *in control* if the structure has the desired electronic properties.

The discipline of process control has been well-developed within the classic chemical-processing industries. However, it is still in the early stages of development in microelectronics processing. Unfortunately, this lack of process control often results in yield problems in the manufacturing line. Yield may be wer than expected; it may be uneconomic; or yield may crash.

Die yield, which is the fraction of working die on a completed wafer, depends on the results of a long sequence of steps. Thus, yield is a complex function of the processing results for the approximately 500 steps. Since each step may require upwards of twenty parameters to describe factory conditions and processing results, yield may depend on 10,000 parameters. Obviously, tracking a yield *faux pas* is challenging to say the least. But it must be done since yield is one the key factors in establishing the economics and productivity of the factory.

The cost of a yield crash is **strongly** related to the cycle time behavior of the fab. Typically, in a yield crash, the problem is not detected until wafer sort. If we assume that the problem is a malfunctioning plasma etcher that is producing one lot per hour, a lot may reach wafer sort thirty hours after completing etching. Thirty lots will be mis-processed by the time the *faux pas* is detected. Those lots must be scrapped. If the lots had a meager potential value of $100,000 each, the yield crash costs three-million dollars.

Since yield is the economic driving force of wafer fabrication, it is useful to understand the types of crashes that commonly occur, as well as the frequency of these costly events.

2.3.1 Taxonomy of Yield Crashes

Class 1: Lost the Process.

This type of crash results in dead wafers, i.e., no good die. Back-tracking from symptoms of a Class 1 crash is complicated due to the non-obvious interactions among the process steps. Linking causes to symptoms can be difficult, if not impossible. Recovery of the process is an open-ended procedure. If successful, it requires one or more weeks.

Class 1 crashes occur approximately once every five years. The value of the lost production may be ten million dollars or more.

Class 2: Classic Yield Crash.

In a Class 2 crash, yield ranges from a few good die to dead wafers. The cause is typically a blunder at one or more process steps. Recovery time ranges from one week to a full cycle time.

Class 2 crashes occur approximately once every two or three years, and may cost the factory eight to fifteen million dollars.

Class 3: Major Yield Loss.

In a Class 3 incident, yield drops below seventy-five percent of the standard yield for a mature product. One or more process steps may be out of control. Recovery time is on the order of a few days to a full cycle time.

Each quarter of operation may experience a major yield loss at a cost of one to ten million dollars.

Class 4: Yield Loss.

In this type of episode, yield drops to between seventy-five and ninety percent of standard for a mature product. One or more process steps may be out of control. Recovery time is on the order of a few days to a full cycle time.

These processing blunders occur more than once per quarter, at cost a few hundred thousand dollars each occurrence.

Class 5: Line Down.

When the line is down, a single process step can not be made to work. The problem may result from mis-processing one or two steps back. Because the nature of the problem is obvious, the downstream process steps usually do not become contaminated by bad wafers. Recovery time is one day to one week.

Having the line down may occur more than once per quarter, and cost a few hundred thousand dollars each incident.

2.3.2 Process Control

The overall subject of yield control and process control in wafer fabrication is vast. Control of the entire fabrication line is especially difficult; there are encyclopedic treatises [2.4] dealing with various aspects of the problem. Areas of concern for process control (Table 2.8) include problem diagnostics and defect classification, as well as monitoring processes, equipment, and contamination.

TABLE 2.8: PROCESS CONTROL CONCERNS
• Manufacturing Yield
• Problem Diagnostics
• Manufacturing Defect Classification
• Product Dimensional Metrology
• Pattern Defect Inspection
• Process and Tool Monitoring
• Contamination Monitoring
• Repair and Rework
• Test Sites and Vehicles for Yield and Process Monitoring
• In-LIne Electrical Test
• Final Test
• Traceability
• Failure Analysis of Semiconductor Devices
• Material and Chemical Analysis of Devices
• Statistical Quality Control
• Reliability/Defect Severity
• Defect Prevention

Of particular reference to semiconductors are: diagnostic test structures for process monitoring, analysis of device structures for chemical and material properties, and failure analysis of devices. This large range of specialties reflects the vast scope of operating parameters that affect yield.

There are stringent requirements on each step of a process operation. Control of individual process steps like photoresist processing is a major task. As Ruska [2.2] points out, constant attention to many small details of processing is required. Particular attention must be paid to direct variables that are both measurable and controllable.

For photoresist processing, such a variable is uniform film thickness of the polymer layer. After the development step, the pattern in the layer must be clear, distinct, and meet certain dimensional tolerances. The inspection for pattern quality is develop-check or after-develop inspect (Table 2.6). However, this inspection is difficult, and does not provide a reliable basis for process control. Rather, control is achieved through direct measurement of film properties.

Lack of process control in photolithography results in a phenomena called *rework*. If the developed resist fails its quality test, its is chemically removed, e.g. *stripped*, and resist is then re-applied and re-patterned. Reworked wafers in effect follow a separate process flow, and thus have their own set of yield problems associated with them.

As should be apparent from this discussion of yield and process control, the wafer fabrication industry has a long road ahead of it to achieve the same degree and sophistication of manufacturing control seen in older process industries. But fortunately classic chemical process control has many potential applications in wafer processing. Blunder prevention, for example, is a necessary first step. In the traditional process industries, blunder prevention is a natural byproduct of plant and operating safety practices [2.5].

It should be pointed out that the industry's shortcomings in area of process control are not the result of disinterest or neglect, but simply due to the youth of the industry and its rapid pace of process innovation. The speed of process innovation in wafer fabrication is unprecedented in industrial history.

2.4 Fallacies in Wafer Fabrication

The speed of process innovation in wafer fabrication has also led to some shortcomings in design and operating policies as well. Many of the current policies have not been re-thought since the invention of the transistor.

Standard lot sizes of 25 wafers, filling 150-200 wafer furnaces, carrying excessive WIP, not planning for engineering lots, etc., are policies that have been around for 20-30 years and are still being practiced. In many fabs, these archaic policies are neither questioned nor evaluated in the context of modern manufacturing. Perhaps even more surprising, the origins of many of them are lost in the collective memory of the industry.

As a means to understand wafer fab performance, as it exists today, we will discuss some of the more questionable operational procedures, along with their origins, and how they differ from reality. The purpose of this fallacy section is to highlight performance issues that will be covered in detail in Chapters 5 and 6.

2.4.1 Infinite Capacity

Fallacy: To increase output, merely increase starts.

Reality: If the production plan exceeds the capacity of the factory, then following the plan will not increase output. It will merely build inventory.

Origin: Many wafer fabs will have something known as a burst capacity, capacity that is attainable for a short period of time. The source of this capacity comes from operating expedients such as postponing maintenance, running with more WIP, minimizing setup, shifting process or lot priorities, etc. As much as 5% additional capacity may be accessed through such manipulations of operating policies. But these 'doctored' operating policies can not be maintained indefinitely, and when procedures are restored to normal, capacity will return to normal. Unfortunately, at this point, it may be necessary to 'drain' the fab of built-up WIP. During this draining (or re-stabilization of the fab), performance may suffer.

2.4.2 High Inventory and High Equipment Utilization

Fallacy: High inventory and high equipment utilization are always desireable goals because they lead to higher throughput.

Reality: High inventory and high equipment utilization usually lead to:

- longer cycle times,
- lower device yields, and, depending on conditions,
- decreased throughput.

Excessively-high inventories lead to congestion and require time to manage the excess lots.

Origin: Having a queue of lots at each process step does insure that equipment stays busy, which looks good on the books as far as cost-of-ownership goes. Furthermore, up until the time that the fab becomes 'clogged' with inventory, the queues do make for higher throughput, which is still the dominant performance indicator in wafer fabs. But, in reality, these are merely justifications to continue operating with high inventory. High inventory generally means stable factory operation and reliable, on-time product shipping.

2.4.3 Idle Time on Equipment

Fallacy: There is idle time on a machine. Therefore, the fab is not operating at capacity.

Reality: Policies to maximize use of available equipment up-time (minimize idle time) require a large queue of work-in-process so that sporadic machine failures do not result in idle equipment downstream. A large queue at each process step dictates that each wafer spends a substantial amount of time waiting. Not only may capacity be lost as a result of excessive-inventory management, but yield (and thus production) will degrade due to the contamination associated with the long cycle times. Long cycle times also have other penalties, such as:

- high cost of carrying inventory,
- less flexible response to market, and
- longer development cycles (fewer turns per year).

Origin: Due to the extremely-long process sequences in wafer fabrication, it is rarely looked upon as a total system. Instead, segments of the flow are segregated (usually by equipment type) and put under the auspices of an area manager. His performance is then evaluated on how well his operators and equipment meet certain yardsticks. Equipment utilization makes an easy yardstick because it may be rigorously tracked, albeit without reference to the rest of the fab. Furthermore, it makes intuitive sense to keep high-priced equipment as busy as possible.

But the natural line imbalance resulting from mismatched load sizes, granularity, and loading and batching rules makes it impossible to keep all equipment busy all the time, especially in an inventory-lean fab. Not only is it an impractical goal, it is not necessarily desireable either because of the high inventory it necessitates. In an efficient fab, only the bottleneck or constraining workstations will achieve equipment utilizations of 95%+ for extended periods. If too many workstations have utilizations over 85%, then the fab is carrying too much inventory.

2.4.4 Engineering Lots

Fallacy: Engineering lots are basically free.

Reality: Engineering lots degrade production disproportionately to the number of lots run. These lots will typically be more advanced and/or more experimental than production lots and may, therefore, use some of the metals/interconnect workstations more heavily. It is not unusual to find fabs with 20-30% of their capacity being consumed by engineering lots. In leading-edge fabs, this number may rise to 50% or more.

Origin: If engineering lots were truly rare occurrences, then the burst capacity of most fabs could handle an occasional extra lot, over and above regular production. However, by the mid 1980's, engineering lots were not only common, but frequent happenings. And regardless of size, an engineering lot is still a load, tying up a machine at each process step (i.e., tying up fab resources in *out-of-production* service). Some industry mindset continues to believe that capacity applies only to production, and that burst capacity will accommodate all other requirements.

2.4.5 Process Creep

Fallacy: A little extra time on one or two process steps is no big deal, especially if it may improve yield. It may cost a little in cycle time, but throughput should not be affected.

Reality: Over the lifetime of a fab, process creep in the range of 10% is not unusual. Longer process steps mean longer cycle times, regardless of where the creep occurs. If the creep occurs at a bottleneck, however, then not only will cycle times lengthen, but throughput will degrade as well. If the creep is at a constraining or high-utilization workstation, a bottleneck shift may result. Such a shift may cause havoc, over and above a simple degradation in throughput. When a bottleneck shifts, all accumulated intuition and operating knowledge about factory performance may no longer be valid. In fact, factory behavior could be so markedly different as to perform like a different plant.

Is yield improvement sufficient to make up for the impact on factory operations? The answer is most likely not known; it is doubtful whether the data, if it exist, has ever been evaluated.

Origin: Back in the old days (1975-1980), long cycles and high inventories disguised many of the consequences of process creep. Now that plants are being operated better, the impact is more noticeable. This is especially true since creep is most likely to happen in the complex interconnect area, at a constraining or high utilization workstation (such as etch).

2.4.6 Process Migration

Fallacy: An extra process step will not impact anything except cycle time, if the new step can be performed at an existing workstation.

Reality: Process migration, like process creep, puts an additional burden on the fab. When an extra step reuses a workstation, the hub order of the work center increases, and the demand increases accordingly. If the additional demand causes the workstation to bottleneck (i.e., insufficient capacity for demand), then throughput will suffer.

Origin: Because of the discrete nature of fabs, and fundamental mismatches in machine load size and granularity, many work centers in the fab have utilization ratios in the 0.40-0.60 range. Machines with utilizations this low can indeed often accommodate another process step without bottlenecking. However, most process migration takes place while pursuing the next generation of chip, and thus happens in the metals/interconnect area. Utilization of equipment in this area is typically in the 0.60-0.80 range. At a ratio of 0.60, an extra step may cause a bottleneck; at a ratio of 0.80, an extra step very likely **will** cause a bottleneck.

2.4.7 Equipment Capacity Mismatches

Fallacy: Filling large-batch furnaces is cost effective.

Reality: In general, waiting for a full load (four or more lots) before running a large furnace leads to degraded performance (longer cycle times, lower throughputs, etc.). Only a large, high-inventory fab may escape the poorer performance.

Origin: Large-batch furnaces are another holdover from the early days of wafer fabrication. Along about the time IC technology was on the verge of becoming the IC industry, the engineers casts their eyes about to see what equipment was available for high-temperature heating. In their line of sight was the standard apparatus for metals heating, a cavernous heated furnace. The engineers reshaped the furnace's large, heated core into something better suited to accommodating wafers: a large horizontal (or vertical) tube. Add to that boats to load wafers, and controls to insure proper firing, and a new industry's standard was successfully appropriated from an old industry's standard. Over the years, furnace technology has changed relatively little.

Because the original furnaces were huge, those developed for the IC industry were huge. Unfortunately, 'huge' is not in keeping with a balanced fab line. In fact, through trial and error, most fabs now operate their furnaces at half to three-quarter loads (2 or 3 lots). Waiting for 4 or more lots to arrive at a furnace workstation before processing can begin is costly in terms of performance. On the other hand, half to three-quarter loads look bad on the books as far as equipment utilization is concerned. Fortunately, furnaces are relatively inexpensive, and most fab managers are willing to trade capital depreciation for throughput.

2.4.8 Spreadsheet Analysis

Fallacy: A spreadsheet model will effectively track dynamic capacity.

Reality: An experienced, first-class engineer can probably do an acceptable job of sizing a fab using a spreadsheet model. But tracking a plant's dynamic capacity, with up to 40,000 changing variables, requires a validated, factory-specific dynamic model.

Origin: Spreadsheets are a powerful calculation tool. They have an awesome ability to generate a large volume of charts and numbers for a relatively modest investment in modeling effort. As a result, spreadsheets have been heavily used for static sizing studies for new fabs. When the results of such calculations are combined with the judgment of an experienced engineer, adequate equipment sizings can result.

However, tracking the dynamic capacity of an operating fab is a performance analysis problem that is two orders of magnitude more difficult than the static sizing problem. Spreadsheets can not deal with the interactions of lots and machines, nor can they deal with the complex structure of fabs. Furthermore, identifying performance parameters like availability is a challenge to most spreadsheet users.

2.4.9 Equipment Availability Ratios

Fallacy: Longer mean-time-to-failure (MTTF) availability is worth the tradeoff of longer mean-time-to-repair (MTTR) maintenance.

Reality: Equipment availability ratios are not as important as the actual lengths of MTTF and MTTR. Machines with MTTF of 70 hours and MTTR of 30 hours degrade overall fab performance **much** more severely than machines with MTTF of 21 hours and MTTR of 9 hours, despite the fact that the average up ratio is 0.7 for both cases and the average down ratio is 0.3.

Origin: Operations Research, of course. Theories developed for 4-variable models quickly lose their 'umph' when applied to a 40,000-variable model. Not only that, they are usually wrong when some of those variables are linearly dependent, others are contra-dependent, and some are both, depending on the dynamics of the system.

Further Reading and Notes

1. BiCMOS processing is described in Alvarez [2.1]. An overview of interconnect processing and metallization is given in Chapter 3.

2. Unit process operations are discussed in Chapter 3.

3. The underlying structure of the flows in Tables 2.1 to 2.7, using the technique of fab graphs, is the subject of Exercises 4.1 to 4.5 at the end of Chapter 4.

4. Additional background on process flows is given in [2.6], [2.7], and [2.11], as well as in Ruska [2.2].

5. Standard texts in classic process control are [2.8], [2.9], and [2.10].

References

[2.1] Alvarez, Antonio R. (ed.), **BiCMOS Technology and Applications**, Kluwer Academic Publishers, Boston, 1989.

[2.2] Ruska, W. Scott, **Microelectronic Processing: An Introduction to the Manufacture of Integrated Circuits**, McGraw-Hill, New York, 1987.

[2.3] Wolf, S., and R. N. Tauber, **Silicon Processing for the VLSI Era: Volume 1 Process Technology**, Lattice Press, Sunset Beach, California, 1986.

[2.4] Landzberg, A. H. (ed.), **Microelectronics Manufacturing Diagnostics Handbook**, Van Nostrand Reinhold, New York, 1993.

[2.5] Roffel, Brian, and John E. Rijnsdorp, **Process Dynamics, Control and Protection**, Ann Arbor Science Publishers, Ann Arbor, MI, 1983.

[2.6] Colclaser, Roy A., **Microelectronics Processing and Device Design**, Wiley, New York, 1980.

[2.7] Glaser, A. B., and G. E. Subak-Sharpe, **Integrated Circuit Engineering: Design, Fabrication, and Applications**, Addison-Wesley Publishing Company, 1977.

[2.8] Himmelblau, David M., **Process Analysis by Statistical Methods**, Wiley, New York, 1970.

[2.9] Ott, Ellis R., and E.G. Schilling, **Process Quality Control: Trouble Shooting and the Interpretation of Data**, McGraw-Hill, New York, 1990.

[2.10] Seborg, Dale E., T.F. Edgar, and D.A. Mellichamp, **Process Dynamics and Control**, Wiley, New York, 1989.

[2.11] Till, William C., and J. T. Luxon, **Integrated Circuits: Materials, Devices, and Fabrication**, Prentice Hall, Englewood Cliffs, N.J., 1982.

Exercises

1. By inserting the photolithography steps (Table 2.6), explode the flow of Table 2.5.

2. Further explode the detailed flow of exercise 1 by inserting the photoresist processing steps (Table 2.7).

3. In Table 2.2, explode steps 10 through 18 by inserting the operations in Fundamental Sequence I, Table 2.3.

4. Continue the explosion of exercise 3 by inserting the patterning steps of Table 2.6.

5. Continue the explosion of exercise 4 by inserting the photoresist processing steps of Table 2.7.

6. Consider a wafer subjected to rework in photolithography. List the full process flow for patterning steps, showing stripping and rework.

7. A process flow with five masking levels has a rework rate of five percent. The rework rate may also be considered the probability of rework at that step. In such a factory, a wafer may follow many possible paths through photolithography, including no rework, one rework, or rework at all five levels. List all possible paths.

8. Compute the probability of each path listed in exercise 7.

9. Repeat the analysis of exercises 7 and 8 for the case where the number of masking levels is ten.

10. Repeat the analysis of exercises 7 and 8 for the case where the number of masking levels is twenty.

11. In light of the results from exercise 10, and the realities of process control, comment on rework as an operations policy.

12. If the required throughput per masking level is 30 wafers per hour and there are twenty masking levels, what is the required throughput of a single photolithography workstation?

13. In exercise 12, if there are two masking workstations and workstation one performs the first twelve mask operations, what is the required throughput of each workstation?

14. The bottleneck of the fab is metal etch (plasma) which can process 36 wafers per hour for a single layer. What is the maximum throughput of a second flow which requires three metal etch steps?

15. A 200-step flow requires 100 hours of processing and an average transport time of 12 minutes per operation. What is the minimum cycle time required to transit the factory? If the cycle time ratio is four, what cycle time is realized and what is the total wait time?

16. Review the yield taxonomy. Compute the expected annual cost of yield crashes.

17. In problem 13, assume each piece of lithography equipment can process 30 wafers an hour. How many units of lithography equipment will be required at each workstation?

Project

1. Exploding the technology sequence into a process flow. For a specific technology, develop a computer program that takes as input a technology flow in terms of structures and gives as output a list of process steps and appropriate equipment.

If you think too hard about all that can go wrong on a wafer, you'll never start one.

-M. S. Buynoski

Chapter 3
Processes for Wafer Fabrication

3.0 Introduction

The physics and chemistry of the processing involved in microelectronics fabrication is a very large subject requiring volumes for a comprehensive treatment (see *Further Reading and Notes*). But since processing considerations play such a dominant role in manufacturing, the subject requires some treatment here.

Much of the literature on the practice and technology of microelectronics begins with treatments of solid-state physics and semiconductor device physics. These treatments then go on to deal with the electronic behavior of P-N junction devices, interfaces, and thin-film devices. Clearly, a significant amount of background material is required to understand active device structures and their close relationship with the manufacturing process.

Fortunately, microstructures involving multilevel interconnect technology are easier to describe. Interconnect processing is, in fact, a useful application for a discussion of silicon process integration. Integration is the key to turning a series of individual process steps into a fabrication line. Even when the individual operations are performed correctly, unless their interaction is successful, the fabrication of working microprocessors will fail. But to succeed at integration, we must first understand the process steps themselves.

One further reason for including a discussion of processing in this chapter is that it is a prerequisite to understanding the manufacturing equipment that performs the various types of process steps. But we will not recapitulate a library here; rather we will emphasize only those aspects of processing that most impact overall factory performance and analysis.

Particular attention will be given in the following discussion to plasma etching, the most common method for transferring the photoresist patterns into silicon or other material layers. This type of etching is required to form structural features that are a half micron or smaller. Because of the complexities and uncertainties of the chemical processes in the environment of an electric discharge, plasma-etch equipment is notoriously difficult to stabilize in manufacturing operations. This step frequently determines yield and device performance. Of specific interest in fab performance, plasma etchers are often the unplanned bottleneck of the factory.

3.1 Semiconductor Process Technologies

One of the unique features of wafer fabrication is the extent that the process flow dominates operational issues like capacity, cycle-time, throughput, and economics. The complexity of the flow, and the equipment involved, makes semiconductor wafer fabrication perhaps the most complex of modern manufacturing processes. A major flow in a semiconductor-device factory may contain over 500 separate steps or operations. Examples of these steps include patterning (photolithography), thin-film etching, thin-film deposition, and oxidation. Table 3.1 lists the process steps that are used to transform the surface of the silicon wafer.

Because electronic structures are built from thin films on the surface of a wafer, another name applied to wafer fabrication is thin-film processing. The sequence of process operations used during the fabrication is determined by the need to build specific structures. Hence, a *technology sequence* lists the major structures in order of fabrication. In Chapter 2, we discussed exploding the technology sequence so that a process flow in terms of process steps and actual machines is given. It is the process flow that establishes the dynamics of wafer flow through the factory.

TABLE 3.1: WAFER TRANSFORMATION PROCESSES

Equipment Type	Thin-Film Application	Physico-Chemical Effect
Deposition	Deposit thin-film layer on wafer surface	Chemical reaction Evaporation Glow discharge
Etching	Remove thin-film layer from wafer surface	Chemical reactions at interface of gas or liquid phase
Photoresist	Apply photoresistive organic-polymer film	Liquid coating on solid surface
Photolithography	Inscribe pattern on photosensitive layer	Precision optics Photochemistry
Resist development	Remove negative or positive image from patterned coating	Liquid etching of organic layer
Wafer clean	Remove unwanted layer or particles	Etching Scrubbing
Plasma etching	Etch underlying layer according to pattern resist	Glow discharge Plasma chemistry
Resist stripping	Remove photoresist layer of hardened organic film	Liquid etching of organic layer
Ion implantation	Implant dopants (ions) into exposed layer	Plasma physics Ion-beam optics
Oxidation	Grow silicon dioxide layer	Chemical reactions Fluid dynamics Diffusion
Diffusion	Introduce dopants into exposed layer	Diffusion Fluid dynamics
Chemical-vapor deposition (CVD)	Deposit layers of dielectrics, metals, semiconductors	Chemical reactions Transport phenomena
Sputtering	Deposit layers of dielectrics, metals	Glow discharge Chemical reactions
Chemical-mechanical polishing (CMP)	Smooth or remove a patterned layer	Abrasion Chemical reaction

Process flows in wafer fabrication have grown so long and complicated that it has become convenient to describe two segments of the flow: the front end of the line (FEOL) and the back end of the line (BEOL). The front end of the line is characterized by transistor formation, high temperatures, and silicon processing. The back end is concerned with metallization, dielectrics, and interconnects. In BEOL processing, temperatures are necessarily lower.

3.1.1 Transistor Types

Semiconductor technology sequences are classified according to the underlying transistor device and the type of electronic circuit that is being implemented. Device technologies include bipolar, CMOS, and BiCMOS. Circuit applications include memories, logic devices like microprocessors, and analog circuits. In describing technology sequences, electronics professionals thus speak of CMOS memory processes, a process for bipolar microprocessors, or of a bipolar analog process. CMOS is currently the favored process for making advanced microprocessors.

Bipolar technology provided the basis for the original growth of the integrated-circuit fabrication industry. In bipolar technology, transistors are formed from P-N metallurgical junctions in silicon. *(P* and *N* refer to the two different types of atoms used to dope single-crystal silicon.)

MOS (metal-oxide-silicon) has since become the dominant technology for fabricating transistors. The transistor is formed using the MOS capacitor. The letter 'C', when added to MOS, stands for complementary.

In CMOS technologies, both P-channel and N-channel MOS transistors are used in the same device. Each transistor is controlled by a gate structure between two doped areas. These doped areas are called the source and drain. Depending upon impurity type, the source and drain are either N-regions or P-regions.

Finally, in BiCMOS technologies, bipolar transistors as well as MOS transistors are used in building integrated systems of millions of devices. An example of a technology sequence for a BiCMOS flow to build microprocessors is listed in Table 2.1.

3.1.2 Feature Size

In specifying a process technology, a dimension is often appended, as in '0.3-micron CMOS. This dimension is the size of a typical feature of the transistor. The magnitude of this dimension is extremely important because it affects chip size, device performance, circuit economics, and the choice of processing equipment.

For example, consider a microprocessor that was originally brought to market using 1-micron technology. Suppose the same essential circuits are now fabricated using 0.3-micron technology. The resulting microprocessor chips will be much smaller, perhaps 10% of the original size. Hence, many more will fit on a wafer, and more good die should result. The new microprocessors will be substantially cheaper to fabricate than the older devices, and, being denser, will be dramatically faster in speed, perhaps by a factor of 3.

Reducing the dimensions of an existing design is called a *die-shrink*. This procedure has been the traditional method to increase yield and lower costs. The shrink described here actually occurred over three generations:

1.0 micron --> 0.8 --> 0.5 --> 0.3 micron

The impact of a reduction in feature size on process steps and equipment can be major. Wafer fabrication machinery must be capable of creating structures with the new, smaller feature size. In the past, shifts in feature size have caused designs of all types of fabrication equipment to become obsolete. The trend continues; and at some point, perhaps five years in the future, contemporary designs will no longer be adequate for new fabs.

At times, entirely new processing technologies develop in order to fabricate the smaller, more demanding features. For example, when device fabrication technology crossed the magic '2-micron' barrier, plasma etching became imperative, and largely displaced wet etching.

A dimensional shift can impose difficult choices on an existing wafer fab. In order to fabricate wafers using a new process technology with smaller dimensions, every required type of equipment in the factory must be capable of making the new smaller features. If one or more equipment types can not function to the new requirements, then that equipment becomes a *technology bottleneck*. Such technology bottlenecks must be removed before the fab can transition to the new process. For example, a plasma etcher that is adequate for a feature size of 0.8-micron may not be able to fabricate 0.5-micron features in metal layers.

3.1.3 Equipment and Processes

Ultimately a factory implementing a semiconductor technology must be filled with correct fabrication machinery, both quantity and type. The factory performance (throughput) will be determined by the number of machines of each type; the device performance will be determined by the process sequence and the equipment selected to perform that sequence.

3.2 Multilevel Interconnect

For two very fundamental reasons, multilevel interconnect structures are now a major technology bottleneck in the evolution of microelectronics processing. Firstly, it is not certain whether additional levels of interconnect can be successfully fabricated. Secondly, even if the structures are fabricated, their electrical performance may not provide acceptable time delays. An even more significant electronic malfunction, *crosstalk*, may arise with additional levels of interconnect.

Rao's book [3.1] reviews developments in the multilevel interconnect area.

In this section, we will use interconnect structures as a vehicle to discuss microstructures, process steps, and their interactions with manufacturing. The focus of the discussion is on building a very specific microstructure. But at the same time, we will illustrate the art of process integration: combining process steps to build a structure that has the desired electronic properties. We will also clarify why wafer fabs are willing to add processing steps to achieve the desired structures.

3.2.1 Overview

An example of the microstructures required for multilevel interconnect is shown in Fig. 3.1. In terms of device physics, the electronic purpose of this structure is relatively simple. Layers of wires (conductors) and insulators (dielectrics) are being fabricated. Because of the circuit complexity of transistors in the underlying layers, several layers of conductors are required.

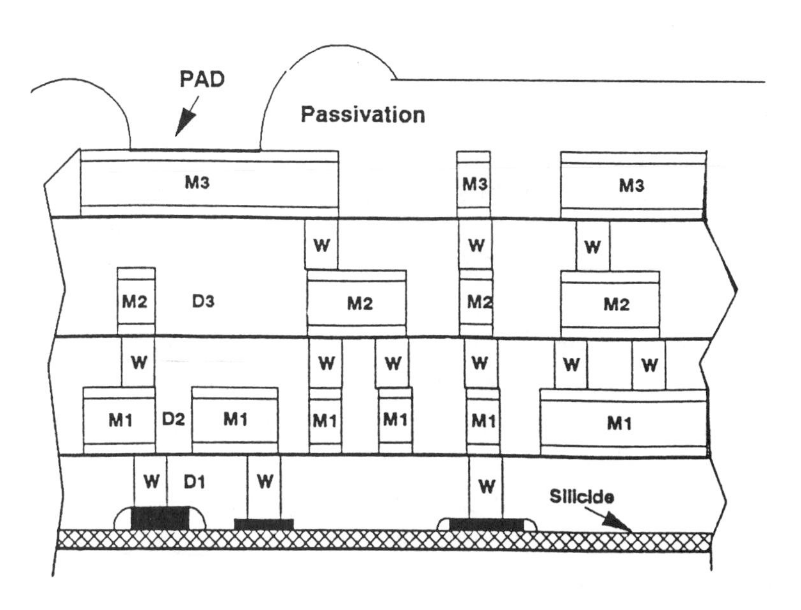

Figure 3.1: Example of Microstructure

Each layer is patterned (Fig. 3.2) to form an array of conductors. Dielectric layers insulate the conducting lines from each other and from the layer below. Vertical conductors or vias connect each layer to the ones below.

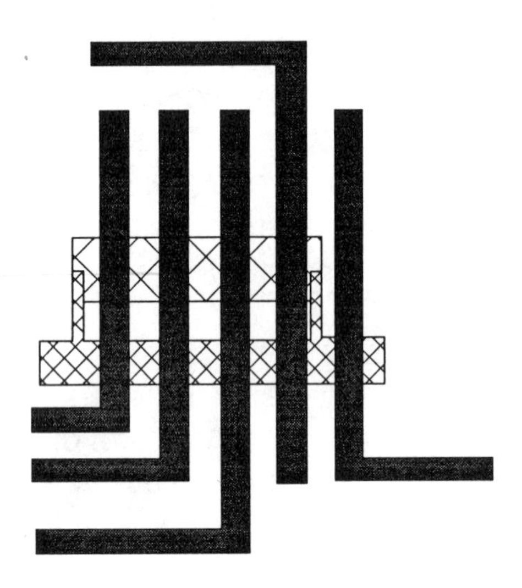

Figure 3.2: Example of Patterned Layers

Despite its apparent simplicity, the interconnect structure presents a highly-complex topography, connecting the millions of transistors of a microprocessor. Performance constraints on this complex structure, in terms of capacitance and time delay, are severe. Effective circuit performance is strongly dependent on the fabrication process. The set of steps involved in building the features of this microstructure are: dielectric deposition, dielectric planarization, contact/via formation, and metallization.

Thus, the beginning point for this discussion will be to make a pattern of metal lines on a dielectric surface. The cross section of this pattern reveals steps and holes with a steep aspect ratio. The formation of the pattern is covered below, in Sections 3.2.7 and 3.2.8 (line metallization and metal patterning, respectively).

3.2.2 Dielectric Deposition

The purpose of the dielectric layer is to separate one conductor from another. This separation is accomplished in three dimensions: lines are separated within a layer, and layers are separated from one another. In successful devices, the dielectric layer must satisfy a list of characteristics including step coverage, physical properties, and electrical properties. In present practice, the material of the dielectric layer consists of forms of amorphous silicon dioxide.

Step coverage is a requirement that results from the need to cover a three dimensional surface, where the surface has mountains and valleys, or, more technically, steps, holes, and narrow gaps. Step coverage is characterized primarily in terms of the uniformity of film thickness on horizontal and vertical surfaces. An additional requirement concerns voids that may form as steep-walled holes; these should be filled during step coverage. Void formation must be minimized, if not eliminated.

The deposited film should have controlled chemical composition and physical properties. Physical continuity of the layer can be interrupted by defects such as particles or pinholes. Measurements of the refractive index and dielectric constant provide indications of film quality and composition.

Electrical properties of the dielectric layer include capacitance, breakdown voltage, leakage, amount of included charge, and the propensity to acquire further electronic charge over time. A function of the film thickness and dielectric constant, the capacitance of the dielectric layer affects the switching speed of the device or microprocessor.

During electrical operation of the integrated circuit, the insulating dielectric layer **must** provide isolation between two conducting layers at different voltages. If the voltage difference is sufficiently high, current could channel irreversibly through the dielectric layer and physically disrupts the material. This disruption is called *breakdown*, and the minimum voltage at which it occurs is called the *breakdown voltage*. Obviously, an integrated circuit will cease functioning once an operating voltage exceeds the breakdown threshold of an integral layer. Current leakage through the layer will detrimentally affect device operation.

3.2.3 Dielectric Planarization

In terms of device structure, the purpose of this set of steps is to cover and smooth over mountains and valleys in the topography of the wafer surface. A single dielectric deposition does not remove the radical steps and valleys of a patterned layer of conductors. If allowed to remain, such irregular topology presents almost insurmountable problems to subsequent processing steps and layers of metallization.

Depositing a second layer of metals on an irregular topography makes step coverage problems similar to those cited in dielectric deposition. Metal layers are even more sensitive to differences in film thickness because of current crowding in thin regions. Increases in current density escalate electromigration and lead to conductor failure. The deposited metal layer must still be patterned, and this leads to a second issue. That is, patterning a seriously non-planar surface adds many difficulties to photolithography processing. Enter planarization.

Planarization is accomplished by a series of depositions followed by smoothing or etching. It may use layers of several deposition types of silicon oxides. The smoothing is often accomplished by a process step known as chemical mechanical polishing (CMP).

The total dielectric layer is a result of planarization processing as well as the original deposition. Consequently, the properties of the total layer determine its functioning in the device structure. Planarization processing is a clear example of the addition of process steps to achieve a desired microstructure.

3.2.4 Via Formation

The next required feature of the microstructure are the holes created in the dielectric layer for the placement of vertical conductors. (The vertical conductors will connect to the next layer of metals.) The holes in the inter-metal dielectrics are called *vias*. A single layer of a microprocessor can require millions of vias.

The requirements on via formation are primarily geometric. The vias should be located accurately, and their formation should be uniform across the die and the wafer. The shape of the via should be correct, and the underlying metal layer should be clear of dielectric. The process steps involved in via formation include patterning using photolithography and plasma etching of silicon dioxide.

3.2.5 Via-plug Metallization

The vias fabricated in the dielectric layer are now to be filled with conducting material. In state-of-the-art practice, via-plugs are formed by depositing tungsten (W) using the chemical-vapor deposition (CVD) process. The deposited metal must fill the hole without leaving voids. In an example of complex metallization issues, tungsten does not adhere well to silicon dioxide. Consequently, an adhesion layer such as titanium nitride (TiN) or titanium tungsten (TiW) must be deposited first. These films are typically deposited using the glow discharge process know as sputtering.

3.2.6 Metal Planarization

The metal used for via filling is typically not used for the main conducting lines. Consequently, the excess metal that is deposited over the dielectric layer must be removed. The surface is replanarized by plasma etching (tungsten etch-back) or by chemical mechanical polishing.

3.2.7 Line Metallization

Line metallization is concerned with depositing the main conducting layer and associated materials. The first layer is the main conductor which is likely to be an aluminum alloy. A second layer is a backup conductor, or *shunt*. The shunt materials can be titanium, tungsten, an alloy of the two, or titanium nitride. The final layer is an anti-reflective coating. This thin film is necessary so that subsequent photolithography processing can successfully pattern the layer. One choice for the anti-reflective coating is titanium nitride (TiN).

All three line-metallization layers are deposited using sputtering.

3.2.8 Metal Patterning

Finally, the process steps involved in forming metal lines include patterning using photolithography, and plasma etching of the metal stack of three or more materials. Plasma etching and lithography are discussed in more detail below.

3.3 The Chemical Reactors of Wafer Fabrication

The previous section has given an overview of the procedures for the fabrication of a microelectronic structure. To carry out each of the procedures requires loading a wafer into a type of highly-specialized machinery that effects material change on the surface of the wafer. Thus, each machine functions as a chemical reactor. These reactors are discussed in more detail below. (Also, see Table 3.1.)

The collection of processes and equipment used in wafer fabrication is heterogeneous and disparate. The major common characteristic defining the set is that each process is useful in making a microelectronic device. The diversity of the processes, however, makes it difficult to organize their intellectual content so that microelectronic processing can be viewed as a coherent whole. But processing overviews are essential for effective process integration; and process integration is the key to building working structures and integrated circuits.

In the older chemical processing industries, this need for intellectual structure was provided by the concept of unit operations. A *unit operation* is a fundamental process operation, and its analysis and design are based upon concepts from chemistry and physics.

Unit operations in wafer fabrication include oxidation, diffusion (both gas and solid state), chemical-vapor deposition, physical deposition, and plasma etching. Fundamental physico-chemical concepts include fluid mechanics, diffusion, heat transfer, and chemical-reaction kinetics. To this standard list can be added glow discharge fundamentals, plasma phenomena, and the characteristics of ion flows.

In analyzing a unit operation like plasma etching, the similarities between etching metals and etching dielectrics are emphasized. Glow-discharge fundamentals and chemical-reaction schemes provide a basis for dealing with each process and its associated equipment. While such an analysis emphasizes the similarities among metal etchers and silicon-oxide etchers, it also helps explain the differences between such equipment.

A more specific paradigm than unit operation for most wafer fab equipment is provided by the chemical reactor. Recall the stringent requirements on via etching described above. Meeting these requirements demands design of a sophisticated chemical reactor, including control of heat transfer, fluid flow, plasma potential, and chemical reactions in the neighborhood of the wafer surface. As the discussion of the fabrication of interconnect structures indicates, control of reactions on the scale of device features ($<<$ 1 micron) is also required.

In general, reactor design and control of process effects occur on four length scales. First is the length scale of the equipment or the reactor chamber. Second is the length scale of the wafer. Third is the length scale of the chip or integrated circuit. Finally, as noted above, is the scale of the device features.

In dealing with chemical reactors, the desired reaction schemes are of paramount importance. Photoresist processing is concerned with photosensitive polymer chemistry. The chemistry of silicon and its compounds is important for a wide variety of process steps. For dopant operations, the chemistry of the compounds arsenic, phosphorus, and boron is required. Developing processes involving the deposition and etching of metals requires a knowledge of their appropriate reaction schemes.

To a large degree, the chemical reactors of wafer fabrication deal with chemically-reacting flows. The range of fluids, the range of pressures, and the types of flow phenomena are quite extensive. For example, liquids-processing is used in cleaning and photoresist coating. Continuum-gas flows, at atmospheric pressure, at high pressures (20 atm.), and at reduced pressures, are used in a variety of deposition processes. Vacuum processing for plasmas and sputtering involve both slip flow and molecular flow.

3.4 Wafer Cleaning

Wafer cleaning steps as a class are one of the most common operations in wafer fabrication. Thus, wafer cleaning processes account for a significant portion of the total required processing time. Essentially, wafers are cleaned before each block of fabrication steps.

A variety of cleaning operations are used depending upon the material layer and structure. The chemical processes used in cleaning also vary with the compositions of the surfaces. Since material is often removed from the surface of the wafer, cleans can also be viewed as chemical-etching processes. The rate and uniformity of the process depends upon chemical reactions at the interface between the wafer and a fluid phase such as a liquid, gas, or plasma.

The purpose of the cleaning operations is to remove a wide variety of impurities and residues. However, the cleaning must be precise for the transistor properties depend upon the introduction of controlled impurities, *dopants*, into precise areas. Unwanted dopants, i.e., *contaminants*, alter device properties, often catastrophically.

More generally, the proper functioning of device structures depends upon control of the chemical composition of individual layers. Impurities that should be removed include metal ionic contaminants, atomic and molecular species, and organics. Many process steps can leave behind unwanted residues.

There is also the issue of particles. Particles can disrupt structures; and disrupted structures do not work.

In more recent designs, cleaning equipment accepts a cassette of wafers, and the wafers are removed and cleaned on an individual-wafer basis. A widely-used standard cleaning sequence for early in a process flow is as follows: (1) Remove wafer from cassette; (2) Preliminary clean; (3) Remove residual organic contaminants and certain metals; (4) Strip hydrous film; (5) Desorb remaining atomic and ionic contaminants; (6) Dry wafer; (7) Return wafer to cassette.

3.5 Photoresist Processing

Photoresist processing is another of the most frequently used process operations in wafer fabrication, and hence a major contributor to the total processing time. Because of its importance in wafer fabrication, a large body of detailed technical knowledge on successful recipes has been accumulated. This discussion highlights the role of photoresist in circuit manufacturing.

Photoresist layers provide the means of transferring the mask image into the solid-state silicon structure. The mask image is first formed in the photoresist, then etching transfers the image to the underlying layer. Thus, the structure is modified according to the circuit design.

The typical photoresist material is an organic polymer containing a photo-activator catalyst. The need to function as a medium of image transfer places many requirements on the photoresist material. The primary requirements are as follows. First, the resist layer must accept and develop an image. Second, it must hold the image while the underlying layer is etched in a corrosive chemical environment; this requirement is the origin of the name resist. Finally, the indurate resist resulting from etching must be susceptible to removal or stripping.

As is frequently the case in wafer fabrication, equipment for photoresist accepts cassettes of wafers, then processes wafers individually. In other words, it uses *single-wafer processing*.

The photoresist processing steps typically include the following: cleaning, baking, priming, coating, softbake, exposure, and development.

In wafer coating, a solution containing solvent and photoresist is spin-coated onto the wafer surface. After drying removes the solvent, the resulting thin film must have uniform thickness and properties. After baking, the resist layer is ready to receive the mask image.

In present day technology, ultra-violent radiation is used to activate the photosensitive chemical groups in the resist. The resulting chemical reactions change the solubility of the exposed areas of the photoresist. Exposure is accomplished using a complex and expensive optical system.

Photo tools are typically one of the most expensive types of equipment in the fab. The most common photolithographic system is the wafer stepper. The stepper contains a master image of the layer for the integrated circuit. This image is projected onto the resist layer on the wafer. The lithography tool then *steps* to the next chip and repeats the exposure operation. Correct placement of the image on the wafer and on the chip is crucial to making a working device.

The final step in transferring the image to the photoresist layer is development. In development, the more soluble portion of the resist is removed by special solutions. This process may be considered as an example of liquid-phase etching, or solvent dissolution, of an organic layer.

At this point, the resist is not yet ready for the rigors of plasma etching. Additional processing (baking) is required to harden the resist so that it can continue to hold the image in a plasma environment.

3.6 Plasma Etching

The use of plasma etching in microelectronics fabrication is remarkable in many ways. Its very introduction illustrates how fabs can develop technology bottlenecks, and overcome them with the subsequent evolution of new fabrication technology. In this instance, when feature sizes were greater than 2-micron, liquid-phase etching was dominant. As feature sizes crossed the 2-micron barrier, liquid-phase etching could no longer build satisfactory structures.

The shortcomings of liquid-phase or wet etching may be characterized in terms of *isotropic etching* and *undercutting*. Prior to the etching, openings in the resist make available portions of the underlying layer, for example silicon dioxide. Solutions containing HF (hydrogen fluoride) will etch any exposed oxide at an equal rate; this is isotropic etching.

As the oxide is etched, vertical surfaces are exposed and also begin to etch. Silicon dioxide is removed from beneath the resist. This phenomena is called undercutting.

After the resist is removed, openings in the oxide are notably larger than the area exposed by the resist. Also, the sidewalls of the etched feature show a large angle. In terms of process integration this etching result is not acceptable for sub-micron device structures. Hence, the transition from wet etching to plasma or dry etching. Despite considerably-higher costs per step, plasma etching has largely replaced liquid-phase etching in pattern reproduction.

This evolution in etching epitomized the importance of molecular-scale phenomena in wafer-fabrication processes. The underlying physics and chemistry of liquid etching had limited usefulness at the small feature sizes, and thus it **had** to be replaced.

The transition to plasma etching was driven by the pragmatic needs of technology, not scientific elegance. New physical processes were required to achieve anisotropic etching. In *anisotropic etching*, etching in the horizontal direction is suppressed or eliminated. Plasma etching was developed to achieve this result. To this day, it remains largely an empirical endeavor.

3.6.1 Plasma State

A *plasma* is a partially-ionized gas. In addition to atoms, molecules, and free radicals, a plasma also contains positive ions, negative ions, and electrons. Sometimes called the fourth state of matter, plasmas are not uncommon. On the scale of the solar system, the high atmosphere of the earth and the sun are in the plasma state. A more prosaic example of a plasma is found in some gas flames. Extrapolating from known examples, plasma has been called the most common state of matter in the universe.

The usefulness of plasmas in microelectronics processing is their ability to generate a wide variety of reactive species and ions. There are a number of ways to generate a plasma for industrial use, and plasma sources are yet another specialty in wafer fabrication. Common types of plasma sources include microwave plasma, inductive discharge, and glow discharge. The glow discharge, common in neon lights, is discussed below.

An operating plasma discharge is a remarkably versatile chemical reactor. Introduce a chemical molecule of only moderate complexity, and a complex *chemical soup* rapidly results.

As an example, consider carbon tetrafluoride, CF_4, used in plasma processes to etch silicon dioxide. In a glow discharge plasma, this molecule rapidly breaks down to yield F, CF, CF_2,

CF_3, and charged versions of these radicals. Further, the free radicals react to form oligomers and polymers of fluorocarbon. Add a second or third compound, and the chemical soup becomes richer yet.

3.6.2 Glow Discharge

The glow discharge is itself an electronic apparatus through which current flows. It has a long, rich history in technology. In its simplest form, a reaction chamber contains two electrodes separated by a low-pressure gas. A voltage source is applied to one electrode. At a sufficiently high voltage, ionization occurs and electrons leave the atoms and molecules; currents of ions and electrons flow between the two electrodes. Electron-impact reactions create additional reactive molecular fragments, and the chemical soup is formed. Highly energetic species emit light, and this light gives the apparatus its name: glow discharge.

It is important to remember that, in microelectronic fabrication, a wafer covered with proto-microprocessors sits in this cauldron. Needless to say, plenty of things can go wrong (Table 3.3).

TABLE 3.2: PROCESS FAULTS IN PLASMA ETCHING	
• Non-uniform etching	• Radiation damage
• Over-etching	• Ionic impurity
• Under-etching	• Charge buildup
• Poor profiles	• Material residue
• Poor selectivity	• Loading effects

In plasma etching, the radio-frequency (rf) discharge is the dominant choice for processing. In this type of glow discharge, the region between the electrodes is not homogeneous. The discharge has considerable internal structure, and there are regions of space with electrical charge where ions dominate. In oxide etching, the wafer is typically mounted on the powered electrode where it is subject to ionic bombardment.

3.6.3 Etching Processes

It should be clear from the foregoing discussion that the details of silicon-dioxide etching in fluorocarbon glow discharges can fill many volumes.

From the viewpoint of wafer fabrication, we seek anisotropic etching and vertical sidewalls for the resulting features. This effect is provided by fluorocarbon polymers. These polymers form on the sidewalls of the etch feature, thus passivating these surfaces. Removing the polymer residues can be tricky, and is the subject of another discussion.

The desired etching process ensues from reactions of fluorine with silicon dioxide to form silicon tetrafluoride, SiF_4. This compound is volatile and is pumped away.

In the glow discharge, positive ions of CF_3 bombard the bottom of the etch feature. This bombardment by ions prevents polymer formation and provides a source of fluorine atoms for the etching reactions. This process is often called *reactive-ion etching* (RIE).

The requirements for a successful etching process are long and detailed, as are the potential hazards (Table 3.2). Nevertheless, etch processes are made to work.

Often, successful etching requires longer processing times. But longer processing times result in less machine throughput and less workstation capacity. If the workstation is a bottleneck or constraining work center, fab performance may be degraded.

Furthermore, the technical difficulty of the plasma-etch process may create maintenance problems which lead to lower machine availability. Lower availability also results in lower workstation capacity and throughput.

But it is the intrinsic complexity of plasma processing that may most impact overall factory performance. Such complexity provides abundant opportunities for blunders and mis-processing. Mis-processing can clearly result in yield crashes.

3.7 Other Processes

Because of their considerable impact on fab operations and performance, wafer cleaning, photoresist processing and photolithography, and plasma etching were singled out and discussed above. But these are by no means the only processes involved in wafer fabrication. Others are discussed briefly below. For reference, Table 3.3 provides equipment-related estimates on approximate process times, equipment costs, and availabilities.

TABLE 3.3: REPRESENTATIVE EQUIPMENT SPECS

Equipment Type	Process Time (hrs)	Machine Cost ($mm)	Availability Ratio
Photoresist	1.10	1.00	0.80
Photolithography	1.00	3.00-4.00	0.85
Plasma Etch	0.80	1.50-3.00	0.70
Ion Implant	0.75	3.00-5.00	0.85
Oxidation Furnace	2.00	0.75-2.00	0.90
Diffusion Furnace	2.00	0.75-2.00	0.90
CVD	0.80	1.50-2.00	0.80
Sputter	0.80	2.00	0.80
Wafer Clean	0.50	0.50-2.00	0.90
CMP	1.00	1.00	0.80
Resist Strip	0.50	0.75	0.85
Metrology	0.50	0.50	0.90
Checker	0.50	0.75	0.90

3.7.1 Sputtering

Sputtering is a glow-discharge process used primarily for deposition of metals in building interconnect structures. However, since it is a versatile deposition process, it is also used to build layers of dielectrics, such as silicon dioxide, and for etching and/or cleaning.

Sputtering is a high-vacuum process where the dominant physical mechanism is ionic bombardment of a substrate, such as aluminum. The bombardment of the substrate causes ejection of atoms from the target material. The ejection process leads to the name 'sputtering.'

The sputter-deposition process is completed by condensation of ejected material onto the wafer surface to form a thin film (of aluminum, for example). Process success is determined by the quality of the deposited metal layer.

Less energetic ionic bombardment of a surface provides the basis for etching and wafer-cleaning processes, where the wafer surface is the target of the ions. In this type of sputtering application, process success hinges on achieving the desired effect without damaging the existing structures.

3.7.2 Plasma Deposition

As discussed above, glow discharge provides the ability to generate a complex chemical soup. If different reagents are chosen, the result of the process on the wafer surface is not etching, but deposition.

Plasma deposition is used primarily to deposit dielectrics such as silicon dioxide and silicon nitride. The composition and properties of plasma-deposited materials can vary widely. Plasma deposition can also be used to deposit organic polymers. The success of the process is determined by uniformity, composition, and properties of the deposited film.

3.7.3 Chemical-Vapor Deposition (CVD)

In contrast to plasma deposition, chemical-vapor deposition is a classic chemical process involving thermal decomposition of a carrier molecule on a wafer surface. A multi-component gas mixture is used to deposit a solid film, following a sequence of reactions on the wafer surface.

CVD is a multipurpose deposition process that can be used to deposit both metals and dielectrics. The material deposited is governed by the systems of chemical reagents and operating temperature. In building interconnect structures, CVD has been used to deposit both tungsten and silicon dioxide.

In comparing CVD to plasma deposition, CVD typically provides a material with more controlled composition and film properties. Related plasma-deposited films can usually be obtained at a lower deposition temperature. Plasma-deposited films, conversely, provide a wider range of film properties that can be used in developing processes to build structures.

3.7.4 Ion Implantation

Ion implanters are small-scale particle accelerators. Within the equipment, the glow discharge is used to generate ions such as arsenic or boron. These ions are segregated and then accelerated toward the wafer surface for doping purposes.

On the wafer surface, a masking layer, typically photoresist, causes the doping to be performed selectively. Thus, ion implantation allows precise control of impurity introduction in forming transistors.

In terms of fab operations, a major impact of ion implanters is the relatively long setup time required to change from one dopant to another. Like most conventional gear, ion-implantation equipment accepts cassettes of wafers.

3.7.5 Diffusion

Diffusion is an older, less precise process for the introduction of dopants in transistor formation. A patterned layer of silicon dioxide on the wafer selects the regions of silicon to be exposed. The wafers are loaded into a high-temperature furnace, and a carrier gas like nitrogen is introduced along with a dopant compound like arsine. Arsenic is carried to the surface of the exposed silicon, and it then diffuses into the silicon crystal.

Diffusion is typically a cheaper process step than ion implantation. Operationally, however, the large load size of furnaces presents potential fab-performance problems.

3.7.6 Oxidation

Like doping, oxidation is a key process in transistor formation. In oxidation, a thin film of silicon dioxide is grown on the single-crystal silicon by direct oxidation. The process is carried out in a furnace at high temperature. Oxidation is a long, slow process.

3.7.7 Chemical-Mechanical Polishing (CMP)

Chemical-mechanical polishing represents a new use in high-technology of an old manufacturing process. Its purpose is to remove or smooth (planarize) layers of metals or dielectrics. The polishing is effected by a slurry of particles and chemical reagents.

3.8 Summary

The uniqueness of wafer fabs as a factory comes from the complex chemical processes and precision machinery necessary to make a modern microprocessor. A further uniqueness relates to how processing considerations often dominate operational dynamics.

Thus, this chapter has provided an overview of the variety of chemical processes involved in building microstructures on the scale of 0.5-micron. Since each process can encompass the career of one or more specialists, the treatment here has been necessarily brief. The notes below provide entries into the literature so that any of these topics may be addressed in more detail.

Unquestionably, wafer fabrication is a chemical-processing industry. Furthermore, it is now a **major** chemical-processing industry in terms of the number of fabs and the technical effort involved. In evaluating the relative scale, wafer fabs can now cost more than petroleum refineries.

Further Reading and Notes

1. Solid state physics and device physics are covered in a number of texts. The usual references are Ghandhi [3.2], Grove [3.3], and Sze [3.4]. The book by Glaser and Subak-Sharpe [3.5] carries device physics into circuit design. Tsividis [3.6] provides a good treatment of transistor physics. See also Nicollian and Brews [3.7].

2. The thin-film processing view of wafer fabrication is given in the handbook of Maissel and Glang [3.8], and in the edited volume of Vossen and Kern [3.9].

3. Detailed discussions of process steps are given in Elliot [3.10], Ruska [3.11], and the volume of Wolf and Tauber [3.12].

4. Process flows and details of MOS and bipolar processing are given in the second volume of Wolf's treatise [3.13]. Background information is provided by Ruska [3.12] and the elementary texts of Colclaser [3.14], and Till and Luxon [3.15].

5. The physical and organic chemistry involved in microelectronic fabrication is discussed in a wide variety of texts. Atkins [3.16] on physical chemistry and Roberts and Caserio [3.17] on organic chemistry are good places to start. Adamson [3.18] gives details on the physical chemistry of surfaces.

6. Chemical engineering rigor is applied to wafer fabrication processes in Middleman and Hochberg [3.19]. See also Grove [3.3]. The classic treatment for diffusion, fluid mechanics, and heat transfer in chemical processes is Bird, Stewart, and Lightfoot

[3.20]. Panchenkov and Lebedev [3.21] treat surface reactions and chemical kinetics. Carberry [3.22] extends these subjects to reactor design. See also Sharma [3.23].

7. Plasma etching, sputtering, and glow discharges in general are treated in Chapman [3.24]. Vossen and Kern [3.9] provide surveys of detailed application to microelectronics. For a view of some of the non-equilibrium processes in plasmas, see Vincenti and Kruger [3.25].

8. Process integration, in general, and for applications in interconnect is discussed in Wolf, Vol. 2 [3.13]. Multilevel interconnect technology is specifically the subject of Rao [3.1].

9. Journals covering topics related to wafer fabrication process development include the **Journal of the Electrochemical Society**, **AIChE Journal**, and the **Journal of Vacuum Science and Technology.**

10. Modeling and numerical simulation of fabrication processes is treated in Middleman and Hochberg [3.19] and in Sharma [3.23]. Aspects of device-level simulation are sometimes called technology CAD or TCAD; TCAD is reviewed in Wolf, Vol. 2 [3.13].

References

[3.1] Rao, G. K., **Multilevel Interconnect Technology**, McGraw-Hill, New York, 1993.

[3.2] Ghandhi, S. K., **The Theory and Practice of Microelectronics**, Wiley, New York, 1968.

[3.3] Grove, A. S., **Physics and Technology of Semiconductor Devices**, Wiley, New York, 1967.

[3.4] Sze, S. M., **Physics of Semiconductor Devices**, Wiley, New York, 1969.

[3.5] Glaser, A. B. and G. E. Subak-Sharpe, **Integrated Circuit Engineering: Design, Fabrication, and Applications**, Addison-Wesley Publishing Company, 1977.

[3.6] Tsividis, Y., **Operation and Modeling of the MOS Transistor**, McGraw-Hill, New York, 1987.

[3.7] Nicollian, E. H., and J.R. Brews, **MOS Physics and Technology**, Wiley, New York, 1982.

[3.8] Maissel, L.I., and R. Glang, (eds.), **Handbook of Thin Film Technology**, McGraw-Hill, New York, 1970.

[3.9] Vossen, J.L., and W. Kern, (eds.), **Thin Film Processes**, Academic Press, New York, 1978.

[3.10] Elliot, David J. **Integrated Circuit Fabrication Technology**, McGraw-Hill, New York, 1989.

[3.11] Ruska, W. Scott, **Microelectronic Processing: An Introduction to the Manufacture of Integrated Circuits**, McGraw-Hill, New York, 1987.

[3.12] Wolf, S., and R. N. Tauber, **Silicon Processing for the VLSI Era: Volume 1 Process Technology**, Lattice Press, Sunset Beach, California, 1986.

[3.13] Wolf, S., **Silicon Processing for the VLSI Era: Volume 2 Process Integration**, Lattice Press, Sunset Beach, California, 1990.

[3.14] Colclaser, Roy A., **Microelectronics Processing and Device Design**, Wiley, New York, 1980.

[3.15] Till, William C., and J.T. Luxon, **Integrated Circuits: Materials, Devices, and Fabrication**, Prentice Hall, Englewood Cliffs, N.J., 1982.

[3.16] Atkins, P.W., **Physical Chemistry**, Freeman, San Francisco, 1978.

[3.17] Roberts, John D., and M. C. Caserio, **Basic Principles of Organic Chemistry**, Benjamin, 1964.

[3.18] Adamson, Arthur, **Physical Chemistry of Surfaces**, (3rd ed.), Wiley, New York, 1976.

[3.19] Middleman, Stanley, and A. K. Hochberg, **Process Engineering Analysis in Semiconductor Device Fabrication**, McGraw-Hill, New York 1993.

[3.20] Bird, R.B., W. Stewart, and E.N. Lightfoot, **Transport Phenomena**, Wiley, New York, 1965.

[3.21] Panchenkov, G.M., and V.P. Lebedev, **Chemical Kinetics and Catalysis**, Mir Publishers, Moscow, 1976.

[3.22] Carberry, James J., **Chemical and Catalytic Reaction Engineering**, McGraw-Hill, New York, 1976.

[3.23] Sharma, Ajay, "Modeling for Manufacturing Diagnostics," in Landzberg [3.26], 1993.

[3.24] Chapman, Brian, **Glow Discharge Processes: Sputtering and Plasma Etching**, Wiley, New York, 1980.

[3.25] Vincenti, W.G., and C.H. Kruger, **Introduction to Physical Gas Dynamics**, Krieger, New York, (1965) 1977.

[3.26] Landzberg, A. H. (ed.), **Microelectronics Manufacturing Diagnostics Handbook**, Van Nostrand Reinhold, New York, 1993.

Exercises

1. Develop a detailed process flow for one level of interconnect metallization described in Section 3.2. Specify each process step and equipment type used.

2. Take the flows and fundamental sequences given in Tables 2.3-2.7 and assign each process step to a type of equipment.

3. Using the typical process times given in Table 3.3, determine the total minimum processing time for the flows and sequences in Tables 2.3-2.7.

4. Assuming the fab is a well run operation, determine the cycle time and cycle-time-to-process-time ratios for the flows and sequences given in Tables 2.3-2.7. (Hint: In well run operations, cycle times are generally a multiple of 2 to 4 times minimum processing time.)

5. Assuming the fab is a poorly run operation, determine the cycle time and cycle-time-to-process-time ratios for the flows and sequences given in Tables 2.3-2.7. (Hint: In poorly run operations, cycle times are generally a multiple of 8 to 10 times minimum processing time.)

6. Determine the number of machines needed for the flows and sequences in Tables 2.3-2.7. To do this, consider the process times for each step, and assume a normalized load size going through each machine. To normalize the flow or sequence, assume the shortest process step requires exactly one machine. (Hint: You can not have a portion of a machine.)

7. Recalculate the number of machines needed in Exercise #6 using the equipment availability numbers in Table 3.3. (In effect, Exercise #6 assumed 100% availability.)

8. Using the cost figures given in Table 3.3 and the sizing from Exercise #6, calculate the capital expenditures for the flows and sequences in Tables 2.3-2.7.

One foot up, the other foot down;
That's the way it is to London town.
- Anon. Ballad

Chapter 4
Fab Graphs and Static Analyses

4.0 Introduction

In designing, analyzing, and/or operating a wafer fab, where does art end and technology begin? It is a difficult question because wafer fabrication is without a doubt one of the most complex and highly-advanced forms of manufacturing in existence today. Not only will it remain so well into the next century, but it is likely to become even more formidable with each new generation of development.

When semiconductor chips were first manufactured, along about 1960, the number of process steps used in their fabrication was measured in multiples of ten. In 1995, the process sequence for a state-of-the-art CMOS microprocessor is 500+ steps, and growing. Over this same brief industrial time span, wafer size has increased from 1 to a whopping 8-inches, with 12-inches on the horizon; the number of metal layers has gone from one to five; and hub orders have gone from two to twenty-something.

Surprisingly, the concept of *hub order* is relatively unknown, even among experienced semiconductor engineers. Yet understanding the number of times a process sequence returns to the various work centers is one of the key measures of fab complexity; and therein lies the crux of this chapter. In order to advance wafer fabrication from art to technology, we need analytical tools that enable us to understand and analyze fab performance, regardless of the inherent complexities and rapid advancements of semiconductor manufacturing.

Throughout this chapter, and the next two, we will develop such a set of tools. We will define and discuss the various measures that determine performance, the operational practices that contribute to both good and bad performance, and the trade-offs involved in making informed decisions on how to achieve good results.

The first of the analytical tools to be explored, and perhaps the simplest, is the *fab graph*, a mapping of the process flow (or flows) onto the factory floor.

4.1 Factory Representation Using Fab Graphs

The fab graph provides a visual, first-level understanding of process and resource interactions. In wafer fabrication, as in any manufacturing operation, there are three fundamental entities involved: the material being processed, the sequence of processing steps leading to a finished product, and the equipment or machines performing those steps. To illustrate the complex interactions of these fundamental entities, consider the simplified fab shown in Fig. 4.1.

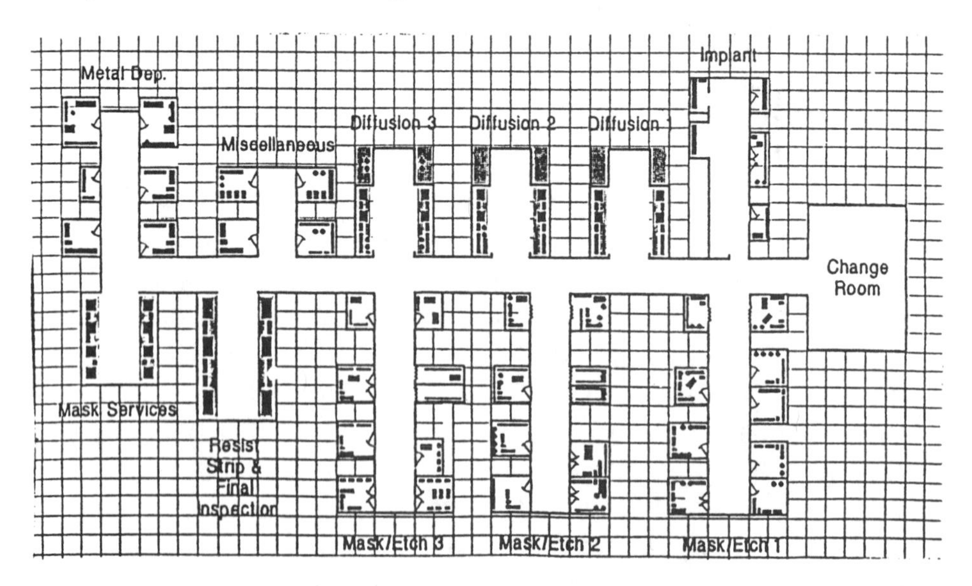

Figure 4.1: Simplified CMOS Fab

In CMOS-1, bare-silicon wafers enter the factory in lots, and then follow specific process flows, moving from work area to work area. In each work area, the lots will doubtless spend time in a queue waiting for processing. Eventually, a lot will make its way to the front of the queue, and then be loaded into a specific machine. The loading will proceed according to the loading, batching, and dispatching rules of the equipment being used to perform the step. When the step is completed, the lot will be unloaded from the machine and moved on to another work area to await the next step in the processing sequence.

TABLE 4.1: SIMPLIFIED CMOS FACTORY

STEP #	PROCESS STEP	TIME	WORKSTATION	WORKSTATION #
1	LASER MARK	0.8	MISC.	1
2	INITIAL OX	2.0	DIFF 1	2
3	P-TUB MASK	1.0	MASK/ETCH 1	3
4	P-TUB IM	0.5	IMPLANT	4
5	NITRIDE DEP	2.0	DIFF 2	5
6	FIELD MASK	1.0	MASK/ETCH 2	6
7	FIELD DIFF/OX	2.0	DIFF 2	5
8	SOURCE/DRAIN MSK 1	1.0	MASK/ETCH 2	6
9	S/D IM 1	0.5	IMPLANT	4
10	S/D OX 1	2.0	DIFF 1	2
11	S/D MASK 2	1.0	MASK/ETCH 2	6
12	S/D IM 2	0.5	IMPLANT	4
13	S/D OX 2	2.0	DIFF 1	2
14	NITRIDE STRIP	1.0	MASK/ETCH 2	6
15	GATE OX	2.0	DIFF 1	2
16	POLY DEP/DOPE 1	2.0	DIFF 3	7
17	POLY MASK 1	1.0	MASK/ETCH 1	3
18	POLY DEP/DOPE 2	2.0	DIFF 3	7
19	POLY MASK 2	1.0	MASK/ETCH 1	3
20	CONTACT MASK	1.0	MASK/ETCH 1	3
21	METAL DEP	1.1	METAL DEP	8
22	METAL MASK	1.0	MASK/ETCH 3	9
23	TOPSIDE DEP	2.0	DIFF 3	7
24	TOPSIDE MASK	1.0	MASK/ETCH 1	3

In the hypothetical processing sequence for CMOS-1, given in Table 4.1, each lot of wafers will first undergo a laser masking at Workstation #1 (Miscellaneous) and then will be moved on to Workstation #2 (Diffusion 1) for the second step of processing, an initial oxidation. Upon completion of the second step, the lot will be moved to Workstation #3 (Mask/Etch 1) for the third process step, P-tub masking. Each lot will continue through the process sequence until all twenty-four steps have been performed. Each wafer in the lot is then a completed product, and is ready to be shipped to the next electronics factory for sort, followed by assembly, then final test.

To visualize what is physically occurring on the factory floor, we could draw the processing sequence for CMOS-1 (i.e., play 'connect the dots'). The result of this exercise, shown in Fig. 4.2, is a mapping of the process flow onto the factory floor. Such a mapping could loosely be called a *fab graph.*

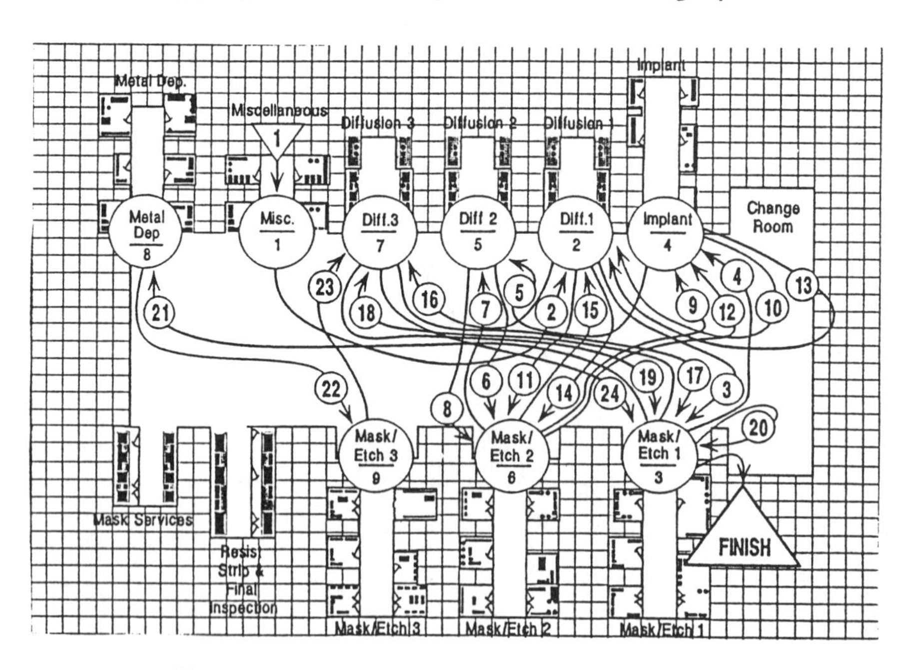

Figure 4.2: Mapping of CMOS-1 onto Factory Floor

Why use a Fab Graph?

About now, one should be asking, "Of what earthly good is this spaghetti representation of a wafer fab?"

Considering how little effort went into drawing it, the answer is, inestimable good. This crude graphical picture provides us with a first-level visual model for fab analysis.

Immediately, we can see which work areas are busiest (Mask/Etch 1, Mask/Etch 2, etc.), how these work areas interact with other work areas (Mask/Etch 1 with Diffusion 3, Mask/Etch 2 with Diffusion 2, etc.), which aisles may need to be wider to accommodate heavy AGV (automated guided vehicles) or HGV (human-guided vehicle) traffic, which work areas may need to be split into two or more work area for better fab performance, how capital expenditures may be saved, and so on.

Concisely, the fab graph has mapped the process flow onto the factory's resources, and in so doing has:

- provided a traffic analysis of the fab,
- supplied a means to study design options,
- produced a template for visualizing the impact of new process flows and/or engineering lots,
- furnished a basis for use of empirical cycle-time data,
- provided a tool to study the impact of bottlenecks, and
- identified the fab's hub structure.

What is hub structure?

We have used the terms hub order and hub structure, but have yet to define them precisely, much less their underlying fundamental concept, the hub.

According to Mr. Webster, a hub is a center of activity, a focal point. In a wafer fab, this definition is still surprisingly accurate. A *hub* is a workstation or work area where more than one step of a process sequence is performed. The *hub order* of a workstation is the number of repeat visits a flow makes there, i.e., the number of steps being performed there. Two processing steps makes the workstation a second-order hub; three steps a third-order hub; etc.

It is one of the unique features of wafer fabrication that there are many hubs in the fab, some of surprisingly high order. In fact, it is the existence and quantity of these high-order hubs that make these factories so inherently complex and difficult to analyze. Masking, for example. How many times do wafers return to masking in a 1995 state-of-the-art fab? Five? Try twenty... or more.

Given the fact that multiple returns do occur in wafer fabrication, why do they cause so much complexity? To explore this question, we will use the analytical tool known as the fab graph.

4.2 Drawing Fab Graphs

The 'spaghetti' fab graph shown in Fig. 4.2 was drawn without forethought and without regard to the types of information one might garner from such an effort. But considering that it was drawn blind, with no rules nor knowledge of how to tailor the graph for different scenarios, it was not a bad first effort. Just consider the power of the tool, though, when it is used with forethought and skill. Hence, the next few sections of this chapter will be spent on developing and honing fab-graphing expertise.

Although CMOS-1 is radically simplified compared to a real wafer fab, the twenty-four step flow is still rather complicated for the uninitiated, first-time grapher. Therefore, to begin the learning and exploration of fab graphs, we will use some very simple examples, basically portions of a process flow.

Example 1: Photolithography

Table 4.2 shows the process sequence for a six-step photolithography flow. The information contained in this table provides what is known as a *traffic analysis* of the flow. It assigns each process step to a workstation containing equipment capable of performing that step. Thus, in this example, six process steps are assigned to five workstations.

TABLE 4.2: PHOTOLITHOGRAPHY FLOW

Process Step		Workstation	
1	Coat	1	Coat
2	Expose	2	Align
3	Develop	3	Develop
4	Inspect	4	Scope
5	Resist Hard	5	Bake
6	Inspect	4	Scope *

Always, the first step in drawing a fab graph is to review the traffic analysis for hub structure. In other words, we need to: (1) determine which workstations are used more than once during the process sequence, (2) determine the number of times each hub is used, and (3) rank-order the hubs, from the highest usage to the lowest.

In this example, only one workstation is used more than once; Scope, Workstation #4 (also designated Wkst 4), is visited twice and is, therefore, a second order hub.

The second step in drawing a fab graph is to identify the workstations that interact with the higher-order hubs. Scope interacts with both Develop and Bake, Workstations #3 and #5, respectively.

Now, having reviewed the traffic analysis, we are ready to put pencil to paper and draw. As a standard rule for graphing, we start with the busiest hub or hubs.

Using circles to represent workstations, place the major hubs near the center of the page, then locate interacting work centers nearby. Using straight, parallel arrows, 'hook-up' the interacting workstations according to the process sequence in Table 4.2. The circles/nodes should be numbered and labeled according to the Workstation list, and the arrows should carry the number of the appropriate process step in the flow's sequence.

Thus, the hub structure for the fragment of flow is shown in Fig. 4.3(a). To be a complete fab graph, however, the remaining workstations and process steps must be added.

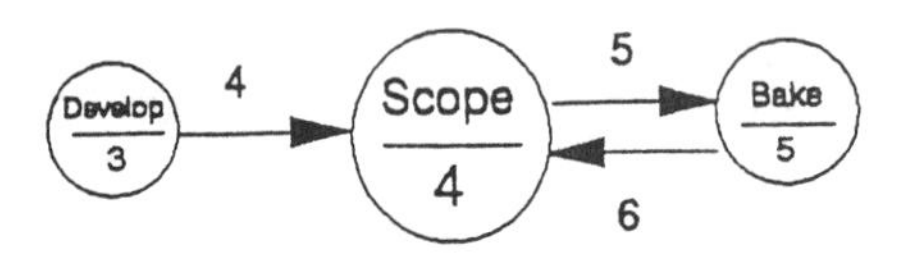

Figure 4.3(a): Hub-Structure Fragment of Photo Flow

Non-hubs are simply located in a logical, linear sequence, heading towards the hub (or hubs) in their respective portion of the process flow. For this example, Align feeds into Develop, and Coat feeds into Align. Thus, they are drawn in the correct order heading towards Develop. Non-hub workstations are linked up according to the total process sequence. The resulting graphical representation of Example 1 is shown in Fig. 4.3(b). The fab graph has five nodes (workstations) and six arrows (process steps).

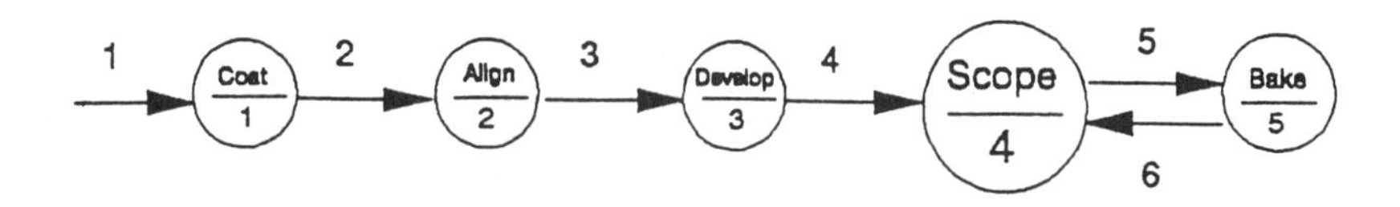

Figure 4.3(b): Fab Graph for Example 1

Strictly speaking, the use of straight arrows to represent process steps and sized circles to represent workstations is an arbitrary choice. There is no exact right or wrong way to draw a fab graph__other than that the result should provide a clear and illuminating picture of how the process fits onto the equipment. Having said as much, then why is Fig. 4.3(b) preferable and more insightful than other representations?

Its first advantage is the use of sized circles to represent hub order. Because Scope is a 2nd-order hub, it will see twice as much traffic as the other four workstations; i.e., it will be visited twice for every one visit to the other work centers. By drawing a larger circle for Workstation #4, it is easier to tell at a glance that it is a hub, and that it is being used more than the other workstations. Although more usage may not necessarily translate into more equipment, it will nonetheless mean more traffic and possibly more congestion.

The next advantage is the use of straight, parallel, individually-numbered arrows to provide an unambiguous and clear representation of how the lots move through the factory. Unnumbered arrows, curved lines, multiply-labeled lines, etc. are imprecise and/or may clutter the fab graph. While using other representations is no mortal sin, imagine the surrealistic effect that could result when drawing a 200+ step flow. The simplified CMOS flow in Fig. 4.2 derived it spaghetti-like characteristics from the use of curved lines.

Example 2: Etch & Clean

Now that we are ready for a more complex fab graph, Table 4.3 provides the traffic analysis for a six-step Etch and Clean operation. The six steps are performed at four different workstations.

TABLE 4.3: ETCH & CLEAN

Process Step		Workstation	
1	Inspect	1	Scope
2	Etch	2	Etch
3	Inspect	1	Scope
4	Ash Resist	3	Asher
5	Clean	4	Ash Resist
6	Inspect	1	Scope

Review of the traffic analysis indicates that Workstation #1, Scope, is used three times; hence, it is a 3rd-order hub. Further review indicates that, while Scope interacts with both Wksts 3 and 4, most of its interactions are with Wkst 2, Etch. Furthermore, Wkst 2 only interacts with Wkst 1.

Having made these observations, several plausible fab-graph representations are shown in Fig. 4.4; these graphs are interesting for various reasons.

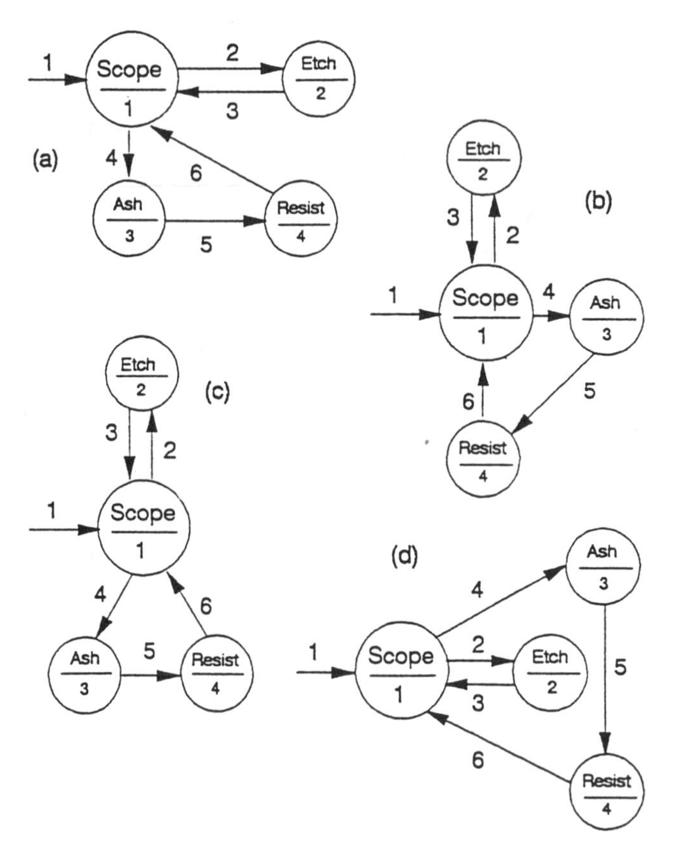

Figure 4.4: Fab Graphs for Example 2

Fig. 4.4(a) is typical of a non-complicated flow. The drawing starts out as a tunnel-type of graph from Workstation #1 to Workstation #2, and only after the return to Wkst 1 do we veer south for the subsequent loop to Wksts 3 and 4. Such an approach to drawing a fab graph might well be termed 'one foot in front of the other'; i.e., branching from the straight vector occurs only when it becomes necessary, and, when it does, is given little or no deliberation. For all but the most simplistic flows, this approach will generally lead to excessive crossovers, and may necessitate iterative attempts before an acceptable graph is finally drawn.

By contrast, a cursory analysis of the process steps and their workstation assignments might lead to Fig. 4.4(b). In this graph, the three interacting workstations (and the incoming flow) are equally spaced around the busy hub. Although this graph is only a slight variation of Fig. 4.4(a), it does nonetheless provide the maximum distance between the traffic paths. Such a graph might have some physical relevance to an area where AGVs or HGVs are being used.

Fig. 4.4(c) is an interesting graph because it basically segregates the two spokes of the flow onto opposite sides of the fab. In fab-graph theory, a *spoke* is a process sequence or sub-sequence that leaves and returns to a hub. Thus, in this example, process steps #2 and #3 make up one spoke, and process steps #4, #5, and #6 make up a second spoke. In the world of manufacturing, wafer fabs are unique because of their hub-and-spoke structure. Understanding the implications and ramifications associated hub-and-spoke manufacturing can radically change the way fab designers and managers tackle sizing, layout, and day-to-day operations.

For example, Fig. 4.4(c) might suggest that Etch occupy one physical area in the fab, and Ash and Ash Resist another. It also indicates that an Inspect station should be nearby and central to both areas.

Clearly, identifying the spokes leaving and returning to a hub indicates which process steps may be 'grouped' together, and hence the physical location of the equipment to perform those steps. Identifying these groupings also provides a first-pass identification of possible candidates for clustering (see Chapter 8). Finally, delineating the spokes, and where they occur in the process flow, may detect when a hub is being overtaxed, or sub-optimally used. Such insight may lead to splitting a hub into two or more work centers. (This concept will be discussed in more detail later.)

The final fab graph for this example, Fig. 4.4(d), exhibits what is known as an imbedded spoke. An *imbedded spoke* is one which is immured within one or more other spokes. In this example, the process sub-sequence going to the etcher (Steps #2 and #3) is surrounded by the sub-sequence performing the Resist/Clean (Steps #4, #5, and #6). Here, the enclosed spoke has no physical relevance to actual fab layout. But in many instances, especially as fab graphs become more realistic, imbedded spokes may be used to represent such steps as dedicated inspects and/or cleans.

Consider the flow in Fig. 4.5 for examples of imbedded spokes with potential physical meaning. The inspection steps after Mask and Etch would be dedicated operations, whereas the outer spoke (Dep/Mask/Etch) could be physically located all over the fab.

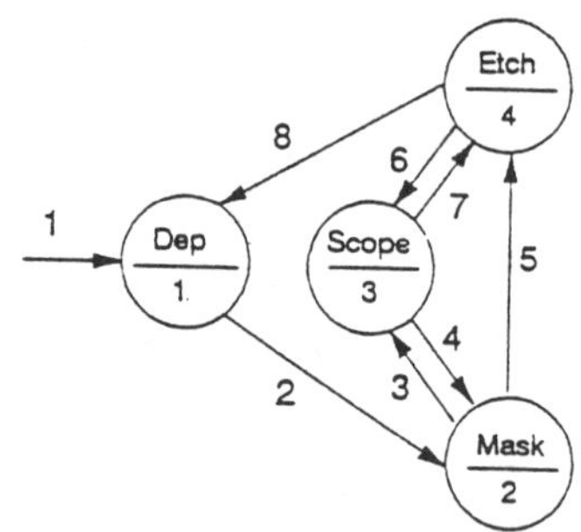

Figure 4.5: Imbedded Spokes with Physical Meaning

This digression on imbedded spokes brings up an interesting point: Should the fab graph represent fab layout?

4.3 Fab Layout

The usefulness of the fab graph as a design or layout tool really depends on the complexity of the manufacturing operation under consideration. Many, if not most, wafer fabs in the U.S., Pacific rim, and Europe still use layout methodologies developed in the early 1980's, despite the dramatic changes in wafer size, product types, flow complexity, equipment advances, etc. (See Fig. 4.6 for a picture of the industry's evolving complexity.)

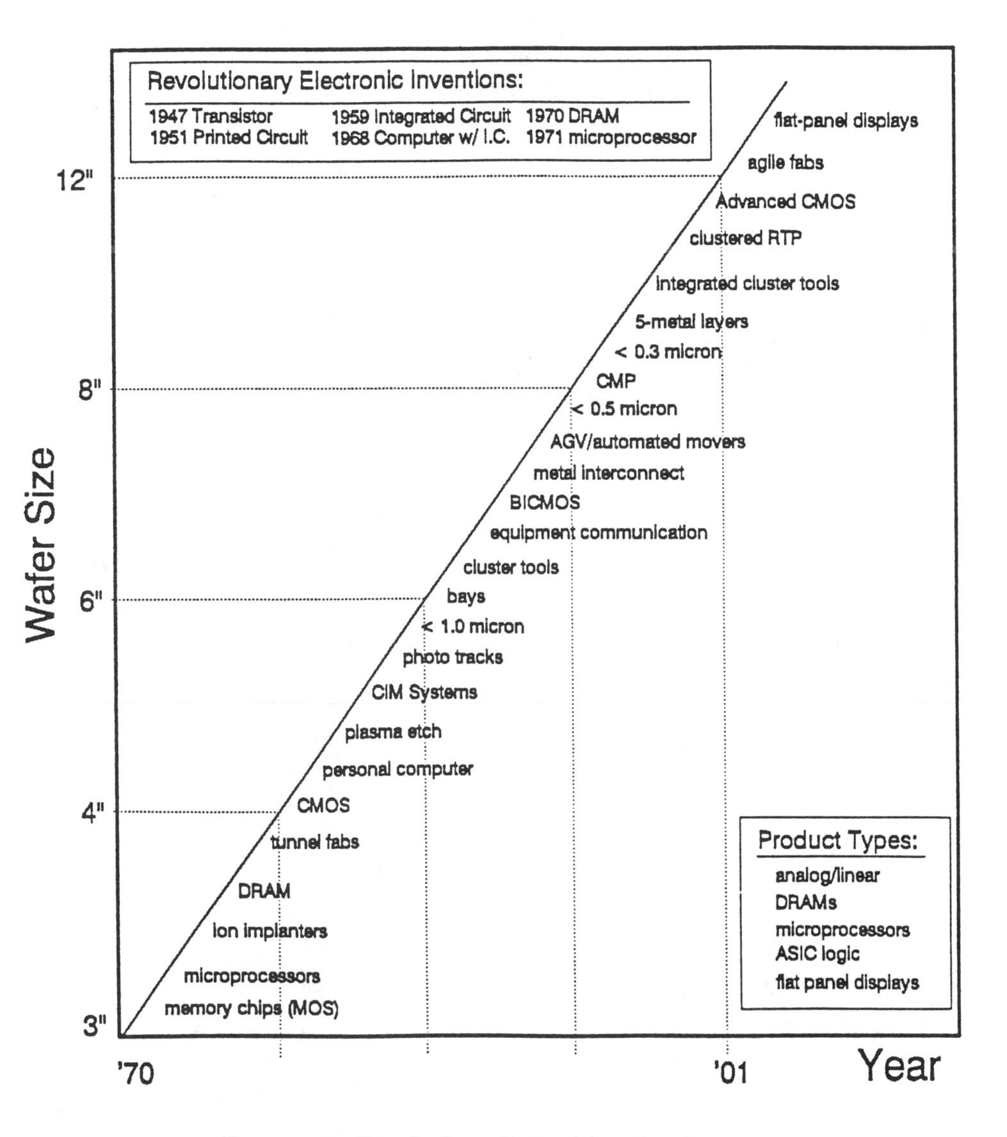

Figure 4.6: The Industry's Evolving Complexity

The persistent methodology for fab layout generally entails dividing a cavernous building, or *ballroom*, into five or six basic function areas (implantation, thin films, furnaces, etc.), and then putting all similar equipment types there under a fab-area manager. When floor space becomes exhausted in one or more of the function areas, equipment is distributed serendipitously around the fab to wherever space is available.

By the mid 1980's, some semiconductor manufacturers began recognizing certain processing and operational issues that were an inherent result of the ballroom-type layout. In an effort to alleviate problems such as lengthening cycle times, lost lots, lot misassignment, etc., the concept of the manufacturing bay was devised. Loosely described, a *manufacturing bay* is a physical subdivision of the factory floor, containing equipment for a sub-sequence of the manufacturing process. A bay **does not** contain equipment for different, unrelated process sub-sequence(s).

With the advent of the contained, dedicated bay, also came the concept of the fully- or partially-automated fab. Layout began looking more like a fab graph. The simplified CMOS fab in Fig. 4.1, for instance, is laid out using bays.

As wafer fabrication moves further up the complexity ladder in Fig. 4.6, past boutique ASIC fabs, single-wafer processing, extensive clustering, and small lot sizes, fab layout will closely approximate or become identical to a good fab graph.

But will the fab still be as flexible?

4.4 Fab Flexibility

The key argument put forth for avoiding layouts that too closely resemble a fab graph is the loss of flexibility over the life cycle of the plant. As processes change and evolve, the fab must be able to add equipment, change equipment usage, add gas and vent lines, and so on. When a fab is tightly constrained along the lines of a fab graph, flexibility will be sacrificed. But the sacrifice in flexibility will undoubtedly buy improved cycle time. And the trade-offs do not end there. (See Chapter 5.)

It should also be pointed out that concerns about flexibility become increasingly less important at the upper end the fab-complexity curve. Highly-clustered fabs offer the promise of ultimate flexibility. Improvements or changes in process flows should no longer cause major reverberations when no physical floor space is available in the plant. Instead, the changes may simply entail adding, replacing, and/or modifying process modules on radial cluster tools.

Not only will the highly-clustered plants of the future be extremely flexible, they will also be considerably smaller than today's facilities. A tool with four or five interchangeable process modules will replace four or five stand-alone machines, and thus will need only a fourth or a fifth of the previously-used space. Further, the associated storage, transport, and operator overhead for those machines will also be eliminated. Suddenly, the question of flexibility changes so radically, one wonders why there is not a determined push to get to the top of the complexity curve. Suffice it to say that in the ultimate case, the layout of the fab will be portrayed by a good fab graph.

But what exactly distinguishes a good fab graph from a bad fab graph? So far, the examples have been too simplistic to see a wide variation in the graphs.

Example 3: Etch, Clean and Metals

Taking the level of difficultly of the examples up a tier, consider the etch, clean and metals flow described in Table 4.4. Thirteen process steps are performed at nine workstations. Analysis of the traffic flow indicates the following hub structure:

Hub	Order
2 Scope	3
5 Spin/Clean	3

TABLE 4.4: ETCH, CLEAN and METALS

Process Step		Workstation	
1	Etch	1	Etch
2	Inspect	2	Scope
3	Ash Resist	3	Asher
4	Clean	4	Resist
5	Inspect	2	Scope
6	Clean	5	Spin
7	Plasma Dep	6	Deposition
8	Clean	5	Spin
9	Inspect	2	Scope
10	Soft Clean	5	Clean
11	PVD	7	Sputter
12	CVD	8	CVD
13	Etchback	9	Plasma Etch

The work areas interacting with Workstation #2 are Etch (#1), Ash (#3), Resist (#4), and Clean (#5). But, note that Clean, Wkst 5, has a total of three interactions with Wkst 2, and is itself a third-order hub. The other work-area interactions for Wkst 5 are Deposition (#6) and Sputter (#7).

From this preliminary analysis, it is obvious that Workstations #2 and #5 should be represented as large nodes on the fab graph, and should be next to each other in the center of the page and appropriately connected (Fig. 4.7(a)).

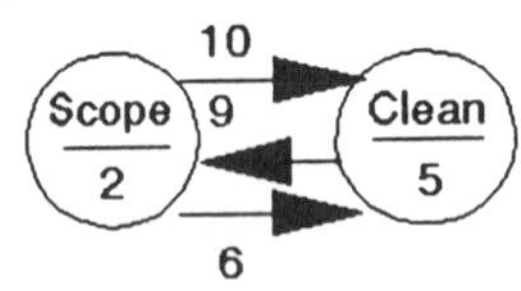

Figure 4.7(a): Interacting Hubs of Example 3

Interacting work areas and spokes should then be distributed around the hubs, which accounts for most of the other workstations. When the sole two outlying work areas (Wksts 8 and 9) are finally added, the fab graph is complete and given in Fig. 4.7(b).

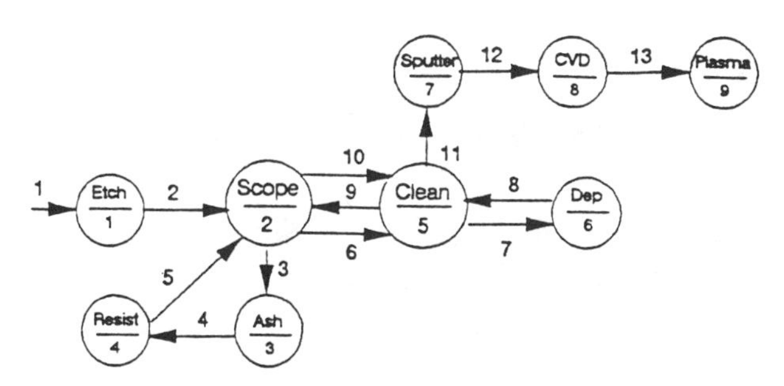

Figure 4.7(b): Fab Graph for Example 3

Are there other, better fab graphs for this particular flow? Analysis of the spoke structure of the hubs will generally provide the answer.

When drawing fab graphs, it is important to remember that there are many possible options. What distinguishes a good graph is whether it accentuates or highlights certain insights. But since any insight is better than no insight, a less-than-optimal fab graph is usually still a good fab graph. Furthermore, every fab graph drawn will make the next one even better.

4.5 Rules of Thumb for Drawing Fab Graphs

Now that we have worked through three simple examples, we may postulate some rules of thumb for drawing fab graphs, Table 4.5. To use these rules on something that approaches a realistic flow, it is time to draw the fab graph for the simplified CMOS factory that was used to begin this chapter.

Review of the traffic analysis in Table 4.1 provides the following hub structure:

Hub	Order	Nearby Wksts
3 Mask/Etch 1	5	7,8,2,4
2 Diffusion 1	4	6,4,3,7,1
6 Mask/Etch 2	4	2,5,4
4 Implant	3	2,6,3,5
7 Diffusion 3	3	3,2,9
5 Diffusion 2	2	6,4

TABLE 4.5: RULES OF THUMB FOR DRAWING FAB GRAPHS
• Review traffic analysis for hubs with the largest number of returns. – Determine number of returns. – Initially assess nearby workstations (i.e., next or previous process step). • Identify hubs with major interactions. • Place major hubs in center of page. Use large nodes (circles). • Place interacting hubs nearby. • Size circles according to hub order. (Singly-used workstationss will have smallest circles, etc.) • Use straight, parallel lines to indicate the process flow. • Minimize cross-overs.

Six of CMOS-1's nine workstations are hubs, some of surprisingly high order considering the simplified nature of the plant. Major hub interactions are as follows:

# of Hub-Hub Interact.	Hubs
4	3,7
3	2,6
3	6,5
2	2,4
2	6,4

Following the rules of thumb for drawing fab graphs, we place the highest order hub, Mask/Etch 1, Wkst 3 in the center of the page. Its major interacting hub, Diffusion 3, Wkst 7, with four interactions, is placed nearby and is sized accordingly. Dedicated spokes and/or stand-alone workstations connected to these two hubs are also drawn. The intermediate result in shown in Fig. 4.8.

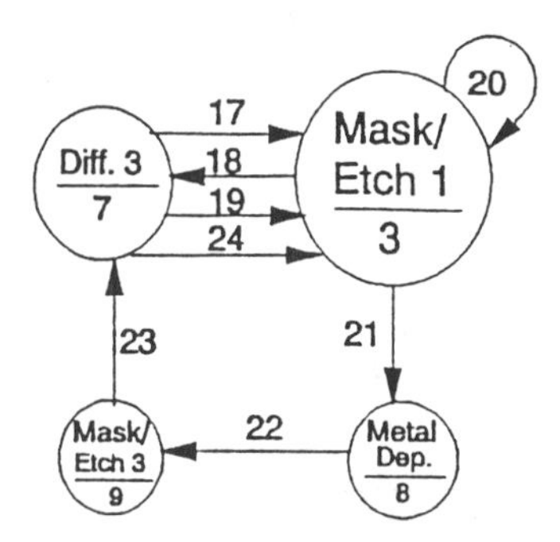

Figure 4.8: Hubs with Major Interactions for CMOS-1

The next highest order hub, Diffusion 1, Wkst 2, has three interactions with Wkst 6, and two interactions with Wkst 4. But note that Wkst 4 and Wkst 6 also have two interactions. To further complicate matters, Wkst 2 and Wkst 4 have dealings with Wkst 3 (which is already tightly adjoined), and Wkst 6 interacts heavily with Wkst 5.

At this point, there are two distinctly different ways to proceed with the fab graph. One option is to add structure around the perimeter of the tightly-drawn inner core of Fig. 4.8, effectively imbedding it. The resulting fab graph is shown in Fig. 4.9. The other option is to build a second tight core of the remaining work areas and segregate them to a different part of the fab graph, connecting the two cores as needed, as shown in Fig. 4.10.

These two fab graphs, Fig. 4.9 and 4.10, are both correct, and very insightful in their own ways. They are also disturbingly different. From an aesthetics point of view, Fig. 4.9 is a tight and compact representation of the mapping of the process onto the equipment. The compactness of the graph actually facilitates study of the fab's spoke structure. For example, Fig. 4.11 shows the spokes of the highest-order hub, Wkst 3, Mask/Etch 1.

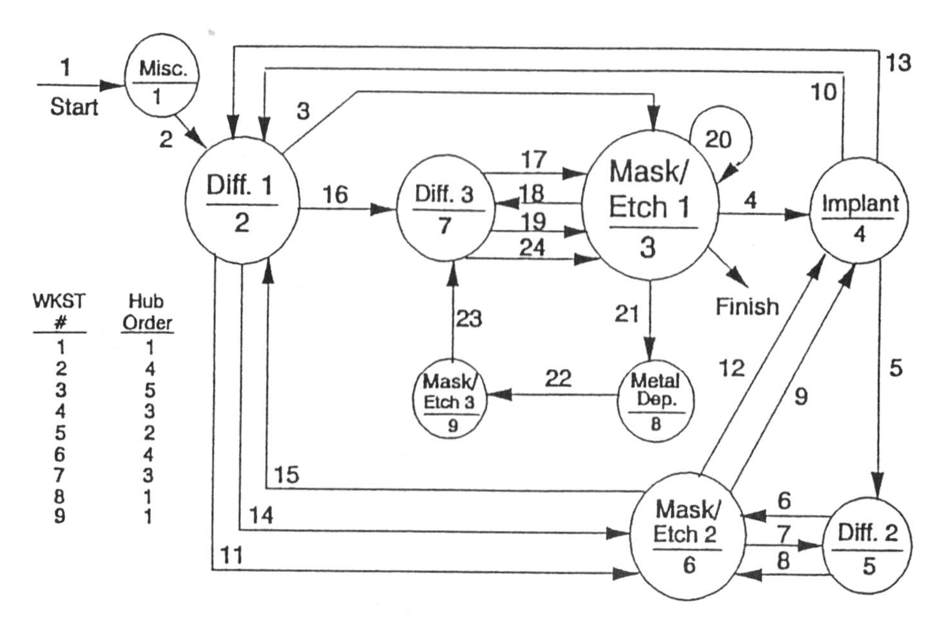

WKST #	Hub Order
1	1
2	4
3	5
4	3
5	2
6	4
7	3
8	1
9	1

Figure 4.9: Fab Graph for CMOS-1

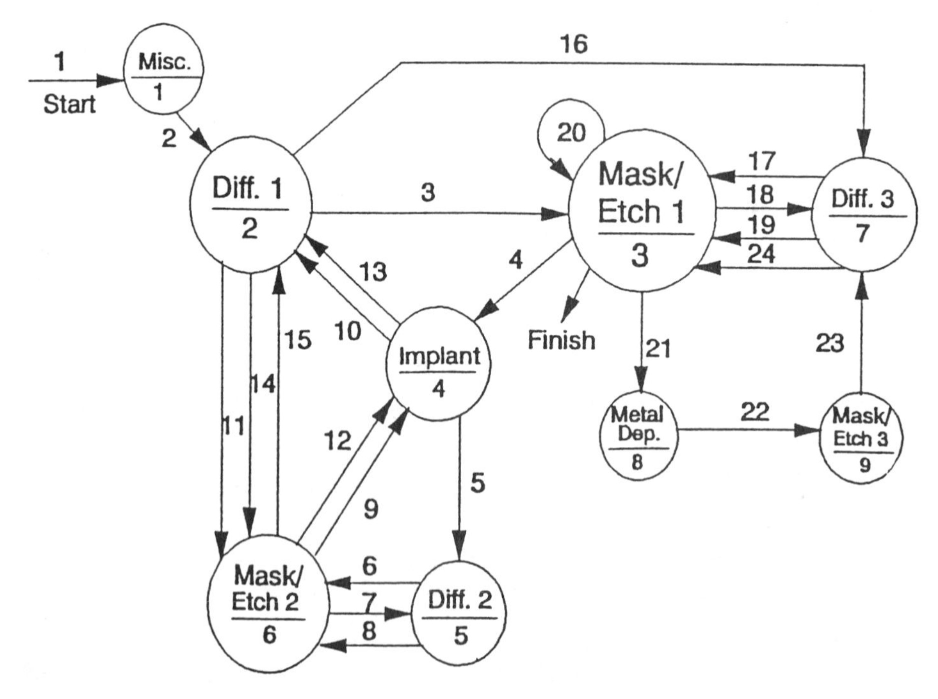

Figure 4.10: Alternate Fab Graph for CMOS-1

Spoke Analysis and Hub Splitting

Not too surprisingly, the number of spokes on any given hub is one less than the hub order. Thus, Wkst 3 has four spokes, Wkst 2 has three, etc. Study of the spoke structure, especially for the higher-order hubs, provides some keen insights which may lead to better fab layout, as well as improved operations from both a manufacturing and a process perspective.

Interestingly, the spoke structure in Fig. 4.11 for Workstation #3 involves every workstation in the fab, except for Wkst 1. The spokes include process steps 4-17, steps 18-19, step 20, and steps 21-24; in other words, all steps except #1 and #2. The most arresting observation gained from this analysis is that Workstation #3 is a critical hub throughout the entire processing sequence. In a real-world fab, this is probably not a wise design.

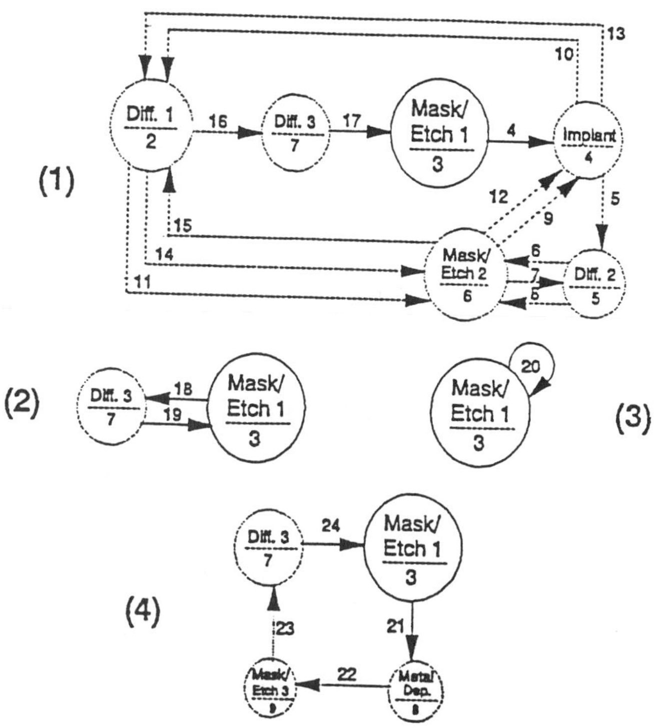

Figure 4.11: Spoke Structure for Workstation #3

Current thinking in wafer fabs is that heavily-used operations, such as masking, should be split into several work centers to deal with different portions of the process sequence (first third, second third, etc.). By splitting a busy hub that is used across the line, it will make for better wafer processing in general, and it will usually insure that the workstation does not become a bottleneck. Furthermore, the splitting makes it easier to concentrate more frequent maintenance and calibration efforts on downstream equipment where the cost of scrapping wafers is horrendously expensive.

Therefore, based upon the spoke analysis for Wkst 3, and the current thoughts on hub splitting, we should consider dividing Mask/Etch 1 into two (or more) work centers. The questions then become: *Where?* and *How?*

Fig. 4.12, the hub-and-spoke analysis for the second highest-order hub, Diffusion 1, Wkst 2, provides the answers.

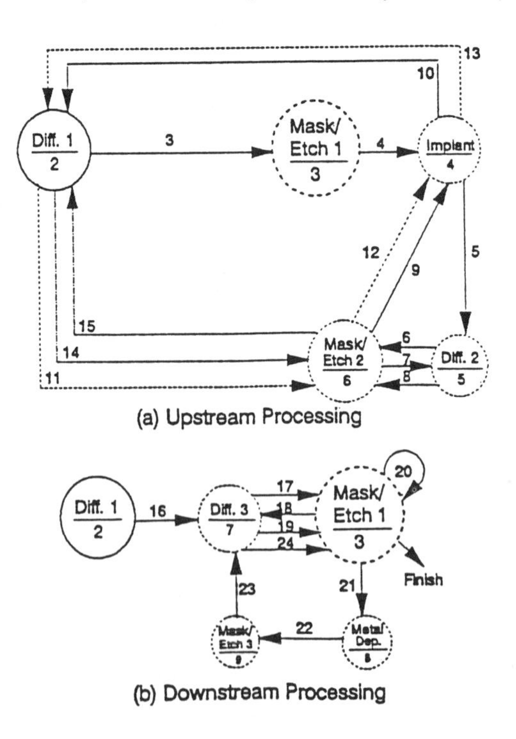

Figure 4.12: Hub and Spoke Analysis for Workstation #2

Wkst 2 is a fourth-order hub with three spokes. Spoke 1 consists of steps 3-10; spoke 2 has steps 11-13; and spoke 3 has steps 14-15. When step 16 leaves Workstation #2 (never to return), it begins the final third of the processing sequence. In the first two-thirds and the final third, there is only one point of overlap: Workstation #3.

Wkst 3 is used only once during the early portion of the processing sequence and four times in the latter third of the flow. If the quantity of machines currently residing at the work center were to be split, using one-fifth for early processing and four-fifths for later processing, the fab could be laid out along the lines shown in Fig. 4.13, with a new Workstation 10 (Mask/Etch 1A).

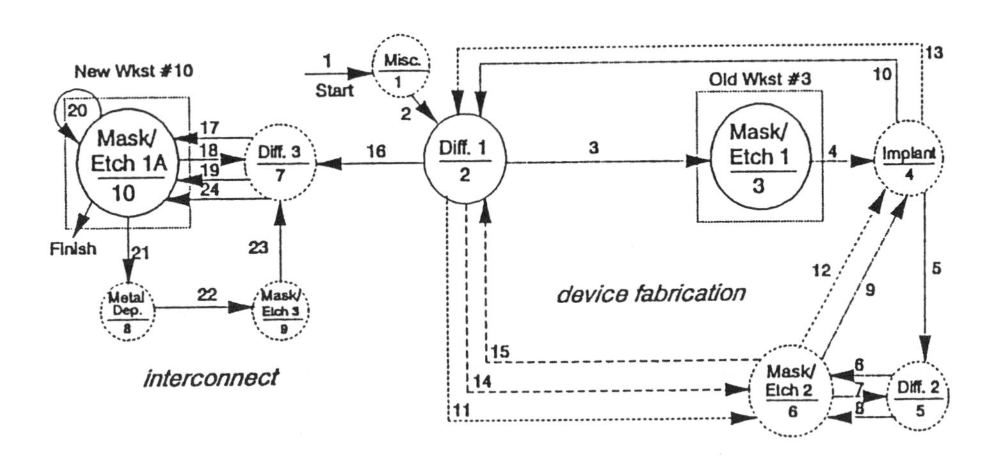

Figure 4.13: Hub Splitting

The beauty of this fab graph is its literalness in representing wafer fabrication. Process steps 1-15 (right side of graph) correspond to device fabrication; steps 16-24 (left side) correspond to the interconnect operations (i.e., wiring up the devices).

One-Step Graphing versus Other Approaches

The fab graph shown in Fig. 4.13 was arrived at through a multi-step approach: drawing a compact, highly-imbedded fab graph (Fig. 4.9), pulling it apart and analyzing it piecemeal (Figs. 4.11 and 4.12), then putting it back together in a real-fab context. But what about the other, one-step approach that gave us the fab graph shown in Fig. 4.10? Are the results as useful?

The answer is: *It depends.*

From the outset, Fig. 4.10 has two obviously segregated, busy portions of the fab. But progressing from this observation to knowing a hub should be split, and which hub, may not be as clear. The primary problem with the fab graph in Fig. 4.10 is that Mask/Etch 1, Wkst 3, appears integral to the total operation; the inner-loop/outer-loop configuration in Fig. 4.9 is not apparent. Without this inner-loop/outer-loop alert, we might not explore beyond the simple spoke analysis for Wkst 2, Fig. 4.14.

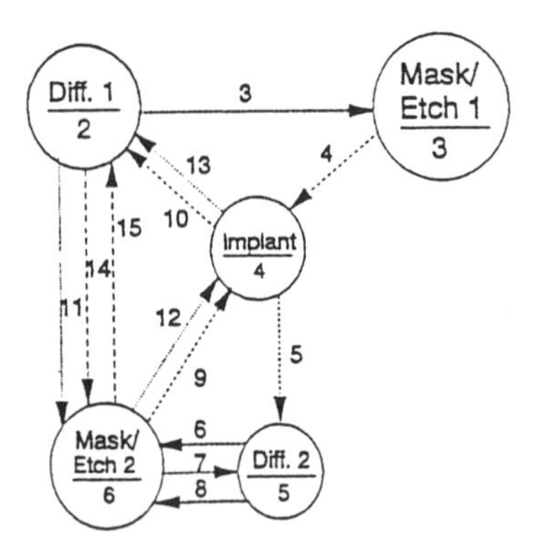

Figure 4.14: Spoke Analysis for Wkst #2, Alternate Fab Graph

However, if we were to do a <u>total</u> hub-and-spoke analysis for this workstation, Figs. 4.15a)-(c), we would in all likelihood still notice that a hub splitting may be in order.

Thus, if we are thorough in our analysis, both paths lead to the same conclusion. Which takes us back to the original question: Is Fig. 4.9 or Fig. 4.10 the better fab graph?

And we supply the standard chicken-or-egg answer: *It depends.* We have seen that Fig. 4.9 may be slightly better for analyzing hub splitting. Conversely, Fig. 4.10 may be better for fab layout when hub splitting is not an issue.

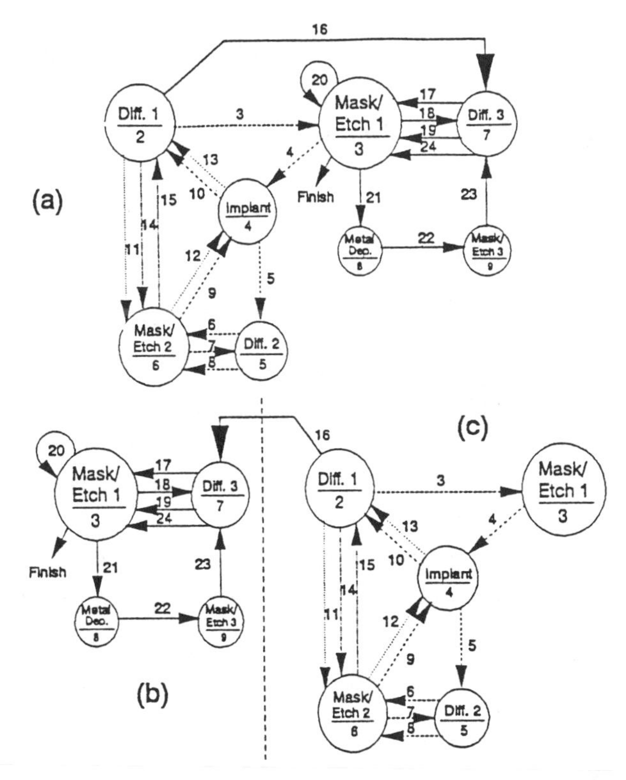

Figure 4.15: Alternate Graph's Hub and Spoke Analysis, Wkst #2

Which brings up the next topic: the usefulness of fab graphs for studying something other than fab layout, such as the impact of a new process or the inveterate engineering lots?

4.6 Multiple Process Flows and Engineering Lots

In wafer fabrication, it is not unusual to have several processes flows within a facility. Added to the multiple production flows, plants are constantly running engineering or development lots.

An *engineering lot* is a lot of wafers produced solely for the purpose of engineering evaluation, not production. The specialized flows that fabricate these lots usually have priority over all production flows.

Though these lots are a well-recognized phenomena in wafer fabrication, they are, nonetheless, seldom taken into account during the design and sizing of a fab. As a result, significant (sometimes substantial) production capacity is never realized. This capacity is sometimes referred to as being 'lost'.

Lost capacity is the difference between a factory's rated capacity and what the plant is actually able to achieve. There are many contributing factors to lost capacity, both processing and operations related. But perhaps the single, most-responsible culprit is the unplanned-for engineering lot. To address this very real industry problem, we will use our newly-acquired fab-graph and hub-analysis skills to assess the operations impact of these lots. (The same analysis technique will work equally well for any multi-process situation.)

The first step in dealing with engineering lots is to recognize that they are, in point of fact, fabricated using most if not all of the fab's resources. Furthermore, their fabrication uses a separate, specialized flow that is usually more complicated than any of the production flows.

For example, the hypothetical engineering flow given in Table 4.6 for CMOS-1 is quite a bit more advanced than the production flow (Table 4.1). The engineering flow is making wafers with two metal layers, as compared to one layer in the production flow. Despite the addition of a new flow with three new process steps, the fab still consists of basically the same equipment set, with the exception of the new Workstation #10, Dielectric.

On a per-lot basis, the demands of the new engineering flow, Fig. 4.16, are noticeably more taxing than the demands of the production flow (Fig. 4.9). When such an unplanned-for, and more-taxing demand is imposed upon the fab's resources, plant throughput will only be met to the extent that there is capacity available. Beyond that, wishing will not grow capacity, and production will generally suffer.

TABLE 4.6: CMOS-1 ENGINEERING SEQUENCE

STEP #	PROCESS STEP	TIME	WORKSTATION	WORKSTATION #
1	LASER MARK	0.8	MISC.	1
2	INITIAL OX	2.0	DIFF 1	2
3	P-TUB MASK	1.0	MASK/ETCH 1	3
4	P-TUB IM	0.5	IMPLANT	4
5	NITRIDE DEP	2.0	DIFF 2	5
6	FIELD MASK	1.0	MASK/ETCH 2	6
7	FIELD DIFF/OX	2.0	DIFF 2	5
8	SOURCE/DRAIN MSK 1	1.0	MASK/ETCH 2	6
9	S/D IM 1	0.5	IMPLANT	4
10	S/D OX 1	2.0	DIFF 1	2
11	S/D MASK 2	1.0	MASK/ETCH 2	6
12	S/D IM 2	0.5	IMPLANT	4
13	S/D OX 2	2.0	DIFF 1	2
14	NITRIDE STRIP	1.0	MASK/ETCH 2	6
15	GATE OX	2.0	DIFF 1	2
16	POLY DEP/DOPE 1	2.0	DIFF 3	7
17	POLY MASK 1	1.0	MASK/ETCH 1	3
18	POLY DEP/DOPE 2	2.0	DIFF 3	7
19	POLY MASK 2	1.0	MASK/ETCH 1	3
20	CONTACT MASK	1.0	MASK/ETCH 1	3
21	METAL DEP	1.1	METAL DEP	8
22	METAL MASK	1.0	MASK/ETCH 3	9
23	DIELECTRIC	2.0	DIEL-FAB	10
24	METAL DEP	1.1	METAL DEP	8
25	METAL MASK	1.0	MASK/ETCH 3	9
26	TOPSIDE DEP	2.0	DIFF 3	7
27	TOPSIDE MASK	1.0	MASK/ETCH 1	3

An even worse case develops when the more-taxing flow has priority over the production flows. The throughput of the taxed workstations may drop disproportionately, causing plant production to fall even further below design.

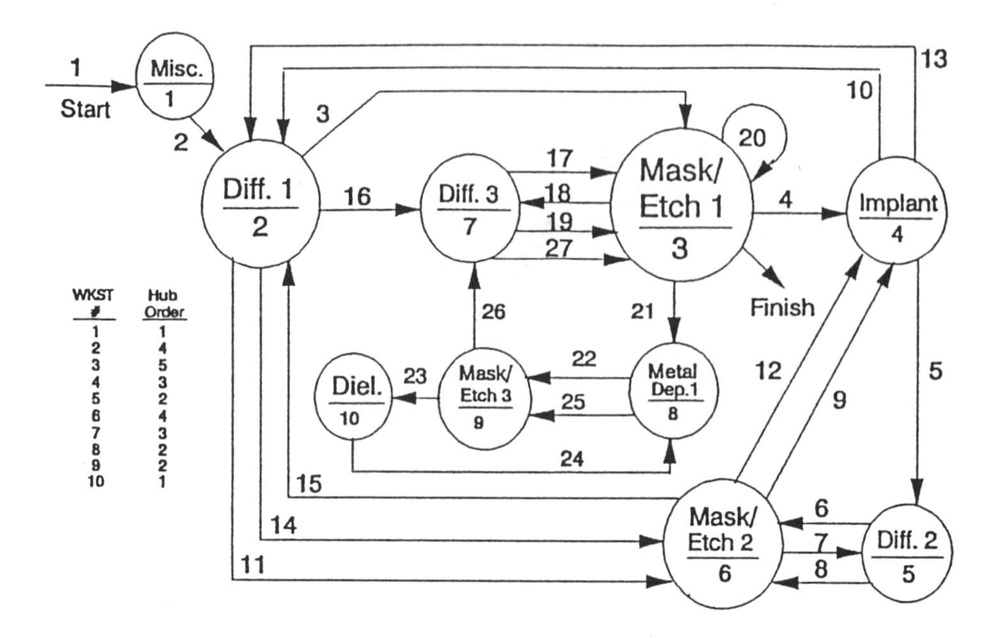

WKST #	Hub Order
1	1
2	4
3	5
4	3
5	2
6	4
7	3
8	2
9	2
10	1

Figure 4.16: Fab Graph for Engineering Flow, CMOS-1

4.6.1 Hub Order and Workstation Throughput

The *throughput of a workstation* is the rate at which lots are processed through that set of equipment, i.e., the number of lots processed over a given time interval. Most wafer fabs collect these data as part of their production reports in order to track equipment performance, as well as schedule product shipments. However, certain factors may distort or discredit such tracking data. One factor is the effective hub order of a workstation when multiple processes are in the fab. The importance of hub order on workstation throughput for a single flow is illustrated below.

Assume for the moment that CMOS-1 has been sized for a production rate of one lot an hour. Consequently, if the plant is expected to produce 1 lot/hr, it will naturally have a lot-start rate of 1 lot/hr. Furthermore, every step of the process sequence will have a throughput rate of approximately 1 lot/hr.

Shifting to a workstation perspective, how do all these process-step rates relate to the throughput of each workstation? Will the measured throughput of each workstation also be 1 lot/hr?

The answer: **Only those workstations that are not hubs.**

Each workstation, in order to keep up with production and not be a factory bottleneck, must have a throughput equal to the lot-start rate multiplied by the number of times the workstation is used in the process sequence, e.g., the workstation's hub order. For instance, Diffusion 1, Wkst 2, which is used four times, would need a throughput rate = 1 lot/hr * 4, or 4 lots/hr, in order not to be a bottleneck. Likewise, Wkst 3, as a fifth order hub, would need a throughput of 5 lots/hr. (See equation 4.7.1.)

Of course, most fabs will not achieve exactly the desired throughput at every workstation, even over long time intervals. But, with good up-front capacity planning and ongoing inventory control, most workstations should operate within 95+% of design throughput. As does CMOS-1 under production-only conditions.

But now enters the nefarious engineering lot, or, to be more realistic, engineering lots. Most wafer fabs will generally be running another 5 to 25% of its production rate as engineering lots. It is not even unheard of for the number of engineering lots to equal the production rate. Needless to say, these lots **will** impact the fab's production performance.

4.6.2 First-Level Performance Analysis

For a first level performance analysis, consider the fab graph shown in Fig. 4.16. For this engineering flow, the hub orders for workstations 1-7 are the same as for the production flow. Workstations #8 and #9, however, are now hubs of order 2 and 3, respectively. The new workstation #10 is not a hub.

With the introduction of engineering lots, there are two distinct process flows utilizing the factory's resources. What then is the effective hub order of the various workstations? Wkst 2, for example, is the hub order 4? Or 8? Or what? And what about Wksts 8 and 9?

The *effective hub order* of a workstation servicing multiple process flows will be a function of how much each flow uses the workstation's resources. More specifically, if h_{ij} is the hub order of workstation j for process i, then the effective hub order for workstation j for all processes in the fab is defined by:

$$\langle \text{Effective Hub Order}_j \rangle = \sum_{i=1}^{\#\text{ of processes}} (w_i * h_{ij})$$

where the fraction of use by each process, w_i, is the start rate for flow i divided by the total start rate for all process flows:

$$w_i = \text{start rate}_i/\text{total start rate} = s_i/S_{TOT}$$

The w_i must sum to 1 for all processes:

$$\sum_{i=1}^{\#\text{ of processes}} w_i = 1$$

Let us now apply this definition to CMOS-1. If the rate of engineering lots is 0.25 lots/hr, and the rate of production lots has been dropped back to 0.75 lots/hr, then the hub order of Wkst 2 is still 4.

$$\langle h_2 \rangle = (0.75 * 4) + (0.25 * 4) = 4$$

Now what about the more probable scenario where 0.25 lots/hr of engineering have simply been added to the existing production rate of 1 lot/hr? What is the effective hub order of Wkst 2 for this case?

$$w_1 = 1/1.25 = 0.80$$

$$w_2 = 0.25/1.25 = 0.20$$

$$\langle h_2 \rangle = (0.80 * 4) + (0.20 * 4) = 4$$

Perhaps, surprisingly, the effective hub order of Wkst 2 is still 4.

The effective hub order for a mix of processes is not dependent on the start rates for the various processes, but rather on how the mix uses the fab's resources. If the use is identical, the effective hub order is the same as the hub order for either flow.

For a case where effective hub order is different, consider a work center in CMOS-1 that is used differently by the two flows, such as Wkst 9.

$$\langle h_9 \rangle = (0.80 * 1) + (0.20 * 2) = 1.2$$

Thus, Wkst 9 is effectively being used 20% more than design. Surely, such impropriety must have an impact on fab performance.

4.6.3 Engineering Lots, Fab Performance, and Bottlenecks

If a factory is sized for a certain production rate, and if that factory is operating at its design capacity, then starting any additional lots (much less 25% more) will derate production. The extent of the derating will depend on the size of the added burden, as well as the factory's operating policies.

But even in the case where the lot-start rate does not exceed the fab's design rate, a shift in product mix can still derate fab performance. To illustrate, consider CMOS-1 where the production start rate is 1 lot/hr and the engineering start rate is

0.25 lots/hr, and the fab was sized for 1.25 lots/hr of production only. Table 4.7 gives the hub orders for the production flow, the hub orders for the engineering flow, and the **effective** hub orders for both flows.

TABLE 4.7: EFFECTIVE HUB ORDER FOR CMOS-1

Workstation #	Design Hub Order Production	Hub Order Engineering	Effective Hub Order Both Flows	Effective/ Design
1	1	1	1.0	1.0
2	4	4	4.0	1.0
3	5	5	5.0	1.0
4	3	3	3.0	1.0
5	2	2	2.0	1.0
6	4	4	4.0	1.0
7	3	3	3.0	1.0
8	1	2	1.2	1.2
9	1	2	1.2	1.2
10	--	1	1.0	1.0

Fab sized only for 1.25 lots/hr prodn; using:
Production start rate = 1 lot/hr
Engineering start rate = 0.25 lots/hr

If the fab was sized for a total start rate of 1.25 lots/hr of production, how does simply substituting engineering lots for production lots impact fab performance? The answer is provided by the effective hub orders for the mix of the two flows. The equipment at Workstations 1 through 7 are presumably doing the same processing, regardless of the flow, and thus will see no derating. Similarly, Wkst 10 will see no derating because it was added for, and is being used exclusively by, the engineering flow. Only Wkst 8 and Wkst 9 are affected by the new mix of product flows. Because of the added complexity of the engineering lots, these two workstations become higher-order hubs than when used in production only.

To quantify the impact of the more-complex engineering lots on fab performance, note the 'Effective' versus the 'Design' hub orders for Workstations #8 and #9. If no additional equipment is added to accommodate the engineering flow, then these two work areas are each undersized by at least 20%. They may be even more undersized if: process times are lengthier in the engineering flow (often the case), if machine setups must be made (ditto), if loading rules differ, etc. But, ignoring these 'ifs', these two workstations ostensively have insufficient capacity for the product mix.

In a wafer fab, the workstation (or equipment type) that has the minimum capacity relative to processing demands is generally the bottleneck. As a bottleneck, the workstation will gate or throttles the throughput for the entire fab. Whether it is something as complex as photolithography or as mundane as an inspection, a bottleneck limits the production of the entire factory to something less than expected.

To determine the best performance achievable in a wafer fab, identify the bottleneck and determine the best performance it can achieve.

For the case presented in Table 4.7, Workstations #8 and #9 are the fab's only seriously undersized workstations, each by 20%. But as a result of these two workstations being undersized by 20%, the entire fab is underdesigned by 20%. If lots are processed as they arrive at each work area (i.e., lot dispatch is *FIFO* or First-in/First-out), then the best the factory will be able to achieve is a production throughput of 0.8333 lots/hr and an engineering throughput of 0.2083 lots/hr, or a total fab throughput of only 1.0416 lots/hr, a far cry from the expected 1.25 lots/hr. These numbers are arrived at by derating the individual-process start rates:

$$Q_{prodn} = (1.0 \text{ lot/hr} / 1.2) = 0.8333 \text{ lots/hr}$$

$$Q_{engg} = (0.25 \text{ lots/hr} / 1.2) = 0.2083 \text{ lots/hr}$$

$$Q_{TOT} = (0.8333 + 0.2083) = 1.0416 \text{ lots/hr}$$

Actual throughputs may be somewhat less than these numbers, and may shift between the two flows. Engineering throughput may very well be lower since it is using the undersized workstations twice, compared to only once by the production flow.

Again, it should be remembered that these throughputs are for a fab using a FIFO lot-dispatch scheme. But, just as engineering lots seem to be a fundamental law of nature, there seems to be a second fundamental law that decrees engineering lots will be hot lots.

By definition, a *hot lot* is one that has a high or absolute priority over the other lots in the fab. If two lots arrive at a work center at the same time, the hot lot will receive attention from the next available operator and be put into the next available machine. When engineering lots have priority over production lots, which is invariably the case, production will suffer by the number of engineering lots started **plus** some additional loss. The extra loss will result because of the increased complexity of the engineering flow, as well as the added overhead of simply dealing with another process flow in the fab (machine setup, exigent operator requirements, increased tracking, etc.).

So, for the simplistic example in Table 4.7, what production rate may we expect if the engineering flow always has priority? To answer this question, we again look to the fab's apparent bottleneck.

Since both Wkst 8, Metal Dep, and Wkst 9, Mask/Etch 3, appear to be undersized by the same amount, we will use our worldly fab experience and assume that Wkst 8 is more likely to be the real bottleneck. There are two reasons behind this guess. First, undersizing photolithography (Wkst 9) is a capital offense, and in modern wafer fabrication is a mistake not often made nowadays. Second, metal-deposition processes will, in general, have lower-than-expected availabilities and thus are more probable candidates. Therefore, we will proceed on the premise the Wkst 8 is the bottleneck.

Before we start, however, remember that Metal Dep, like the fab itself, has been designed and sized for 1.25 lots/hr of capacity.

Since the engineering flow, with a lot-start rate of 0.25 lots/hr, will always have priority, and since Wkst 8 is a second order hub for this flow, engineering will use 0.50 lots/hr (hub order multiplied by rate) of Metal Dep's capacity. Which leaves only 0.75 lots/hr (e.g., 1.25 - 0.50) of capacity available for the production flow. Thus, the best overall rate we can expect from the entire fab is 1.0 lot/hr, <u>80%</u> the amount anticipated.

$$Q_{engg} = 0.25 \text{ lots/hr}$$

$$Q_{prodn} = 0.75 \text{ lots/hr}$$

$$Q_{TOT} = (0.25 + 0.75) = 1.0 \text{ lot/hr}$$

When queuing losses and other dynamic events are factored in, these rates may be even lower.

The bottom line gleaned from these simplistic examples is that starting engineering lots can derate the performance of the entire fab. Giving engineering lots priority can exacerbate the derating, depending on how often the flow uses the bottleneck workstation as a hub.

4.6.4 Summary

It should be obvious by now that fab graphs and hub analyses are useful tools for studying the impact of engineering lots, new process flows, bottlenecks, and other process and resource interactions. Even when fabs are too complex to hand draw a graph, the hub analysis can be performed with nothing more than the process flows and a list of fab equipment. Such information should be readily available from in-house tracking systems.

As a cautionary note to this type of simplistic bottleneck analysis, however, it must be recognized that the effective hub order does not take into account any variation in process times from one equipment use to another. A longer process time, say for a metals step in an engineering flow, will tie up a machine longer than a shorter time, thus putting even more of a burden on the workstation. If such a situation exists, then one should expect even lower throughputs through that workstation.

Also, this type of analysis does not deal in any fashion with equipment availability. (Section 4.8 provides a static-analysis tool that factors in process times and availabilities.)

4.7 Using Fab Graphs with Tracking Data

If one considers all the things that can go wrong during the 500+ steps of wafer processing, it is truly a wonder that wafers are ever started in the first place. Such fertile opportunity for Murphy's Law is a prime reason for the industry-wide installation of data-tracking systems. These systems, sometimes called CAM (Computer-Aided Manufacturing) or CIM (Computer-Integrated Manufacturing), collect an extensive array of data as lots progress through the wafer fabrication sequence. The data provided by these systems often include: WIP tracking, equipment utilization, order entry, machine availability, cum product yields, and equipment throughput reports.

Interpreting tracking data can be difficult. First, if the time interval over which the data are collected is too short, the throughput numbers may not be representative of how the equipment is actually performing. Other misleading contributors include: poor inventory staging, unexpected and/or unreported equipment downtime, upstream bottlenecks, etc. But the most common defect in analyzing tracking reports is the lack of knowledge of the effective hub number of the work centers.

Nevertheless, adjustments to the factory are often made, on a shift-to-shift basis, based on the very same tracking data. Frequently, the changes are made without a clear understanding of the cause-and-effect relationship between the adjustments and the desired outcome. As a case in point, consider some of the possible long-term and/or short-term remedial actions which may be taken for a perceived bottleneck. Adjustments may include:

- adding machines at bottleneck workstation(s)
- slowing lot starts
- altering process priorities
- minimizing setup times
- rescheduling preventative maintenance
- adding personnel
- etc.

Any one of these actions will generally have a high cost associated with it, whether it is a long-term capital action such as adding equipment or a short-term operating action like delaying maintenance. In light of the cost, and the variety of possible side-effects resulting from the remedial action itself, it is wise to verify that a suspected workstation is indeed a true bottleneck.

For instance, in the above simplistic example we arbitrarily chose Wkst 8 as the bottleneck. But it could just as easily have been Wkst 9, or even one of the other work centers, for reasons not considered in the analysis (process times and/or availabilities, etc.). But can such a bottleneck verification be made from a tracking system's reports?

It can if the CAM system's throughput information for the equipment is accurate, and if the effective hub orders are known for the workstations in the fab. There is, however, a caveat: upstream bottlenecks can sometimes hide or disguise the existence of downstream bottlenecks. In such a situation, only a fab-specific, dynamic model is capable of finding **all** undersized work centers regardless of their position in the process sequence.

For the most part, though, fab graphs and the associated hub analyses provide useful tools for interpreting the voluminous data that are monitored and/or collected continuously in the fab. To illustrate how to use these tools with real tracking data, consider the data given in Table 4.8 for CMOS-1.

TABLE 4.8: BOTTLENECK IDENTIFICATION FOR CMOS-1
USING TRACKING DATA AND EFFECTIVE HUB ORDER

Workstation #	Actual Thrupt (lots)	Effective Hub Order	Expected Thrupt (lots)	Bottleneck Indicator Actual/Expected
1	633	1.0	630	1.0000
2	2,364	4.0	2,520	0.9380
3	2,855	5.0	3,150	0.9060
4	1,761	3.0	1,890	0.9320
5	1,200	2.0	1,260	0.9520
6	2,362	4.0	2,520	0.9370
7	1,659	3.0	1,890	0.8778
8	634	1.2	756	0.8386
9	630	1.2	756	0.8333
10	108	1.0	126	0.8571

Fab sized for 1.25 lots/hr production.
Throughput data collected every 504 hrs.
Expected Thruput = 504 x 1.25 x eff. hub order

Again, this small, idealized fab is operating with both a production and an engineering flow; lot dispatch for the period of data collection is FIFO.

As with most tracking reports, the 'Actual Thrupt' data is an aggregate of all the flows in the fab. Generally, CAM systems do not break out equipment performance by process or product. But, in this example, since we know the effective hub order of each workstation, there is enough information to calculate the expected throughput, q_{WKST}.

$$q_{WKST} = S_{TOT} * t * \langle h_{WKST} \rangle \quad (4.7.1)$$

where S_{TOT} is the total lot-start rate (for all processes), t is the report interval in hours, and $\langle h_{WKST} \rangle$ is the effective hub order for that particular workstation.

By comparing the actual throughput at each workstation to that which is expected, we obtain an indicator, B_{WKST}, which may point to potential fab bottlenecks. However, this number is not absolute, nor is it conclusive in identifying true bottleneck work centers. Rather, it is useful, in conjunction with the process-sequence information, for indicating 'choke points' in the flow (or flows).

Again, the reason the indicator may not be conclusive is because any upstream inhibitor(s) will cloud the performance data of downstream equipment. In Table 4.8, for example, Wkst 3 has a bottleneck indicator of 0.9060, instead of the 0.93-0.95 that would be expected for a non-bottleneck workstation. Further examination of the throughput data and the fab graphs for the flows shows that Wkst 7 is actually choking, or throttling, the flow to Wkst 3 by virtue of being used upstream of process step 17.

Just as Wkst 7 throttles the flow to Wkst 3, it also impedes the throughput to workstations that are exclusively downstream of it (i.e., Wksts 8-10). In general, an impedance makes it difficult to ascertain downstream behavior. Thus, a cursory analysis might lead one to conclude that Wkst 7 looks like the best bottleneck candidate.

However, if one looks further, to Wkst 8, one sees that Metal Dep, by far, has the worst throughput and the lowest bottleneck indicator. In fact, studying the fab graphs, one might conclude that since Wkst 8 is in the loop that feeds Wkst 7 (steps #22, 23, 24, 25, and 26), it may well be the true bottleneck, and the one throttling flow to Wkst 7, which in turn throttles the flow to Wkst 3. What makes this argument even more convincing is that the throughput of Wkst 3, which is being fed mostly by Wkst 7, is higher than Wkst 7's.

But at this point, and with this tracking data, it would be impossible to categorically state whether Wkst 7 or Wkst 8 is the true bottleneck. This question could only be answered in one of three way:

- buy a machine for Wkst 7,
- buy a machine for Wkst 8, or
- use a model to simulate a 'pretend' equipment buy.

Whereas, the odds are 50/50 that one of the first two options would succeed, the answer from the model would be definitive, and a correct capital purchase could be made.

Of course, there is one other option, the one that is often most favored. That is to do nothing. Management could simply be satisfied that the fab is only making 1.0414 lots/hr instead of 1.25 lots/hr. After all, that is only 17% less throughput than expected; probably only a few thousand less lots a year. Couldn't be worth more than a couple hundred million dollars. But at least the fab will not have to fork out $4-5mm for a piece of equipment.

4.7.1 Tracking Data and Priority Flows

Table 4.9 presents another set of equipment performance data from the in-house CAM system for CMOS-1. During this tracking period, the engineering lots were handled as hot lots, always given priority over production. As may be seen, the impact at the probable bottlenecks, Workstations #7 and #8, is even more pronounced on B_7 and B_8, and even on B_3 as well.

TABLE 4.9: PRIORITY FLOWS AND BOTTLENECKS
ENGINEERING FLOW HAS PRIORITY OVER PRODUCTION

Workstation #	Actual Thrupt (lots)	Effective Hub Order	Expected Thrupt (lots)	Bottleneck Indicator Actual/Expected
1	633	1.0	630	1.0000
2	2,357	4.0	2,520	0.9353
3	2,670	5.0	3,150	0.8476
4	1,751	3.0	1,890	0.9265
5	1,198	2.0	1,260	0.9501
6	2,353	4.0	2,520	0.9337
7	1,531	3.0	1,890	0.8101
8	613	1.2	756	0.8108
9	646	1.2	756	0.8545
10	114	1.0	126	0.9048

Fab sized only for 1.25 lots/hr.
Throughput data collected for 504 hrs.
Expected Thruput = 504 x 1.25 x eff. hub order

But a major clue to the bottleneck situation is provided by Workstations #9 and #10. These workstations, in addition to

Wkst 8, are used either more heavily or exclusively by the engineering flow. By giving the engineering flow priority over production, throughput performance actually improves at two of the three engineering-intensive work centers. But not at Wkst 8 where it declines even further. This behavior indicates that Metal Dep is definitely a bottleneck.

But is Wkst 7 also a bottleneck?

Very possibly, since Wkst 3, which is mostly fed by Wkst 7, sees a rather precipitous drop in throughput as well. But the argument for it being a bottleneck can not be made quite so emphatically as it can for Wkst 8.

If one were to attempt to remedy the throughput shortfall, modeling would be the preferable first course of action to verify the bottlenecks. But barring that, the next most logical way to proceed is to buy another machine for Wkst 8 and see if the bottleneck goes away. If not, then another diffusion furnace for Wkst 7 may be in order.

But from Chapter 3, we learned that diffusion furnaces are relatively inexpensive. Might it not, therefore, be wise to proceed with a purchase order there as well, considering how much the throughput shortfall is costing a day?

In general, it is indeed better not to wait on capital purchases if it will buy throughput. However, inasmuch as Wkst 7 is comprised of large-batch furnaces which may be run in a variety of ways, it is wiser still to see how the fab is using the machines it already has before buying more. Batching, as well as other fab control rules, can, in and of themselves, be the cause of bottlenecks. (See Chapter 6 for more on this topic.)

Thus, before new equipment purchases are made, it is judicious to understand the fab's batching, dispatching, setup, and other control rules. We must never get so wrapped up in playing with performance tools, or tracking data, that we ignore the physical reality of what is happening on the factory floor. There will be many instances when static analyses and tracking data simply will not be sufficient to make capacity or operational decisions.

4.7.2 Nebulous Tracking Data

Almost invariably in a wafer fab, we can bet our bottom dollar that there is at least one bottleneck. Finding it, however, may be extremely difficult without a coherent strategy for taking apart the tracking data. In practice, such strategies are rare. Tracking data are often collected, stored, and eventually forgotten or destroyed. Which is unfortunate, because there are two possible ways to grapple with the data even when it ostensibly provides no clear-cut answers.

Perhaps the most obvious way to use the data is to march through the major process flow (Table 4.1) and look for the first workstation that takes a major hit in throughput. In Table 4.9, for example, that would be Wkst 3. But the very next workstation used in the process flow, #4, has an increase in throughput, as does the one used after it. Consequently, it is unlikely that Wkst 3 is a true bottleneck.

Looking at the hub splitting that was carried out in Fig. 4.13, we can say that, if Wkst 3 is a bottleneck, it is the downstream portion of the flow (interconnect) that is making it one. Therefore, we should put further interest in Wkst 3 aside until its next use (process step 17).

The next major hit in throughput comes at Wkst 7. Indeed, every workstation after #7 has degraded performance, except for Wkst 10 which is used exclusively for the hot engineering flow. It is safe to say, therefore, that Wkst 7 is a bottleneck. Whether it is a sufficient bottleneck to explain the downstream performance of Wkst 3 is an open question, and one that can only be answered by eliminating #7 as a primary bottleneck (which requires modelling or hardware). We are also limited on the speculations that may be made about Wkst 8, although it looks to be a promising bottleneck candidate.

At this point, the analysis is basically over unless more information is forthcoming. Outstanding candidates for primary fab bottleneck are: Wkst 7, Wkst 8, and Wkst 3 (for interconnect portion of flow). Fab performance now depends upon either:

doing the modeling, buying the equipment, or losing the throughput.

Before we leave this topic, we said that there are two ways to grapple with ambiguous tracking data. The first is marching through the process flow as we have just done. The other is to look for the last equipment set in the process flow that sees acceptable throughput. In Table 4.9, that equipment set is Wkst 5. The 'true' primary candidates for bottleneck would be those equipment sets used exclusively after the last use of Workstation #5: Wksts 7, 8, 9, and 10.

The advantage of this latter method is that it could save a tremendous amount of time and effort for a real fab that has 75 to 125 workstation and is running multiple 400-500 step process flows.

4.7.3 Summary

This brief look at tracking data has provided one very important additional insight. Because of the complex hub structure in wafer fabs, debottlenecking, if done at all, may have to be done in stages, removing upstream constrictions to reveal others downstream.

But, regardless of the method chosen, hopefully we have shown that fab graphs and hub analyses provide tremendous insight into fab operations, as well as helping to make sense out of often otherwise useless data. When used in conjunction with tracking data, these techniques furnish a first-level analysis tool for fab performance.

4.8 Static Capacity Planner

The final use of fab graphs and hub analyses that we will cover here is for determining the number of pieces of equipment required at a workstation for a given throughput. This calculation is sometimes called the *fab sizing* problem. As will be seen, hub analysis is critical in establishing the total processing requirement at a workstation. And it is this total processing requirement, together with equipment availability, which sets a workstation's capacity requirement. That is, the capacity of a workstation is a function of the number of pieces of equipment, total processing times performed there, and availability of the equipment.

A *static-capacity planner* calculates the number of machines needed at a workstation under static conditions (i.e., no dynamics) for each process flow in the fab. The sum of the machines over all the flows in the plant determines the fab's required equipment set for that workstation.

Fab sizing begins by calculating the total processing requirements of a work center. The hub analysis identifies the process steps performed at each workstation, for each process flow. Each of these workstation-specific groupings of process steps is called a *hub set*.

Let P_{ij} be the total processing requirement (hours/lot) for process i at workstation j. Let $pt_{ij}(k)$ be the process time for steps k in the hub set for workstation j and process i. Then P_{ij} is determined by summing over the hub set.

$$P_{ij} = \sum_{k=1}^{\#\text{ of hub-set steps}_j} pt_{ij}(k)$$

Continuing the calculation, let ls_i be the start rate for process i in lots per hour, and let M_{ij} be the number of required machines at workstation j, for process i. The estimated number of required machines is simply the start rate multiplied by the total processing requirement P_{ij}.

To correct for availability, let a_j be the availability for the machines at the work center j. Then dividing the product above by a_j results in the following estimate for M_{ij}.

$$M_{ij} = (P_{ij} * ls_i)/a_j \tag{4.8.1}$$

where

$$a_j = MTTF_j/(MTTF_j + MTTR_j).$$

For the availability, MTTF is mean-time-to-failure, i.e, average up time; and MTTR is mean-time-to-repair, i.e., average down time. For CMOS-1, this information is provided in Table 4.10.

TABLE 4.10: FACTORY INFORMATION, CMOS-1

One Production Flow and One Engineering Flow

Workstation		MTTF (hrs)	MTTR (hrs)	Load (lots)	Batch (lots)	Process (hrs)	
						Prodn	Engg
1	MISC.	99	1	1	1	0.8	0.8
2	DIFFUSION 1	95	5	4	2	8.0	8.0
3	MASK/ETCH 1	97	3	1	1	5.0	5.0
4	IMPLANT	85	15	1	1	1.5	1.5
5	DIFFUSION 2	95	5	4	2	4.0	4.0
6	MASK/ETCH 2	97	3	1	1	4.0	4.0
7	DIFFUSION 3	95	5	4	2	6.0	6.0
8	METAL DEP 1	85	15	1	1	1.1	2.2
9	MASK/ETCH 3	97	3	1	1	1.0	2.0
10	DIEL-FAB	97	7	1	1	0.0	2.0

lots = Standard lots of 25 wafers.

Be aware that the estimated $M_{i\,j}$ are not necessarily integers. For a single flow, fab sizing may require rounding to the next **largest** integer. For multiple flows, the recommended procedure is to sum over all process flows, and then round to the next **largest** integer.

There are two notable exceptions to the capacity formula used here, and both are caused by batching. First is the case where a machine has a large batch size and can take more than a standard load, such as with a 200-wafer furnace. For a workstation with large-batch equipment, equation 4.8.1 is modified according on the fab's batching rules. (See Chapter 6 for a detailed discussion on batching rules.)

Briefly, if a furnace can process up to four lots at a time, but the fab's batching rule says to proceed whenever two or more lots are available, then each time the furnace is run, it may have either two, three, or four lots in it. Because of the factory's dynamics, the actual number going into the machine will only be known at the time of loading.

But as far as the static-capacity planner is concerned, it must be prepared for the worst-case scenario where every load will be a two-lot batch. Thus, the required number of machines at a large-batch work center is:

$$M_{ij} = [P_{ij} * ls_i)/a_j] / BS_j$$

where BS_j is the equipment's batch size in standard lots. (See Table 4.10.)

Of course, using the worst-case scenario for work centers with large-batch equipment does result in some oversizing, by about 15-25%. But considering that this analysis tool totally ignores factory dynamics, what is one more inaccuracy?

The second batching exception covers the situation when the machine load size is less than one lot. Each lot must be processed in two or more batches. Then the estimated M_{ij} must be multiplied by the number of batches. Again, some oversizing may result, though not as much as in the large-batch case.

We can not caution strongly enough about putting too much value in static analyses. Static analyses may provide insight, ballpark estimates, and possibly even rules of thumb, but they do not provide answers. Nor do spreadsheet calculations incorporating a static analysis. Analysis of real factory performance requires dynamic models.

Having said thus, we will use the static capacity planner to look for some insights on CMOS-1 and the three workstations we suspect of being bottleneck contenders.

Table 4.11 shows the results of the static-capacity exercise for the production flow, the engineering flow, and the sum of the two. If this were a perfect world, the calculated total machines would be the correct equipment set for the fab. We could compare these calculated requirements with what is actually on the factory floor (last column in table) and see which workstations are short of capacity.

TABLE 4.11: STATIC CAPACITY PLANNER, CMOS-1

Workstation		Calculated Req'd Machines			Adj. # Mach.	Actual # in Plant
		Prodn	Engg	Total		
1	MISC.	0.81	0.20	1.01	1.21	2
2	DIFFUSION 1	4.21	1.05	5.26	4.21 *	5
3	MASK/ETCH 1	5.15	1.29	6.44	7.73	8
4	IMPLANT	1.76	0.44	2.20	2.64	3
5	DIFFUSION 2	2.11	0.53	2.64	2.11 *	3
6	MASK/ETCH 2	4.12	1.03	5.15	6.18	7
7	DIFFUSION 3	3.16	0.79	3.95	3.16 *	4
8	METAL DEP 1	1.29	0.65	1.94	2.33	2
9	MASK/ETCH 3	1.03	0.52	1.55	1.86	2
10	DIEL-FAB	0.00	0.54	0.54	0.65	1

Adjustments:
+ 20% for factory dynamics
- 20% for large furnace loads (>2 lots) *

But as we have already discussed, this is not a perfect world. The three diffusion workstations (#2, 5, 7) have most likely been oversized, because of fuller batches, and all the other work centers have probably been undersized, because of factory dynamics. Before we compare the calculated requirements with the actual equipment set, therefore, we may want to make some purely arbitrary adjustments, say require 20% less capacity at furnaces, but 20% more capacity everywhere else. Now, we are ready to compare requirements with actual hardware.

With all the caveats and assumptions made, Wkst 8 appears to be the only work center short of capacity, and thus the most probable bottleneck. This result, when used in conjunction with the hub analysis and tracking data, makes a fairly strong case. But it is, once again, by no means conclusive. For conclusive, we need to analyze the fab as it actually operates, dynamically and under a whole spectrum of initial and boundary conditions.

Chapters 5 and 6 will covers factory dynamics in considerable detail.

Further Reading and Notes

1. Some of the mathematical properties of fab graphs are discussed in Chapter 7.

2. Developments and advancements in the technologies of wafer fabrication and related aspects of microelectronics are discussed in chapter one of Glaser and Subak-Sharpe [4.1]. Additional background is given in [4.2] and [4.3]. Rao [4.4] also charts technology trends.

References

[4.1] Glaser, Arthur B. and Gerald E. Subak-Sharpe, **Integrated Circuit Engineering**, Addison-Wesley, Reading, MA, 1979.

[4.2] Gilder, George, **Microcosm**, Simon and Schuster, 1989.

[4.3] Bunch, Bryan, and Alexander Hellemens, (eds.), **The Timetables of Technology: A Chronology of the Most Important People and Events in the History of Technology**, Simon & Schuster, New York, 1993.

[4.4] Rao, G. K., **Multilevel Interconnect Technology**, McGraw-Hill, New York, 1993.

Exercises

1. Draw the spoke structure for all CMOS-1 hubs. (Fig. 4.11 has the spoke structure for Workstation #3.)

2. Draw the fab graph for the process flow of Table 2.1.

3. Draw the fab graph for the process flow of Table 2.2. What are the structural differences between this fab graph the one from Exercise 2?

4. Draw the fab graphs for Tables 2.3, 2.4, and 2.5.

5. Draw the fab graphs for Tables 2.6 and 2.7.

6. Explode the process sequence of Table 2.4 by inserting the photolithography steps of Table 2.6. Draw the fab graph of the resulting process flow.

7. Further explode the detailed flow of Exercise 5 by inserting the photoresist processing steps of Table 2.7. Draw the fab graph of the resulting process flow.

8. Obtain a process listing for a real fabrication process. Draw the fab graph.

9. Perform a comprehensive hub analysis. Redraw the fab graph as necessary.

10. Obtain tracking data for workstation throughput corresponding to this flow. Perform the performance analyses described sections 4.6 and 4.7.

Beware the Jabberwock, my son!
The jaws that bite, the claws that catch.
-Lewis Carroll

Chapter 5
Factory Performance Basics

5.0 Introduction

The complexities of wafer fabrication stem from a number of factors, including: the intricate processing that is performed, the exceptionally-long process sequences involved, and the large variety of high-tech chemical-processing equipment used. In addition, fabs generally manufacture multiple process flows, requiring that the equipment run under diverse operating rules that often change from product to product. To complicate matters further, fabs must meet time-varying demands while carrying large, long-cycle-time inventories. But perhaps the most confounding aspect of all is that wafer fabrication, does not have a sole set of operating conditions that will achieve 'optimum factory performance'. In fact, the definition of 'optimum factory performance' will vary from fab to fab, and possibly even from manager to manager within a fab.

Factory performance is many things to many people. To the fab operations manager, throughput is the quintessential yardstick. To the customer support manager, schedule stability tops the list. To the misguided, nothing counts save cost of ownership. But in reality, perhaps the only objective indicator of factory performance is cost per good die__regardless of how the good die is achieved. This number, by definition, must take into consideration: yield, throughput, schedule stability, cycle time, equipment utilization, and the overall equipment effectiveness of the bottleneck(s).

Whereas some of these performance indicators are mutually and beneficially dependent, others are at cross-purposes and trade-offs will have to be made to realize good die costs.

For example, yield correlates with fast cycle time; that is, as product moves more rapidly through the fab, yield improves. However, a faster cycle time will usually necessitate smaller inventories which, in turn, may lead to lower throughputs. Thus, in an effort to boost yield, cycle time may improve, but throughput may decline.

Unfortunately, not all of the complex interrelationships in the fab are as well understood as the cycle time-yield connection. And trying to gain an understanding of these interrelationships on the factory floor is well nigh impossible. An operating adjustment on the floor, such as a revision in the preventative maintenance plan or a change in furnace loading rules, may not have a detectable impact until one or more cycles of product have completed manufacture. By then, if the operating change has an unexpected or counter-intuitive result, the price tag could be in the millions of dollars.

The costly magnitude of mistakes, and the often long delay in detecting them, makes experimenting on the factory floor *verboten*. Multi-million dollar CIM systems track materials movement, machine status, operator availability, and thousands of other parameters, not to improve manufacturing and not to schedule more reliably, but simply to sustain repeatability [5.1].

Perhaps a better understanding of the complex interactions that occur in wafer fabrication will alleviate some of the fears associated with change, and move the factory away from the leaden, unresponsive monolith of the 1980's and more towards a fast cycle time, highly-flexible factory of the twenty-first century.

5.1 Factory Performance Definitions

In previous chapters, key performance parameters have, on occasion, been used or discussed without precisely defining how they fit into the overall picture of wafer fabrication. These parameters include: things that cause the factory to perform a certain way (control variables), the behavior that results based on how the factory is designed and operated (response variables), and performance that may be both cause and effect (state variables) Table 5.1 lists the various performance parameters.

TABLE 5.1: FACTORY PERFORMANCE PARAMETERS

Control	Response	Control & Response
Equipment Set	Die Yield	Dynamic Capacity
Availability	Throughput	Inventory
Facilities	Cycle Time	Bottlenecks
Product Mix	Schedule Stability	
Process Priority	Equip Utilization	
Control Rules		
Scheduling		

5.1.1 Design and Operating Parameters (Control Variables)

The performance parameters that fall into this category are those which determine how the factory performs. Since either management and/or decision control may be exerted over these parameters, they are primarily voluntary.

Equipment Set: The equipment set consists of all hardware and associated software used during wafer fabrication, including processing equipment, material movers, and automation systems. The more common industry usage, however, applies this term to the processing equipment only, since those machines: make up the bulk of the capital expenditures, require more decision making in their selection, and are more likely to include the culprit bottleneck, if there is one. (Chapter 3 covers processing equipment in more detail.)

Equipment Availability: Equipment in real factories may be unavailable for processing for significant segments of time. Machines may be out of service because of planned maintenance or sporadic equipment failure; or, they may be in service and simply not being used productively (calibration steps, waiting for a full or near-full batch, etc). Machine availability and equipment usage are two sources of major concern for most wafer fabs.

During fab sizing, equipment is purchased based on the availability numbers supplied by the equipment vendors. Such an approach is unfortunate and also naive. A *mean-time-to-failure* specification denotes that, on average, a machine will be available for that length of time before failing. But the spec can make **no** guarantees that the machine will actually spend the time processing; that is dependent on the rest of the system.

Perhaps a good analogy is the pass play in football. The quarterback puts the ball in the air, but what happens afterwards depends on the rest of the team and the opposition. In theory, the pass could lead to a touchdown; the reality is usually much different.

Similarly, in a wafer fab, there is often a vast difference between a machine's theoretical and actual performance. When this shortfall occurs at one or more key workstations, bottlenecks will result and the fab may be undersized by 20%, or more. Without a detailed and accurate dynamic model, it is generally not possible to anticipate bottlenecks during fab sizing. Even after a fab is built and operating, it may still require an expert and a good computer model to interpret fab behavior. With tracking data and a model, it may be possible to 'back out' good equipment availability numbers which, in turn, may then be used to re-size bottleneck workstations. (Equipment availability is covered in detail in Section 6.9.)

Product Mix: The mix of products within a wafer fab is normally a function of the age of the facility. Fabs are designed for one major process, but do not remain that way long. After about twelve to eighteen months, the fab will evolve to a multi-product, sometimes multi-process facility. The evolution comes about because of an essential mismatch in fab life and chip progress.

The life of a wafer fab is about eight to twelve years; the next generation of chip can be expected in as little as a year. Chip development and prototype production are often done in the same facility that is running the previous generation.

Needless to say, introducing new products and/or new processes can lead to some interesting operational problems for a wafer fab. As a note, *process* refers to the flow (500 steps, 4-metals layers, etc.), and *product* refers to the same process flow with changes in one or more of the masks.

The introduction of new products merely means some small additional setup time for the photolithography equipment, to change out masks. So, the addition of new products, per se, is not earthshaking... unless the factory is already operating at capacity. More interesting and severe operational problems, however, can occur when a new process is added to an existing line. The new process will generally have more steps, more hubs, and probably not much more (if any) added equipment. In such a case, overall production could degrade by more than just the capacity undersizing. (Section 6.1 discusses multiple products/processes in detail.)

Process Priority: When a wafer fab is running multiple processes, it is not uncommon that one of them will have priority over all the others. For example, hot lots generally have priority over production lots.

If no one process is assigned priority over the others, then the fab will generally use a first-in/first-out (FIFO) loading rule, i.e., the lot that arrives at a work center first will receive attention before all those that arrive after it.

Control Rules: Because so many things are happening in a wafer fab, and because there are so many options available with regard to the complex interactions of lots, equipment, and processes, fabs will generally have rules governing how the operators deals with lots and the equipment. Typical rules relate to: machine setup, batching, dispatching, and equipment maintenance.

Setup is the machine preparation that must be performed when the next lot to be loaded is from a different process than the one that last used the machine. Generally, the CIM system will tell an operator which lot should next be loaded and whether or not a setup is needed. (See Section 6.2 for more on setup rules.)

Batching rules govern how lots are loaded into nonstandard-size machines. Currently, standard lot size is twenty-five (or twenty-four) wafers, the standard capacity of cassettes. However, some machines have a larger load size, and some a smaller. Furnaces, for example, typically have a load size of 100, 150, or 200 wafers. Thus, for a full load, four, six, or eight lots, respectively, would need to be available for processing. But since waiting may not be advisable, fabs will generally have batching rules to determine when to load a furnace. Options are:

- wait for a full load
- wait for a specified number of lots (2, 3, etc.)
- load whatever is available and process (load & go)

Machines with load sizes smaller than the standard lot size are not too prevalent nowadays, but some are still being sold. Small-load equipment also needs batching rules. After filling a machine with something less than twenty-five wafers, there will be some number of wafers in the lot left over. When the left-over wafers are loaded for processing, will the machine's remaining capacity be filled with wafers from another lot?

In general, the batching rule at most fabs is 'no splitting of lots', i.e., wafers from two lots will not be in the machine at the same time. Thus, if a machine's capacity is 16 wafers, the lot will be processed in two batches, one of 16 wafers and one of 9 wafers, respectively.

Batching rules are covered in more detail in Section 6.3.

But before we move on, since we have brought up the concept of **standard lot size**, it makes an interesting historical footnote to point out the origin of the magical 25-wafer lot. Back in the 1960's, when etching was basically an exercise in wet chemistry, technicians used to wield oversized platypus-looking tweezers and move the then standard 2"-wafers from one sink to the next--for an etch, a clean, another etch, another clean, etc. But even under the best of circumstances, tweezers made poor excuses for wafer handlers and movers.

So along came the concept of a wafer carrier, something that the silicon disks could rest in between steps and be carried in from bench to bench. The chemical-supply catalogs had some small plastic boxes that sort of did the job.

Eventually, though, it was noticed that considerable time and effort went into individually taking the wafers out of the box, dipping them in the sink, and putting them back into the carrier; time and effort could be saved, as well as wafer breakage, if the entire carrier could be immersed for the wet clean and etch steps. As a result of a perceived need, the cassette was born. Sized for twenty-five 2"-wafers, and made of strong, acid-resistant plastic, the entire ensemble was lightweight enough to be manageable yet strong enough to protect the payload and get the job done.

Some time during the reign of the 4"-wafer, dunking of the entire cassette ceased to be practical; wafer uniformity could not be maintained. Also, the increase in wafer size made the cassettes much heavier and more difficult to handle. By the time wafer size had reached 6", cassettes were once again merely carriers and holders, and robotics were needed for a substantial amount of the handling. With 8" wafers, the twenty-five wafer cassettes are so heavy that human handling is dangerous, both for the health of the operators **and** the well-being of the very pricey payload. A dropped cassette of 8"-wafers would be the death of ten-thousand 486 chips.

Amazingly, despite the fact that the 25-wafer cassette has become unmanageable, and despite the fact that cassettes were originally sized with dunking in mind, the magic 25-wafer lot size endures. Equipment is designed and operated to the standard, and profit sheets are balanced around it. Not bad for a historical artifact that has outlived its usefulness. Moving on....

Dispatching rules determine which lot in a queue will be worked on next. Again, this decision will generally be made by the CIM system, based upon some rule and/or calculation. One simplistic rule might be to take the oldest lot in the queue. Another choice might be to take the lot with the earliest due date. But the most commonly used rule is **critical ratio** which selects the lot based on the ratio of the time remaining to the work remaining. The formula [5.2] for the critical ratio, CR_{LOT}, is:

$$CR_{LOT} = \frac{\text{Due date - Now}}{\text{Lead time remaining (incl. setup, process, wait)}}$$

If the ratio is 1.0, the job is on time. A ratio of less than 1.0 means the job is behind schedule, while a ratio greater than 1.0 indicates the job is ahead of schedule. The rule would be to process the lot with the smallest critical ratio. (Dispatching rules are covered in more detail in Section 6.4.)

Equipment maintenance rules govern when machines will be taken out-of-service for maintenance. The most-often used rules include:

- let it fail
- follow PM (preventative maintenance) schedule, completing lot in machine
- bring it down for (%) process drift

Scheduling: Factory scheduling may encompass a set of planning activities ranging from lot dispatching to long-range production planning. The length of the planning horizon provides a natural index for characterizing scheduling tasks. In most wafer fabs, detailed scheduling or lot dispatching is performed as often as once per shift, and provisional schedules may be generated on a daily, weekly, or monthly basis. Long-range production planning, on the other hand, is done on a time horizon that may be as long as a year. As the planning horizon lengthens, the accuracy of the projections decreases. In most wafer fabs, the greatest sources of scheduling uncertainty are customer orders, equipment availability, and yield.

In the smart fab houses, scheduling is done with the assist of a validated, factory-specific dynamic model that fully uses the fab's tracking data. As conditions change, e.g., failed equipment, the CIM system reports the new state of the factory and the model generates new schedules based on the factory's new status. The types of schedules and reports that are generated may include:

- lot-dispatch schedules
- machine schedules, and
- management inventory reports.

In the not-so-smart fab houses, scheduling is done based upon an infinite-loading mentality. Infinite-loading methods [5.2], as the name implies, assume that the fab has an infinite capacity. Even though these infinite-capacity, dispatch-control methods have simplicity on their side, needing only a knowledge of which lots are waiting at which workstation, the factories using these methods are generally plagued with high inventories, large queues, long cycle-times, and high equipment utilization. All of these effects are perfectly understandable once one concedes the irrefutable fact that the **factory is a physical system**.

The inherent advantage of scheduling using a model of the factory as a physical system is that factory capacity is addressed directly. When required production matches factory capacity, achievable production goals are reached, while still maintaining low inventory, small to moderate wait times, and acceptable cycle times.

In addition to dealing directly with finite capacity, a model also allows evaluation of alternative operating scenarios. When one or more feasible operations policies are identified, the model can then generate detailed schedules. (Chapter 7 discusses factory modeling.)

5.1.2 Factory Response Parameters (Response Variables)

The performance parameters that fall into this category are those which **result** from the operations policies exerted on the fab. The parameters are primarily involuntary since, in order to affect them, one or more of the control parameters would have to be altered.

Die Yield: As discussed in Chapter 2, a die or chip is the portion of a wafer that contains one complete integrated circuit, whatever its complexity. Die yield is the fraction of die on a wafer that function to specification as electronic systems. Many factors may impact die yield, some uncontrollable, some controllable [5.3]. In most fabs, control of particulate contamination is considered to be a key to good yield.

Yield loss may result from unplanned plant upsets such as impurities in a gas tank; or it may be the result of poor fab operating practices like carrying excessive inventory, thus leaving wafers sitting in long queues, waiting. In fact, two parameters which are virtually inseparable are yield and cycle time. Long cycle time leads to lower yield; only bad things happen when wafers wait. In fact, excessively long cycle times are usually indicative of sloppy manufacturing practices in general.

Throughput: Throughput may be defined as either the **flux** of material through a physical system over a given time interval, or the output or production rate. In other words, throughput is a gauge of productivity. For a wafer fab, throughput may be measured in terms of lots, wafers, or die (chips) over a period of hours, days, weeks, quarters, etc.

In addition to the throughput of the entire fab, there are also throughputs associated with workstations, processes, and process steps.

Cycle Time: Whereas throughput is the flux of material through a physical system, cycle time is the material's residence time within the system. It is a common fallacy to assume that throughput and cycle time either mean the same thing, or that they at least perform in the same direction. The reasoning behind this fallacy goes something like: if cycle time is markedly shortened, more cycles will be run a year, and thus more wafers will be produced.

This statement is true only if the factory starts empty (i.e., no inventory). If the factory has WIP, then new wafer starts will follow old wafer starts, and the flux of material moving through the factory will be a combination of all lots.

Highway Analogy

One way to visualize the cycle-time/throughput relationship is to draw an analogy with another physical system: the freeway. Analyzing performance and developing accurate intuition for wafer fabs is difficult, in part, because of the long time delay between stimulus and response. Freeways, on the other hand, are simpler systems and respond much faster; understanding their dynamics provides a basis for understanding fab dynamics. Thus, we will use the freeway in a fallacy-busting example to the idea that cycle time and throughput are interchangeable entities.

For this example, assume that the cycle time of interest is the length of time it takes for a car, starting at Exit A, to get to Exit B. If there is only one car on the road, then the cycle time is the 15 minutes it takes to get from A to B. The throughput of the system (in this case the freeway) is 1 vehicle/15 minutes (an average of 4 vehicles/hr).

If instead of being alone on the empty freeway, the vehicle is travelling in a close pack with forty other cars, then the cycle time is still 15 minutes, but the throughput is 41 vehicles/15 minutes (an average 164 vehicles/hr).

The above two scenarios correspond roughly to an empty factory (i. e., no WIP). Now consider the more realistic situation where the highway is free-flowing, but loaded with traffic. As the first vehicle starts at Exit A, there are 280 cars and trucks between it and Exit B. Because the freeway is free flowing during the time period in question, the cycle time is still 15 minutes; the throughput of the system, however, is 281 vehicles/15 minutes or 1124 vehicles/hr, when the lead car arrives. It is easy to see how WIP can make a huge difference in throughput.

Of the scenarios considered so far, the freeway has been free-flowing; the capacity of the system has not been exceeded. Now suppose that the 41 vehicles join the other 280 vehicles on a clogged freeway. Because the freeway is a semi parking lot, it takes the first of the 41 vehicles an hour and 15 minutes to transverse the distance between A and B. The cycle time has lengthened to 1.25 hours. To determine the throughput, we have to decide whether we want it over the original 15 minutes, or over the 1.25 hours that it takes the lead car to cross the distance.

Over the original 15 minute time horizon, none of the 41 vehicles, nor most of the other 280 vehicles (WIP) reaches Exit B. For this example, say only 84 cars make it across the finish line. (In the wafer fab, this would correspond to lots completing processing and leaving the factory.) The throughput for this time horizon, is 84 vehicles/15 minutes or 336 vehicles/hr.

Over the 1.25 hours that it takes the 41 vehicles to reach Exit B, the throughput is 321 vehicles/1.25 hours or 256.8 vehicles/hr. The 321 vehicles include all of the 280 vehicles already on the freeway between Exits A and B, added to the 41 starting at Exit A.

Table 5.2 summarizes how the different scenarios impact cycle time and throughput.

TABLE 5.2: CYCLE-TIME/THRUPT FREEWAY ANALOGY

	without WIP		with WIP	
Vehicle Start Rate	Cycle Time (hrs)	Throughput[1] (vehicles/hr)	Cycle Time (hrs)	Throughput[1] (vehicles/hr)
1 vehicle	0.25	4	0.25	1,124
41 vehicles	0.25	164	1.25[2]	256

1 Averaged over an hour.
2 Freeway is bottlenecked.

Schedule Stability: Schedule stability is the ability to meet production plans and ship scheduled lots on time. The most common method employed by wafer fabs to insure reliable and stable production is to carry a sizeable WIP. If WIP becomes excessive, however, long cycle times will result. Further, if the fab is operating at its capacity, schedule stability becomes a bottleneck control problem. If the fab is operating beyond its capacity, schedule stability is a joke. (See Section 5.3.5.)

Equipment Utilization: At any given point in time, a piece of equipment in the factory can be: in use processing; undergoing setup, calibration, testing, maintenance, etc.; or sitting idle waiting for work. Equipment utilization, or how the gear is being used, is defined in a variety of ways. One school of thought says that it is the fraction of time that the equipment at a workstation spends in production (i.e., processing). Others define utilization as the fraction of time that the equipment is busy doing anything (i.e., not idle). Yet others define it somewhere in between.

In the fab, utilization of equipment can range from 30 to almost 100% at the various workstations. The factors influencing the extent of workstation utilization are: whether it is a bottleneck or not, the availability of the equipment, any special batching rules, and the inventory level at the station.

If a workstation is a bottleneck, the equipment there may be busy virtually all the time that it is up and running. But, low availability may actually be the cause of a bottleneck, as is often the case in a wafer fab. In this situation, not only will availability be low, but equipment utilization will be low as well.

The factor most directly impacting equipment utilization is inventory, and it is the most dominant factor as well. High-inventory fabs generally have queues at each workstation. The lots waiting in these queues insure that equipment does not have to wait for lots to process, and large-batch equipment will be able to run at full or near-full loads.

High equipment utilization sounds like a worthwhile goal in fab operation, but is it really? The answer is both 'yes' and 'no'.

At the bottleneck workstations, equipment utilization should be as high as possible since these work center gate the throughput of the entire fab. At the non-bottleneck workstations, trying to force high equipment utilization is generally counter-productive. It will generally lead to long cycle times, and it may even have a disastrous impact on throughput. Furthermore, trying to force high equipment utilization may not even be sound economically, as will be shown later in the chapter.

5.1.3 Control and Response Parameters (State Variables)

Dynamic capacity, inventory, and bottlenecks are in many ways the most potent entries in Table 5.1. These factory conditions may be deliberately set by design and/or operating policies, or they may simply be allowed to 'just happen' as a result of the same or different policies. More specifically, fab policies may try to change or control these conditions, but the very same conditions are constantly responding to **everything** that is occurring within the fab. Thus, these parameters are both control and response variables. As such, they fall under the broad umbrella of state variables.

As a general definition, *state variables* use a system's past history to enable computation of its future behavior. Thus, in the strictest sense, state variables are not concerned with control so much as they are with defining the cause-and-effect relationships of system behavior.

Nevertheless, well-run wafer fabs do try to operate by setting these state variables. That is, operations managers try to directly control dynamic capacity, bottlenecks, and/or inventory level. But as the remainder of this chapter shows, controlling a fab's state variables is complex and challenging. The first, necessary step towards success is understanding the above-mentioned cause-and-effect relationships of wafer fabrication. For, if key variables like dynamic capacity are not well understood, and properly managed, then the fab will forever be running the operators and engineers, not the other way around.

Dynamic Capacity: Factory capacity is synonymous with the fab itself. It is both the dynamically-changing, physical system that does the fabrication of microelectronic structures, as well as the maximum quantity of structures that can be produced as the physical system changes with time. Thus, dynamic capacity is both the system and its capabilities.

A fab's capacity is inseparable from the equipment set housed within its walls, the process flows being performed with said equipment set, and the material being started and moved through the process sequences which will eventually become the fabricated products.

In an ideal world, a fab's dynamic capacity, its product throughput, and its planned production are identical. To be sure, if the fab has been correctly sized, and if equipment has been replaced or added as demands changed, then under normal operating conditions, capacity, throughput, and planned production do converge. However, if the fab has been mis-sized, the equipment set has not been updated as needed, or there have been unexpected operational problems, then capacity, throughput, and planned production may be vastly different.

Bottlenecks: The bottleneck of a factory is the workstation or workstations that throttle or choke the flow of lots through the fab, causing throughput to drop below design. Bottlenecks generally occur because the factory was under-designed to begin with, or the demands have changed over time. Whatever the source, and opinions may vary, there is generally consensus that bottlenecks are bad things and should be eliminated. They are, after all, the major cause of lost capacity.

The goal of any debottlenecking effort must be to increase throughput through the culprit workstations. This may be done straightforwardly by adding equipment, or indirectly by using the existing equipment more effectively.

One course of action that will definitely impact the behavior of bottlenecks, and may possibly shift or eliminate them, is to alter the inventory-management policies.

Inventory: As it turns out, inventory or WIP is the single, most influential, control variable a fab manager has for regulating overall factory behavior and productivity. A good inventory-management policy, one well-matched to a particular fab, will provide optimum factory performance. Without such a policy, inventory becomes almost entirely a response variable, responding to how the fab is being operated instead of helping to control it. Left uncontrolled, inventory swings fab performance in its erratic wake.

In order to be an effective control variable, inventory must reach a certain level, enough to insure that critical workstations stay busy. But, by the same token, inventory must **not** become excessive, clogging those critical workstations, and hence the fab. Unfortunately, it is this latter mode that many fabs find themselves in. Because large inventories act as a dampener to chaotic behavior, and because they ostensibly lead to higher throughputs, it is an operational policy with many advocates.

Still, progress has been made. By the mid 1980's, fab managers were realizing that cycle times ranging from ten to fourteen times minimum processing time (10 to 14x) were not only excessive, but also undesirable. When they began looking for the cause of the long cycle times, they did not have far to look. For all the apparent good that comes from running with large WIP, managers could not help but notice **what** WIP does most: WIP waits. The problem does not stop there.

Even though the quintessential experiment has not been run to correlate yield loss with long cycle time, it is well recognized that during long waits, bad things happens to the wafers. So in addition to long cycle times, excessive WIP leads to yield loss.

Still, lengthy cycle times were something many fab managers were willing to live with in exchange for stable factory operation. But when American manufacturing, in general, came to recognize that long cycle times meant slower development cycles, wafer fabrication also jumped on the bandwagon and embraced the concept of short-cycle time manufacturing.

With a concerted push to shorten cycle time, it was not uncommon in 1985 to find it in the range of 8 to 10x. By the end of the decade the multiplier had dropped to 5-6x. Now, in the mid 1990's, multipliers of 3 to 4x are standard.

Few fab managers can honestly say that when they set out to reduce cycle time they planned to do it by cutting back inventory. Yet of all the policies implemented in the quest for faster cycles, inventory reduction is the only one that can be pointed to that will categorically succeed. Still, reducing inventory makes fab managers nervous and must be approached gingerly. Although it is now widely recognized that much of the early operating lore was based on fallacies, it is hard to abandon an approach if it has been shown to give stable factory operation.

5.2 Inexorably-Bound Factory Performance Relationships

Of the numerous control and/or response variables described in Section 5.1, some are highly dependent, some are dependent only in certain operating regimes, and some are counter-dependent. When normally dependent or counter-dependent relationships disappear, then fab performance has become dominated by extreme conditions, and operation is generally unstable and/or outside of acceptable limits.

Leaving extreme situations aside for the moment, certain performance relationships are inexorably-bound in manufacturing, and most especially in wafer fabrication. These are:

- higher inventory gives higher throughput
- longer cycle time decreases yield
- higher inventory causes higher equipment utilization
- bottlenecks cost capacity

5.2.1 Inventory and Throughput

For many fabs, the fundamental premise of their operating policy is: If a fab is designed for a certain capacity, and if all equipment is kept busy, then design throughput will be achieved.

The necessary condition behind this premise, besides having a correctly-sized capacity, is that there be sufficient WIP to keep the equipment busy. At the very least, the bottleneck or constraining workstation must stay active.

Historically, to insure that equipment stayed busy, fabs have run with large to excessively-large inventories. And, because meeting production was the only performance yardstick that really mattered, operating with high inventory provided a direct means to achieve the throughput goal. But does higher inventory always result in higher throughput?

The answer is usually yes, but not always.

Fig. 5.1 shows the performance of two real CMOS fabs as a function of various initial inventory levels, distributed across the production lines. The results shown are for average throughput and cycle time for a fixed time interval of operation.

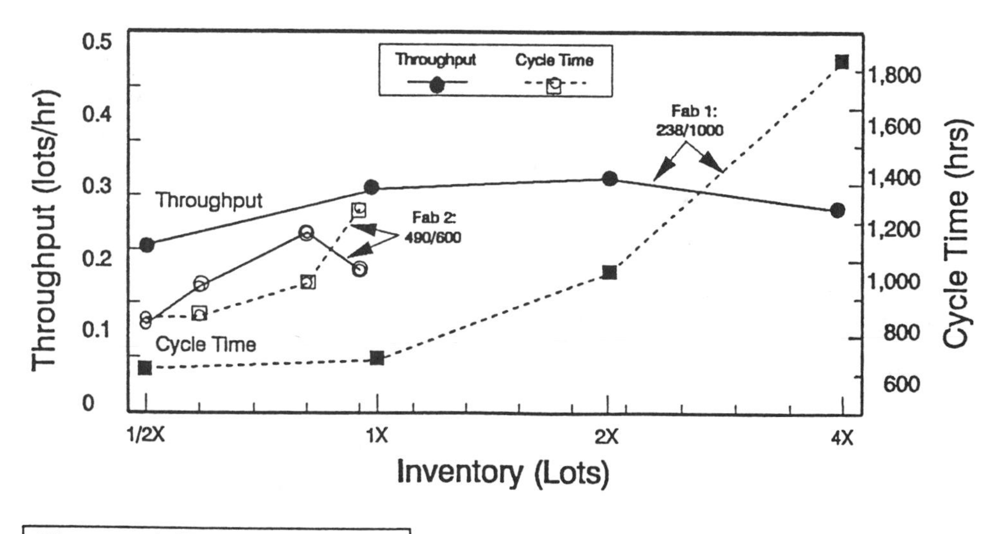

Figure 5.1: Performance vs Inventory, Two CMOS Fabs

Because of when the fabs were built, the two are markedly different in size and complexity. *Fab 1* is an old line, having only one metal layer, 238 process steps, and a moderate capacity of approximately 1000 wafer starts per week. *Fab 2* is a modern 'boutique' line, having four metal layers, 490 process steps, but only 600 approximate wafer starts a week. Both fabs are running at capacity; i.e., inventory is neither being built nor depleted.

The 'X' multiplier on the x-axis is simply a representative WIP for fab capacities. Note that in order to put these two fabs on the same graph, a semi-log plot is used, with the logarithmic x-axis having a base of 2. (On a linear plot, the slopes of the lines would not be so pronounced.)

As may be seen with both of the CMOS fabs, there are different regimes of fab performance, depending on the level of inventory across the line. For low-inventory or just-in-time (JIT) operation, any slight increase in WIP leads to an increase in throughput. For low-to-moderate WIP, increases in inventory still lead to increases in throughput, although the degree of the impact is smaller with larger initial inventories. For high to excessively-high WIP, more inventory generally does not lead to higher throughputs; it can, in fact, cause a drop in throughput by 'clogging' the fab. The congestion resulting from large WIP may exacerbate bottlenecks, and may even shift them from one work center to another.

To demonstrate the relationship between WIP-load, throughput, and equipment activity, Fig. 5.2 shows the Fab-1 workstations with activity ratios above 0.80, i.e., bottlenecks and potential bottlenecks.

At the lowest inventory level, '1/2X', the fab is almost a JIT operation; throughput is low, as is equipment utilization. Only one workstation, Wkst 37, is seeing use (i.e., not idle) more than 90% of the time.

Increasing the inventory level to the '1X' level makes for better fab performance; throughput increases without negatively impacting cycle time. Also, two additional workstations become busy more than 90% of the time (Wksts 23 and 38, in addition to Wkst 37).

At successively higher inventory levels, the equipment in the fab becomes increasingly busier. At the inventory level of

'2X', the fab is unarguably bottlenecked; any additional WIP causes throughput to decrease. The activity ratios indicate that Workstation #37 stays busy all the time (Utln = 1.00); even Workstations #23 and 38 may be a little *too* busy for comfort (Utln = 0.96 and 0.97, respectively).

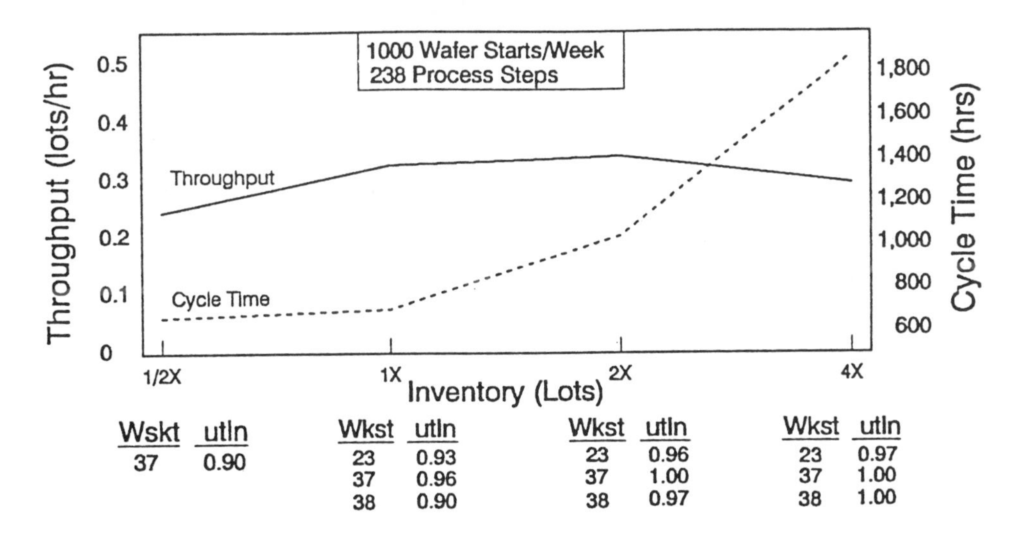

Inventory	Wkst	utln
1/2X	37	0.90
1X	23	0.93
1X	37	0.96
1X	38	0.90
2X	23	0.96
2X	37	1.00
2X	38	0.97
4X	23	0.97
4X	37	1.00
4X	38	1.00

Fig. 5.2: Equipment Activity vs WIP-Load

At the highest inventory level, '4X', the workstations that were busy, get busier. Since the fab was already bottlenecked at the 2X-level, increasing the inventory to 4X did nothing positive for performance; throughput went down, cycle time went through the ceiling, and the congestion in the fab reached such a high level that it can not clear on its own.

From what we now know about the three regimes of performance as a function of WIP load, the one that is of most interest for this fab is the one bracketed by 1X and 2X. Above 2X, performance declines; below 1X, throughput is probably not high enough to meet demand. So which is the better inventory level for this fab?

A cursory examination of Fig. 5.2 would leave one with an unclear idea of which is the better inventory level. The throughputs for the two cases are in the same ballpark at 0.337 and 0.376 lots/hr for the 1X and 2X cases, respectively. Cycle time, however, is markedly in favor of the 1X case (671.4 hours versus 1205.4 hours for the 2X case). But for a truly valid comparison, we must to put this performance information on a basis with an objective yardstick: $/good die.

If the fab is running 25-wafer lots, then the throughput in the two cases is 8.43 wafers/hr and 9.40 wafers/hr, or a difference of 0.97 wafers/hr more in the 2X case. Consequently, over a 168-hour work week, the higher inventory case is producing 163 more wafers than the smaller inventory case. If these wafers are the lower-end MOR (middle-of-the-road) variety that sell for $7000 each and have a production cost of $1500, then the throughput difference is worth $896,500 for the a one week period. If the wafers are the *el primo* $15,000 variety (production cost = $5000 ea.), then the higher throughput is worth $1,630,000 for the week. In other words, this apparently small throughput difference is worth between $11.65mm and $21.19mm for a quarter's operation, depending on whether the wafers contain low- or high-end chips.

This back-of-the-envelope calculation should demonstrate one thing quite emphatically: a small throughput difference can translate into big bucks. From a strictly throughput perspective, the best WIP load for this fab is 2X.

But is throughput the only criterion for choosing the best inventory level for a fab?

5.2.2 Cycle Time and Yield

The other performance parameter presented in Fig. 5.2 is cycle time, the time elapsed between when a wafer starts its processing sequence and when it finishes. As discussed earlier, short cycle times are important for a number of reasons, among them more development cycles per year, as well as higher yields. It is difficult to quantify the value of speeding up development cycles, but not so the value of improved yield.

While wafers are in the fab, they are either being processed or waiting around to be processed. It is this latter activity (or rather inactivity) that is one cause of yield loss. Certainly, there can be yield losses that occur during processing, but then whole wafers are generally affected and the event is more likely to be a yield crash as opposed to a yield loss. But while wafers are sitting around on racks waiting to go into machines, particles are being deposited on their surfaces—even in Class-1 clean rooms. Particles lead to defects, and defects lead to die loss.

There are a number correlations that relate yield to defect density on a wafer, among them the crude, but simple to use, Seed's Model [5.4]:

$$y = \exp(AD)^{-1/2}$$

where **A** is the area of the chip and **D** is the lethal defect density (particles per square centimeter). The lethal defect density is a direct result of the cleanness of the environment where wafers wait. It is a function of particle deposition rate, **p**, and the length of time the wafers spends waiting, **H**:

$$D = Hp$$

The longer a wafer waits, the higher the lethal defect density, and the lower the yield. Thus, a long cycle time, which is the result of long wait times, leads to lower product yields.

To illustrate, consider the cycle times for the 1X and 2X inventory cases, 671.4 and 1205.4 hours, respectively. The minimum processing time for this flow is 174.6 hours. Thus, the wait times for the two cases are 496.8 hours and 1030.8 hours.

The lethal particle deposition rate for the current industry standard, a Class-1 clean room, is 0.00005 particles/sq.cm-hr. Therefore, the lethal defect densities for the two cases are 0.0248 and 0.0515 particles/sq.cm.

With the size of current microprocessors at 2.5 sq.cm, the yields for the 1X and 2X inventory cases are 0.7794 and 0.6984, respectively. The much shorter cycle time in the 1X-WIP load case has superior yield advantages to the 2X-WIP load, making for a very interesting problem. As expected, the 2X case gives better throughput, but the 1X case gives better yield. But just as throughput means money, so does yield. So where do we run the fab?

Table 5.3 puts the two inventory cases on a common basis, '# of Good Die', which should make it possible to determine where to run the fab.

TABLE 5.3: THROUGHPUT vs CYCLE TIME
Fab 1 CMOS Process

	1X WIP Load	2X WIP Load
Thrupt (lots/hr)	0.3373	0.3760
(wafers/hr)	8.43	9.40
(die/hr)	843	940
Cycle Time (hrs)	671.4	1205.4
Yield	0.7794	0.6984
# of Good Die	657.1	656.5

Although the 2X case has notably better throughput, the superior yield in the 1X case makes it a toss as to which is the more economically attractive operating regime; the number of good die in the two cases is virtually the same. Which leaves the fab manager no better off than when he started: does he run with low or high inventory? What additional performance criteria will tip the scales? Will it be the faster development cycles in the low-inventory case, or the better schedule stability in the high-inventory case?

Furthermore, it should be pointed out that the value of **p** is a critical number in these calculations. The actual value of the lethal particle deposition rate will probably vary from fab to fab, depending on clean room practices, the use of SMIF (Standard Mechanical InterFace) boxes, degree of automation, etc. Also, the definition of 'lethal' may be open to interpretation. Table 5.4 shows how variations in **p** can impact yield (according to Seed's Model).

TABLE 5.4: YIELD vs LETHAL PARTICLE DEPOSITION

P (Particles per sq.cm-hr)	0.001	0.0005	0.0001	0.00005	0.00001
1X-WIP Yield	0.3281	0.4547	0.7030	0.7794	0.8945
2X-WIP Yield	0.2008	0.3214	0.6019	0.6984	0.8517

1X Case: 496.8 hrs wait time
2X Case: 1030.8 hrs wait time
Chip Size = 2.5 sq. cm.

Finally, it should be noted that the above comparison, in addition to not considering indeterminate factors such as faster development cycles or degree of schedule stability, also gives no credit for the economics of higher equipment utilization in the higher inventory case.

5.2.3 Inventory and Equipment Utilization

From the utilizations ratios given in Fig. 5.2, it is apparent that the equipment stays much busier in the 2X-inventory case than it does in the 1X case. It is, after all, a fundamental law of manufacturing that large inventories keep equipment busy. High equipment utilization is second only to high throughput as the key reason for fabs carrying large inventories. To understand why, one need look no further than the cost of the equipment.

As may be seen from Table 5.5, semiconductor processing equipment is not cheap. When one machine can cost several million dollars, it is only human nature, and presumably sound economics, to want to keep the tool as busy as possible. Because not only is fab performance rated on the number of good die made, it is also rated on the cost of making those good die. Figured heavily in that expense is the cost of the factory, i.e., the depreciation of the equipment and building.

For example, take Workstation #38 which consist of one plasma etcher costing $1,500,000. If the write-off period is five years, and simple straight-line accounting is used, then the depreciation for this machine is $300,000/year. Assuming the fab can get 360 good days a year, and uses 24-hour days, then the depreciation is $34.72 per hour.

TABLE 5.5: EQUIPMENT PRICING

Equipment Type *	Cost
Stepper	$2,500,000
Coater/Developer	1,000,000
Asher	300,000
Dry Etch	1,352,000
CVD	1,355,000
Metal	2,500,000
LPCVD	1,000,000
Diffusion/Oxidation	750,000
L. Implant	2,100,000
H. Implant	3,000,000
Sputter	2,500,000
Wet Stations	750,000

* Non Cluster Tools.

It stands to reason that if equipment is used more per hour, then cost will be spread over more wafers and more die. It is a logical assumption, one that pervades the industry, and one that can be grossly wrong.

When fabs are designed and sized using spreadsheets, the first line of attack is to determine the throughput of the machines at each work center. This calculation is based on an expected process time for the step or steps that will be performed there, as well as the expected down time of the equipment type. Other than down time, spreadsheet modelers usually assume no idle time at a workstation, disregarding the dynamics of fab operation. In reality, only bottleneck or constraining workstations have 100% utilization. Bottleneck performance will usually swamp anticipated advantages stemming from distributing depreciation over a few more wafers or die.

Again, consider Workstation #38 with a design throughput of 9.7 wafers/hr (970 die/hr). As may be seen by the utilization numbers on Fig. 5.2, in the high-inventory (2X) case, the equipment is active 97% of the time (Utln ratio = 0.97). Thus, the actual throughput is 9.4 wafers/hr (940 die/hr). With a machine depreciation of $34.72/hr, spread over 9.4 wafers/hr throughput, the product depreciation cost is $3.69/wafer, or $0.0369/die.

Using this same simplistic reasoning for the small-inventory (1X) case, the utilization ratio of 0.90 would give the workstation a throughput of 8.73 wafers/hr or 873 die/hr (i.e., 0.90 times the design value of 9.7 wafers/hr). The machine depreciation spread over this throughput would give a product depreciation cost of $3.98/wafer or $0.0398/die. Obviously, because of the inferior equipment utilization in the smaller inventory case, the fixed cost per wafer or die is more.

WRONG!

As is often the case, when attempting to analyze a complex system by analyzing its individual parts, this result is not only erroneous, but also misleading. Product costs, whether fixed or operating, may only be written off against *good* die or wafers. All others are destined for the dust bin. Understanding this basic and inviolable axiom is imperative for making good operating decisions in wafer fabrication. To get a meaningful perspective on the importance of equipment utilization when writing off capital expenditures, it is necessary to take a global or fab perspective, as oppose to a myopic or equipment perspective.

We know from Table 5.3 that the number of good die produced, not just through one step but through the entire fabrication process, is 657/hr for the low-inventory case and 657/hr for the high-inventory case. Therefore, the depreciation cost for Workstation #38, spread over the **saleable** product, is $0.0528/die for either case. Not only do the numbers change by looking at the whole picture (Table 5.6), but also the conclusions on how the fab should be operated. Basing fab operation on equipment utilization would mandate operating with the higher inventory. But the lower inventory case, despite the idle time on equipment, gives just as good an ROI. One could argue that if fab performance is maximized in terms of throughput **and** yield (i.e., # of good die), then equipment utilization should not enter into the cost equation, per se.

TABLE 5.6: EQUIPMENT DEPRECIATION
Two Perspectives

	Throughput (die/hr)	Depreciation Cost ($/die)
Utln Perspective:		
2X	970	0.0369
1X	873	0.0398
Actual Operation:		
2X	657	0.0528
1X	657	0.0528

Workstation #37 is $1.5mm Plasma Etcher.
2X Utilization = 0.97
1X Utilization = 0.90

The only instance where tracking of equipment utilization is critical is for bottleneck control.

5.2.4 Bottlenecks and Dynamic Capacity

Even with the best design tools in the world, more often than not, by the time a fab comes on-line, its capacity is undersized, i.e., the fab can not achieve the design throughput. From there, things only get worse.

When a fab is sized, design engineers are given a process flow which lays out the process steps, the order of those steps, and the length of time required for each. This information, combined with the desired throughput, enables the engineers to determine the type and quantity of machines needed at each work center.

But between the time the fab is sized and when it comes on-line, the process continues to evolve, sometimes quite radically. Product development never stops, not during fab sizing, not during the construction, and not during or after start up. Process times lengthen, steps are added, some steps are clustered, etc. Almost invariably, the new, improved changes to the original process flow require additional resources (hardware), which generally do not make it onto the factory floor. Unfortunately, all it takes is for one workstation to be short of equipment, and it will bottleneck the whole factory. The factory can never do better than its worst bottleneck; it chokes or gates the flow through the entire line. Conversely, removing or easing a bottleneck can expand the capacity of the line.

Even when a bottleneck is removed, and the fab achieves desired production, it will not last. Over time, more bottlenecks will surface, some due to operating practices such as:

- excessive inventory
- more lengthening of steps (process creep)
- unplanned-for engineering lots
- changes in product mix and/or priority
- etc.

and some due to the aging of the plant itself such as:

- poorer equipment availability
- longer and more frequent calibrations
- catastrophic failures on certain equipment
- etc.

Some bottlenecks will persevere until remedial action is taken, such as adding equipment at a workstation. Others will disappear on their own during normal fab operation, such as with the clearing of an inventory bubble. Thus, just as factory capacity is a dynamic variable, so too are the existence and shifting of bottlenecks. The tracking of bottlenecks, and controlling them when warranted, is the single, biggest leverage that can be exerted over a factory's capacity, and hence its production throughput.

To study the relationship between bottlenecks, bottleneck shifts, and capacity, consider again the performance results of Fab-1 (Fig. 5.2). As is often the case when a fab is running full out, at or beyond design capacity, high inventory can cause certain workstations to be busy **all** the time that their equipment is in service. Witness Workstation #37 in the 2X case.

In general, as was the case in this real fab, long cycle times are directly attributable to the bottleneck workstation(s). When this type of performance occurs, there are three options available to the fab manager:

- do nothing,
- back off on inventory, or
- add equipment.

If production **is** being met, many if not most managers will opt for the 'do nothing' approach. But before this option is chosen, an analysis should be made to determine the yield penalty associated with the long cycle times.

The second option, backing-off on inventory, will also cost nothing in terms of capital outlay. But the controlled reduction of WIP would provide the benefits usually associated with reducing cycle time (improved yield, shorter development cycles, greater market responsiveness, etc.).

However, if backing off on WIP causes an unacceptable drop in production, then the third option, adding equipment, may be the only sensible choice. The additional throughput gained, as well as yield improvements, will most likely pay for the added equipment in a matter of weeks.

To illustrate, at the 2X-inventory level, Fab-1 has three workstations with utilization ratios above 0.95: Workstations #23, 37, and 38. For a first-pass debottlenecking, one piece of gear will be added at each of these work centers. Since Workstation #37 has a utilization over 0.95 for the 1X-inventory case, one machine will be added there for comparison. The results of the additions are shown in Fig. 5.3 and given in Table 5.7.

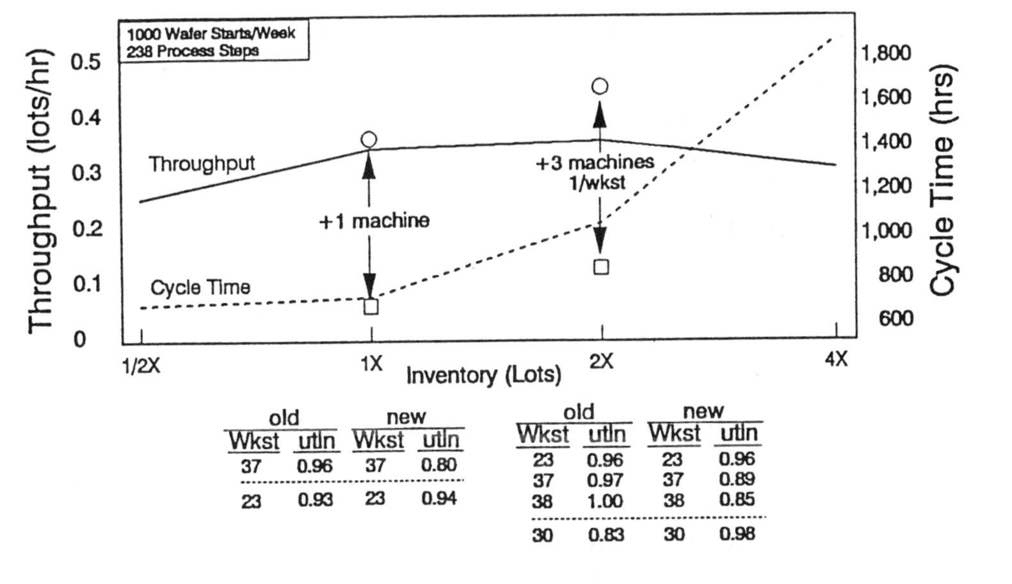

Figure 5.3: Debottlenecking Fab 1

TABLE 5.7: ADDING EQUIPMENT AT BOTTLENECKS

Fab 1 CMOS Process

	1X Case		2X Case	
	Base	+ 1 Machines	Base	+ 3 Machines
Cycle Time (hrs)	671.4	656.8	1205.4	850.4
Thrupt (lots/hr)	.337	.343	.376	.458
Extra Capital Cost $	–	1,500,000	–	3,750,000
Yield	0.7794	0.7823	0.6984	0.7478
# of Good Die per Hr	657	671	657	857
Δ Good Die/Hr	–	+14	–	+ 200
Δ Value/Hr ($75/die)	–	$1050	–	$15,000
Capital Payback	–	8.5 weeks	–	1.5 weeks

238-Step, 1000 Wafer/Week CMOS Process Flow.
Wkst #23: $0.75MM Flo-Ox Furnace
Wksts #37 & 38: $1.5MM Plasma Etchers

For the 2X-inventory case, the addition of three machines (two plasma etchers and one oxidation furnace) raised throughput from 0.376 to 0.458 wafers/hour. The addition also reduced cycle time, from 1205 hours to 850 hours, which coincidentally gave a yield improvement of 5% (from 0.70 to 0.75). The total impact of spending an additional $3.75mm on equipment ($0.75mm for #23 and $1.5mm each for #37 and 38) was to increase the number of good die produced in an hour from 657 to 857. The capital payback period is less than two weeks.

The impact of adding the one plasma etcher in the 1X-inventory case was far less dramatic. There were modest improvements in both throughput and cycle time, and hence yield, which increased the number of good die per hour from 657 to 671. But even with such small performance improvements, the capital payback period for the $1.5mm plasma etcher (Wkst #37) was still short, less than nine weeks. However, the best operating strategy, by far, for this fab is to add the three pieces of equipment and run at the higher 2X inventory level.

Studying Fig. 5.3 brings up an interesting dilemma. When the three pieces of equipment are added in the 2X-WIP case, the utilizations on the two most heavily used workstations, #37 and 38, drop below 90%. These two workstation are no longer limiting fab capacity. A bottleneck shift has occurred, and the new fab bottleneck is Workstation #30, with a utilization of 0.98, followed by Workstation #23, with a utilization of 0.96. Would adding one more piece of gear at each of these workstations give a further boost to fab performance?

Fig. 5.4 and Table 5.8 show the performance impact on the fab of adding another oxidation furnace (Wkst #23) and another metal etcher (Wkst #30).

Whoops! Who would ever have thought adding more equipment would cause performance to crater? Someone shoot the fab manager! Or, at a minimum, find out what went wrong.

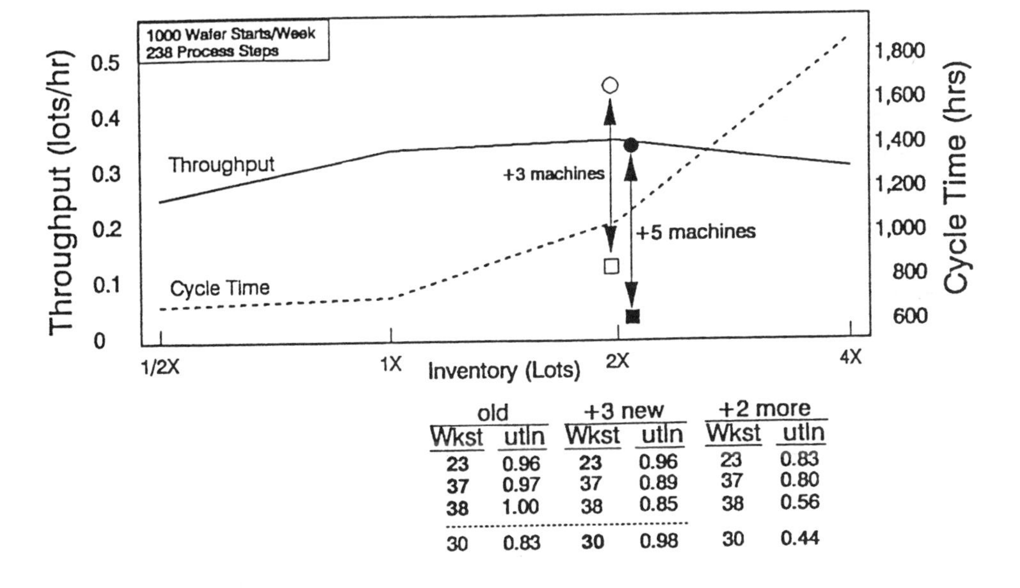

Figure 5.4: Second-Level Debottlenecking of Fab 1

TABLE 5.8: SECOND-LEVEL DEBOTTLENECKING

Fab 1 CMOS Process

	2X Case		
	Base	+ 3 Machines	+ 2 More
Cycle Time (hrs)	1205.4	850.4	602.3
Thrupt (lots/hr)	.376	.458	.350
Extra Capital Cost $	–	3,750,000	6,000,000
Yield	0.6984	0.7478	0.7936
# of Good Die per Hr	657	857	694
ΔGood Die/Hr	–	+200	+37
ΔValue/Hr ($75/die)	–	$15,000	$2,775
Capital Payback	–	1.5 weeks	12.9 weeks

238-Steps, 1000-Wafers/Week CMOS Process Flow.
Wkst #23: $0.75MM Flo-Ox Furnace
Wksts #37 & 38: $1.5MM Plasma Etchers
Wkst #30: $1.5MM M-Etcher

5.2.5 Bottlenecks and Factory Control

The addition of the last two pieces of equipment to Fab-1 effectively eliminated any real or potential bottlenecks. No workstation has a utilization ratio above 0.83. In other words, the entire fab has a burst capacity of 17%. With this much excess capacity, there are no physical constraints to choke or control the flow of lots. As a result, the fab will quickly 'chew up' or clear inventory, bringing the WIP down close to its inherent 1X level. In other words, inventory will consist of lots in machines or shortly to be in machines; queues will disappear.

In general, when excess capacity is added to a fab, and no attempt is made to control lot moves (i.e., not loading machines simply because they are available), performance will approach the 1X case in terms of both throughput and cycle time. (For Fab 1, the 1X case originally had a burst capacity of 4%, which was more than sufficient to keep in-line inventory low.) As we have seen throughout much of this chapter, whereas the 1X (or JIT) operation may be optimal in terms of fast cycle times, it is far from optimal in terms of throughput, schedule stability, and true economics.

In actuality, excess capacity is rarely a problem in real wafer fabs. Rather, the other extreme is usually the case. But understanding the causal chain of phenomena in fab operation is a prerequisite to understanding and analyzing factory performance:

Bottlenecks set plant capacity. Plant capacity, while trying to accommodate the lot-start rate, determines the in-line inventory. In-line inventory bounds factory performance. The cause-and-effect chain reaction of these phenomena may only be changed or disrupted by cognizant scheduling of lot-moves, and a willingness to leave machines idle even though non-scheduled lots may be available for processing.

To illustrate why factory performance always comes back to the existence and handling of bottlenecks, we will reintroduce the simplified CMOS fab that was used extensively in Chapter 4 (see Table 4.1) and shown again in Fig. 5.5. For the following discussion, it will be assumed that CMOS-1 does not have an intelligent move scheduler.

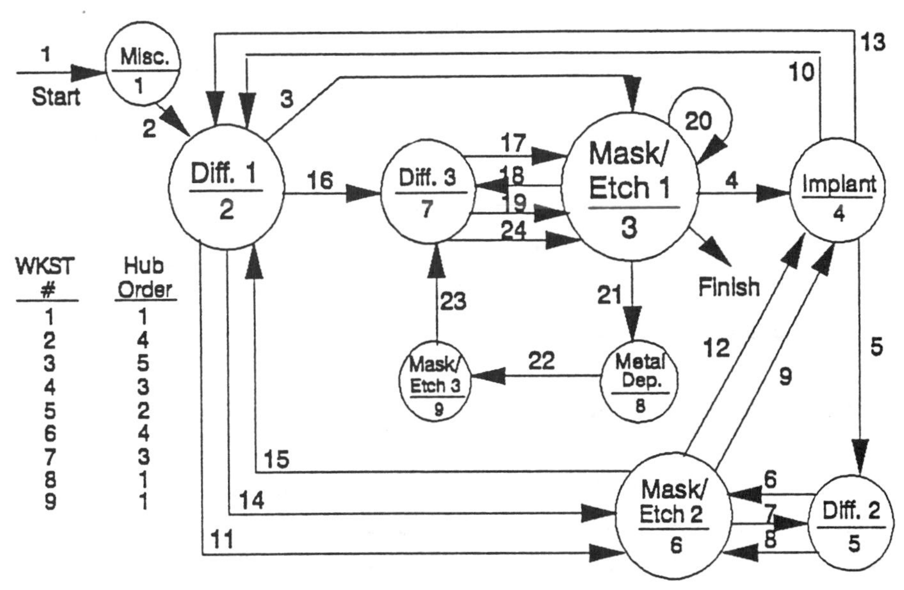

Figure 5.5: Simplified CMOS-1 Fab

So, when is a bottleneck *really* a bottleneck? Is the existence of a bottleneck necessarily a bad thing? How should bottleneck(s) be handled?

By definition, the bottleneck of a fab is the workstation (or workstations) that throttle or choke the flow of lots through the entire factory. Generally, the bottleneck workstation(s) will be the one(s) with the minimum capacity relative to processing needs. But, what if one of these minimum-capacity workstations has a utilization ratio of only 0.9, is it truly a bottleneck?

The answer to this question is: probably not. If the highest-utilized workstation still has a burst capacity of 10%, then it is unlikely to be choking flow through the fab.

What if the utilization ratio is 0.97? Is the workstation then a bottleneck?

This is a much tougher question, and the answer is: maybe, from time to time, depending on the dynamic conditions of the fab.

Then what if the utilization ratio is 1.0? Is the workstation then a bottleneck?

The answer is: Definitely.

Thus, we have three different possibilities for the most heavily used workstation in the plant, ranging from 'probably not' to 'definitely is' a bottleneck. Perhaps it is time to redefine the term bottleneck.

The true measure of whether a workstation is a bottleneck or not is if it is *limiting the capacity of the plant below the desired start rate.* So, with this criterion in mind, does the simplified CMOS-1 have a bottleneck that is limiting factory performance?

5.3 Operations Policy and Fab Performance

To explore performance as a function of operations policy, we will observe CMOS-1's behavior over a range of lot-start rates. That is, the physical capacity of the plant will remain unchanged, but the start rate will be varied from 1 lot/hr to 1 lot/0.7 hrs, in effect varying capacity demand.

Furthermore, since we know inventory can play a pivotal role in fab performance, each start rate will be investigated with initial WIP ranging from small to large quantities (an average of 1, 3, 8, and 15 lots per process step).

5.3.1 Excess Capacity

For the case where CMOS-1 has a design capacity, and hence a lot-start rate, of 1 lot/hr, Fig. 5.6(a) shows the factory's throughput behavior over time with various levels of initial WIP. In the case where initial inventory is small (1X), throughput is initially lower than design but gradually works its way up.

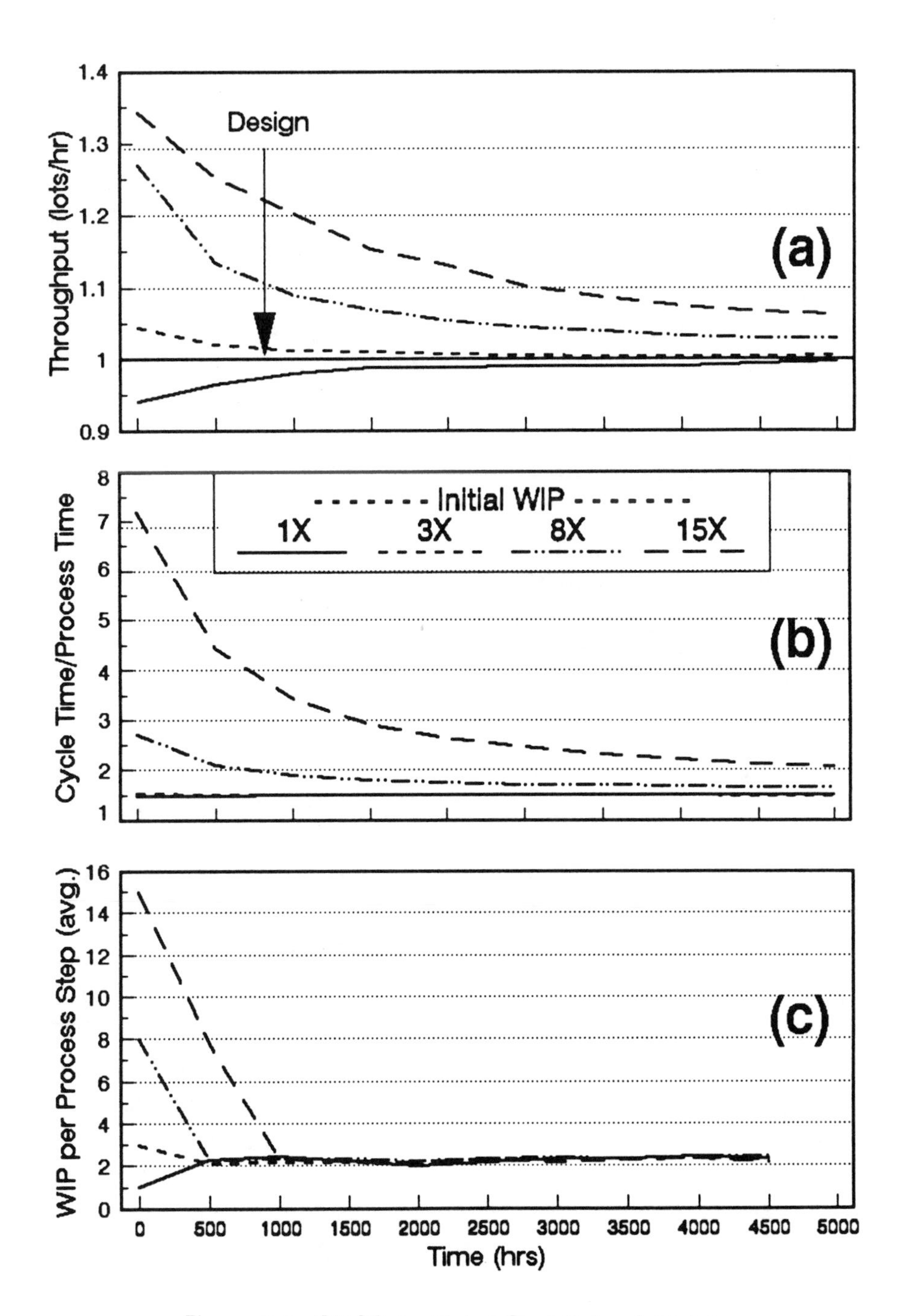

Figure 5.6: CMOS-1 with Lot-Start Rate of 1 lot/hr

With the close-to-optimal WIP (3X), throughput starts out a little high, but quickly settles to the design value. In the high inventory cases (8X and 15X), throughput starts out quite high, falls off precipitously to a still-high value, then gradually declines over the months, converging on design. Looking at the total throughput picture for this lot-start rate, the fab apparently converges to its design value in every case, regardless of starting inventory.

Analogously, cycle time behavior for this lot-start rate, Fig. 5.6(b), seems always to settle down to a very low, JIT-type of operation, even when there is a high starting inventory.

A clue as to how the factory's performance is adjusting itself over time is provided by the WIP profile, or WIP stability, of the fab, Fig. 5.6(c). Regardless of whether the fab starts with very high or very low inventory, the WIP level quickly adjusts itself to an average, low value of two lots per process step. The actual value of this 'intrinsic' WIP is factory-specific, but the trends observed in Fig. 5.6(c) are typical of real fabs.

Intrinsic WIP is the small amount of work-in-progress that is associated with a JIT operation. It consists primarily of lots being processed in machines, lots being loaded into or unloaded from machines, lots being moved from workstation to workstation, or lots experiencing a very small wait while the lot in front of it is wrapping up a step or a move. The term has no relevance in situations where there are queue waits.

The explanation for the performance of CMOS-1 at the lot-start rate of 1 lot/hr is that the fab has considerable extra capacity; all workstations have 20% or more idle time. With 20+% burst capacity, the fab quickly clear excess WIP. Low WIP (i.e., no queues) means low throughputs and fast cycle times. It would take a catastrophic set of events, or deliberate operational changes, to constrain a fab with this much excess capacity.

Moving to a situation with a greater demand on the factory, Figs. 5.7(a)-(c) show the performance behavior of CMOS-1 with a lot-start rate of 1 lot/0.9 hrs.

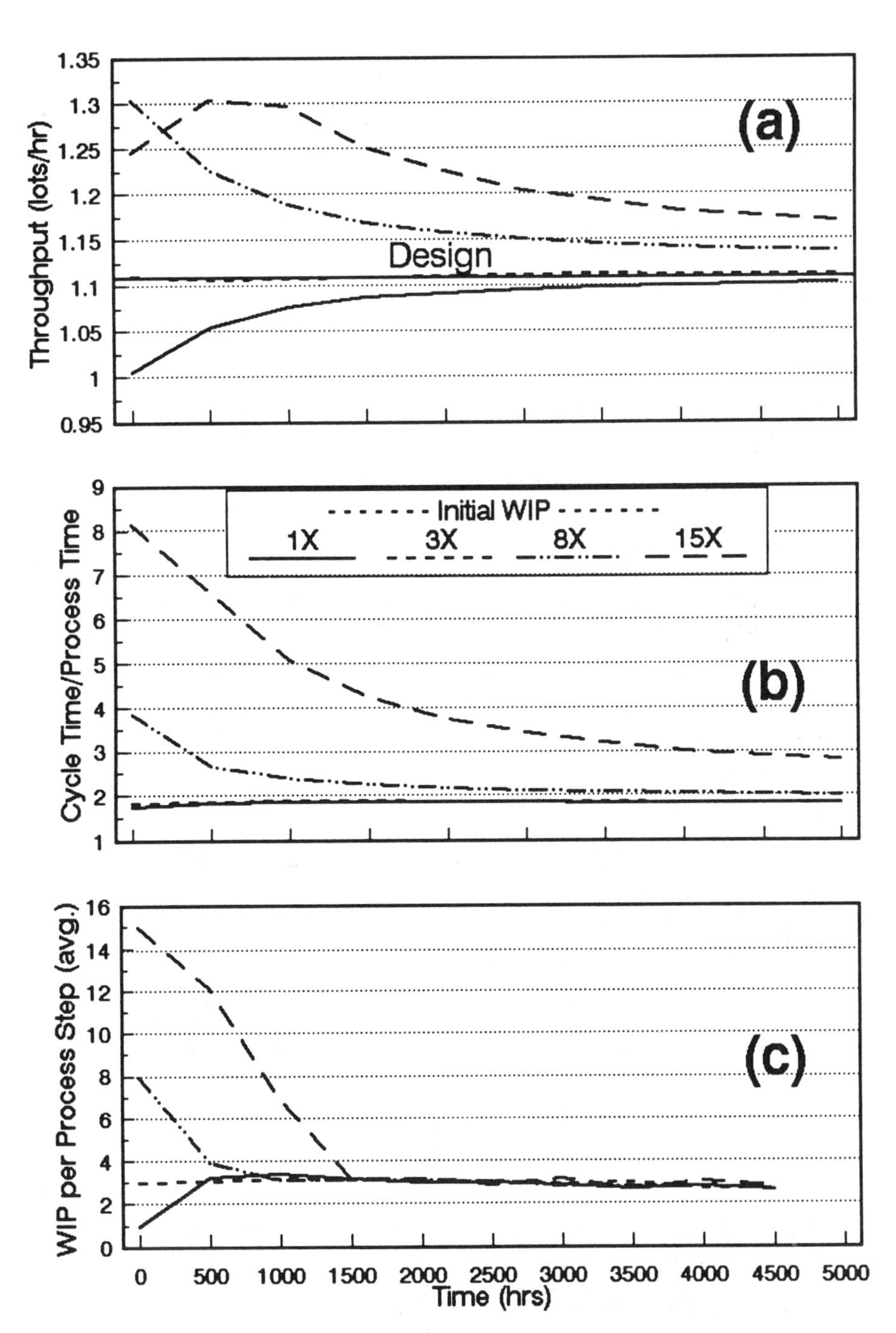

Figure 5.7: CMOS-1 with Lot-Start Rate of 1 lot/0.9 hrs

At 1 lot every 0.9 hrs, lots are being started more frequently than in the previous case, thus using up some portion of the excess capacity. But there is still enough spare capacity to enable CMOS-1 to control itself. It takes the fab a little longer to 'settle down', or converge to its inherent WIP, but it still gets there rather quickly, regardless of starting inventory. For this start rate and this fab, the inherent WIP is about 3 lots per process step.

Likewise, at the faster start rate, it takes throughput and cycle time performance longer to converge to their steady-state values (especially for the high 15X inventory case), but they will eventually get there. Even in the 15X-WIP case, the highest equipment utilization ratio (Wkst #3) is only 0.92. With a burst capacity of 8% or more, convergence is inevitable.

The only unusual result for this lot-start interval occurs in the very early stages with the 15X WIP case. Even though the initial inventory is staged, it is staged by process step and not by burst capacity. When the fab starts up operation, the work center with the smallest burst capacity, Workstation #3, will also have a very large bubble of inventory in front of it because it is also a fifth order hub. The inventory waiting in front of Workstation #3 will actually 'starve' other portions of the flow for a while until the bubble is redistributed. Hence, when operation begins, the utilization ratio of Workstation #3 is 1.0, and throughput will be lower than expected until the ratio drops off, indicating the bubble has cleared. The improved throughput is only temporary, however; once the bubble has cleared, the fab has burst capacity and it can begin to drain WIP, leading to lower throughputs.

So far, we have seen two cases where the fab has some, or a lot, of excess capacity. In both cases, barring human intervention or catastrophic events, the fab will basically adjust its WIP to its intrinsic level. More inventory than that level and the fab will devour the excess; less inventory and the fab will build it up until it reaches the comfort zone.

For a single-process fab, the intrinsic WIP level is a function of:

- number of process steps, workstations, and granularity,
- equipment availability and load-size mismatches,
- idle time at bottleneck(s), if any, and
- lot start rate.

For a multi-process fab, process mix and product priority also become factors.

Determining the intrinsic WIP level of a fab may be one of the few engineering problems better solved by practice than theory. Without proactive control, and assuming there is at least some small idle time at every work center, over time a fab will adjust its inventory to its intrinsic or average steady-state level.

Ironically, a fab's comfort zone for inventory level may be diametrically opposed to a fab manager's comfort zone, since the intrinsic level is usually small (only large enough to balance out the line). Two to three lots per process step will not provide that warm, cozy feeling when equipment unexpectedly fails. Nor will such a low level keep equipment utilization high enough for most managers. Also, the smaller the inventory, the lower the throughput. So, regardless of the value of the inventory state variable, the incentives of production stability, equipment utilization, and throughput beckon the fab managers toward higher inventory levels.

But, if a fab is always adjusting itself back to its intrinsic WIP, how can higher inventories be achieved and maintained?

The most obvious method for keeping inventory higher is to adjust the capacity of the fab, which usually happens with time anyway.

The capacity of the fab may be adjusted by:

- starting more lots (production or engineering)
- adding process steps or suffering process creep
- experiencing lower availability at workstations
- holding up lot moves

The long-term effect of these strategies, if not done judiciously, is that they will either clog the fab, create one or more bottlenecks, or otherwise artificially hamstring moves. Whatever the outcome, the bottom line is that capacity will be adjusted in such a way that the *burst* capacity either disappears or becomes controllable.

In the real world this discussion is somewhat academic. Excess capacity is not a common problem for a fab. By the time a fab comes on-line, numerous process changes have usually taken place since the original sizing. Since re-sizing is not an option, it is fortuitous luck if there is even sufficient capacity, much less any burst.

5.3.2 At Capacity

Figs. 5.8(a)-(c) show the performance of CMOS-1 when the start-rate interval is further cut back, to 1 lot/0.8 hrs. As opposed to the previous two excess-capacity cases, the shapes of these curves have a different story to tell. Perhaps of greatest interest, fab behavior now seems to be influenceable by the starting conditions, in particular, the starting WIP. At this lot-start rate, fab capacity is constrained, but not bottlenecked.

CMOS-1 goes through an initial WIP redistribution period, where bubbles of inventory at high-utilization workstations are smoothed out. Until this smoothing out takes place, the throughput is always below design, regardless of initial WIP. But once the inventory has redistributed, the throughput works its way towards the design value, even in the very low WIP case. So, from a throughput perspective, the manager of this fab will be able to meet production, yet still feel free to pick the inventory level at which he wants to run the fab.

Cycle time, on the other hand, is very much determined by the WIP level run in the fab. For the low inventory cases, 1X and 3X, the cycle-time-to-process-time ratio is just over 2, a JIT-type of operation. For the 8X case, the CT/P ratio settles at about 4, a value that is reflective of 1990's operation. For the 15X case, the ratio is about 8.5, a fairly common value in the not too distant past.

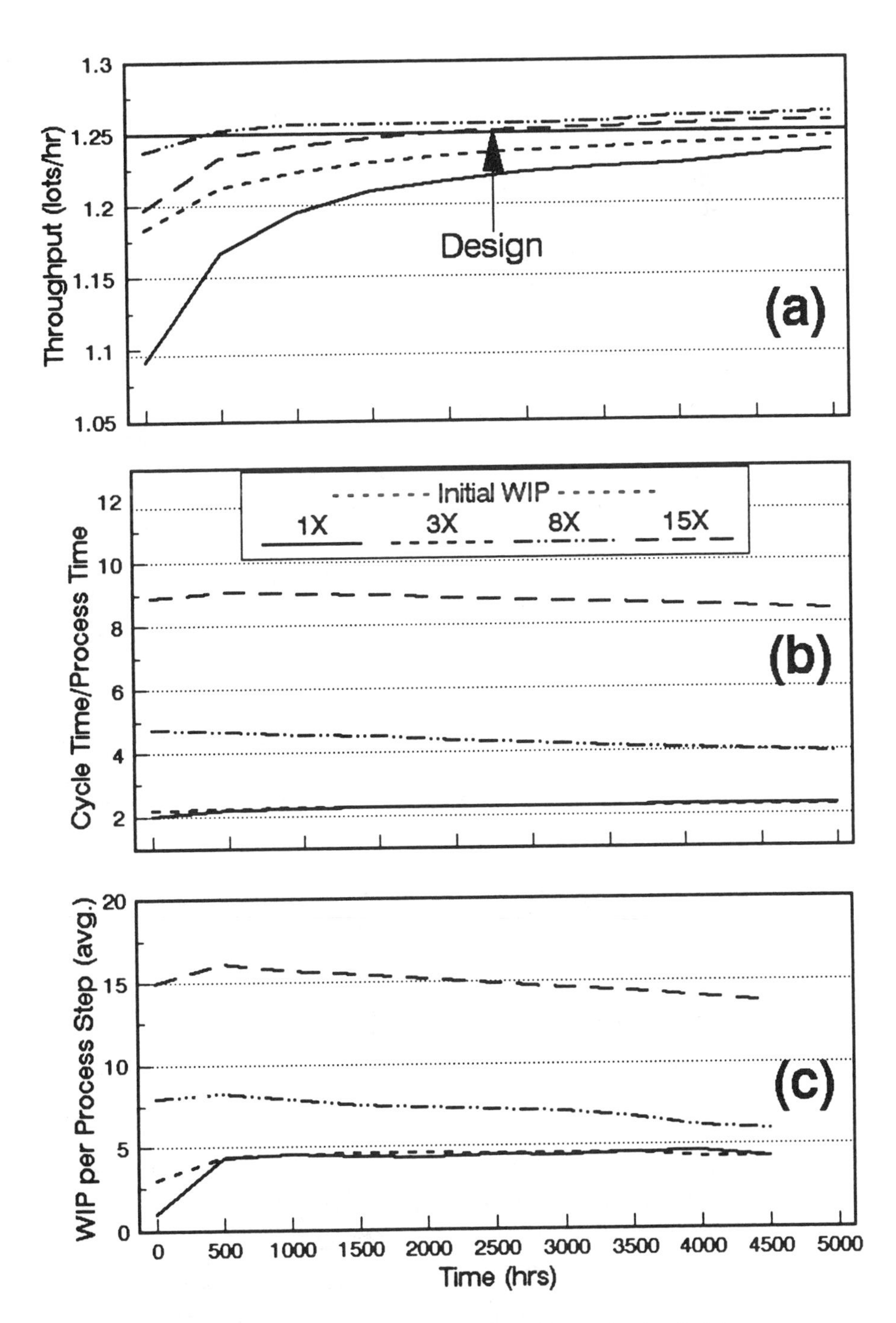

Figure 5.8: CMOS-1 with Lot-Start Rate of 1 lot/0.8 hrs

The tight relationship between CT/P ratio and initial WIP, for a properly-designed fab, elucidates the direction the industry has been taking over the last decade. Inventories have been getting smaller, and CT/P ratios have been getting smaller. The beauty of operating at, or very near, design capacity is that production will generally be met, and an inventory level may be selected that will give a reasonable cycle time, and yet provide good operational and schedule stability. Once the inventory level is decided, it will remain reasonably constant over time, barring serious fab upsets or operational changes.

So, why at this start rate, and regardless of starting WIP, is this fab operating at capacity? A look at Table 5.9 provides the necessary insight.

TABLE 5.9: HEAVILY-UTILIZED WORKSTATION PROFILE

Start Interval	1X		3X		8X		15X	
	wkst	utln	wkst	utln	wkst	utln	wkst	utln
1.0	--	--	--	--	--	--	--	--
0.9	--	--	--	--	3	0.900	3	0.920
0.8	3	0.986	2	0.913	3	1.000	3	1.000
	6	0.928	3	0.991			6	0.940
			6	0.943				
0.7	2	0.916	2	0.914	2	0.911	3	1.000
	3	0.995	3	0.999	3	1.000	6	1.000
	4	0.926	4	0.927	4	0.927		
	6	0.995	6	0.998	6	1.000		
	8	0.934						

Workstations w/ Utilization Ratios > 0.90, only.
Lot-Start Rate = 1 lot/Intv-hr (i.e., 1 lot/0.9

At the low inventory levels, 1X and 3X, only Workstation #3 has an especially high equipment utilization ratio. But even with high utilization, there is still a small amount of burst capacity. This burst capacity gives the fab some slight maneuverability to adjust itself. The adjustment that it makes is to establish the intrinsic WIP level at a comfortable 4.5 lots per process step.

Once the WIP level has adjusted to this level, spread across the fab, production approaches design, and cycle times are reflective of a constant, low-inventory operation.

At the high inventory levels, 8X and 15X, Workstation #3 is a true bottleneck; its equipment is in use all the time that it is in-service. With a utilization ratio of 1.0, there is no question that this work center is gating the flow of material through the entire fab. Why then is there not more of an adverse effect on throughput and cycle time performance?

The reason is that even with, or more accurately because of the bottleneck, the capacity of the fab can handle exactly the number of lots started: 1 every 0.8 hours. Because the start rate matches capacity, the only impact of having a larger starting WIP is that inventory stays high and it takes product longer to get through the process sequence, hence longer cycle times. The ideal place to run this accurately-sized fab is with a starting WIP of 4.5 lots per process step.

Running a fab at its exact capacity is both very enticing, as well as being somewhat chancy. The fab is obviously very stable, and able to meet production demands over the long haul. But should a mishap occur that causes inventory to build, there is no excess capacity to clear it. Hence, the standard one or two 'unexpected' interrupts a week due to any variety of reasons could produce a significant increase in WIP, as well as the associated lengthening of cycle time. But Section 5.4 covers methods to handle unexpected occurrences.

5.3.3 Insufficient Capacity

Figs. 5.9(a)-(c) show the performance of CMOS-1 when the start-rate interval is cut back to 1 lot/0.7 hrs. Of the cases presented so far in this section, this one is perhaps most representative of industry practice. This start rate is too high for the capacity of the fab under any, and all, circumstances.

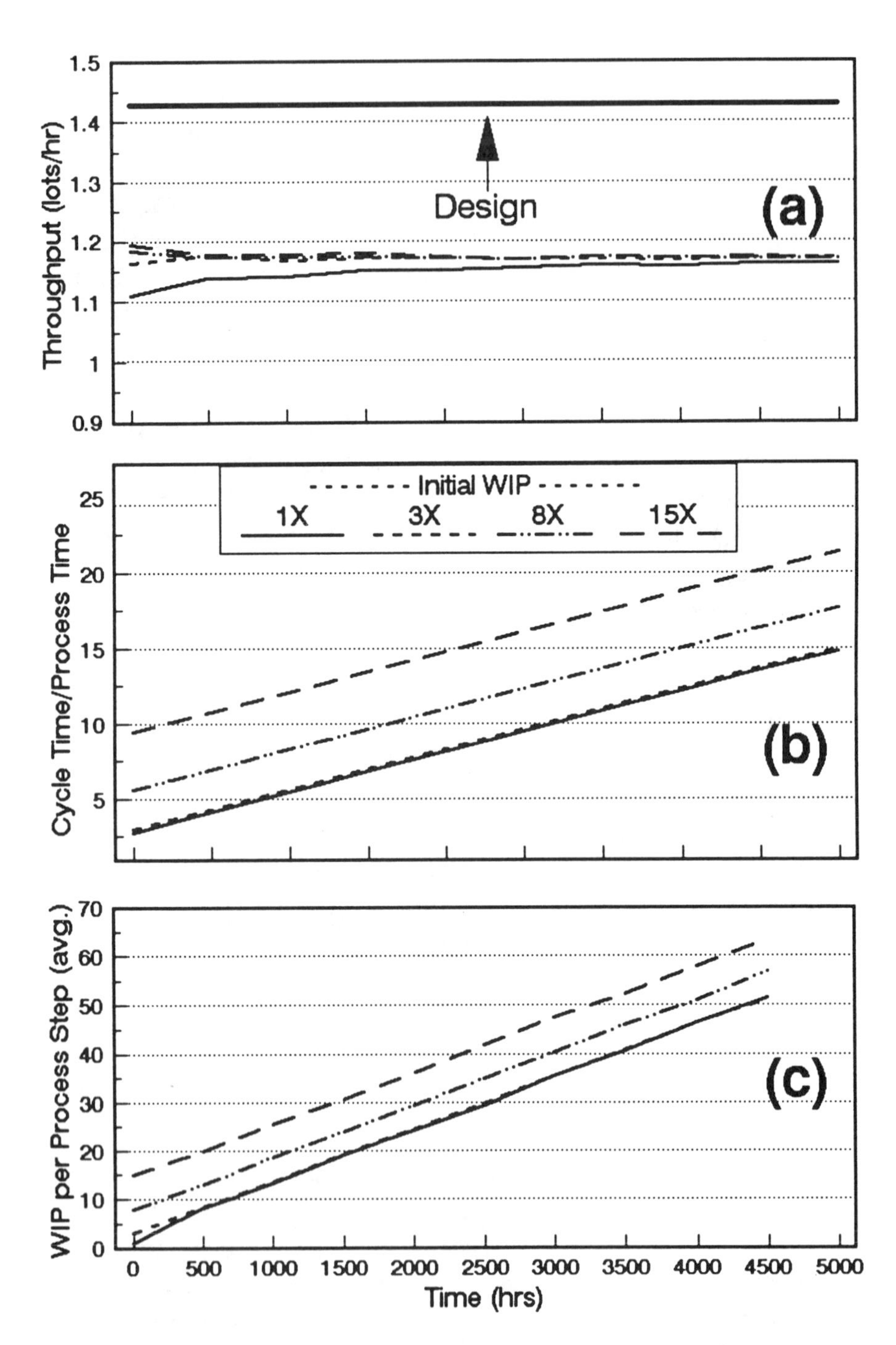

Figure 5.9: CMOS-1 with Lot-Start Rate of 1 lot/0.7 hrs

A start rate of 1 lot every 0.7 hours implies a production rate of 1.43 lots/hr. As may be seen by Fig. 5.9(a), the fab can not even break 1.2 lots/hr—under any scenario. In addition to never being able to meet production, cycle times simply get longer and longer, as seen in Fig. 5.9(b). Which is not too surprising since WIP simply builds...and builds...and builds (Fig. 5.9(c)).

The only way CMOS-1 can operate at this lot-start rate is to use a standard binge-purge operating policy: lots will be started at the specified rate for a length of time determined by how long it takes to fill the racks in front of the first bottleneck; then no lots will be started until the racks have been cleared out to a certain degree. Thus, on paper, there will be periods when it looks like production is being met, and other periods when meeting production is totally hopeless—but explainable. Usually some equipment vendor's fault. Needless to say, this is a very stressful fab to run.

For a fab in this position, there are two remedial courses of action: (1) back off on lot starts, or (2) buy more equipment.

As was seen in Figs. 5.8(a)-(c), running the fab at the slower start rate of 1 lot every 0.8 hours gives much superior performance in terms of throughput, cycle time, and WIP stability. It may seem counter to intuition that starting more produces less, but at the faster rate, demand far exceeds capacity. The fallacy of infinite capacity, and the poor results that may arise from following such a doctrine, were discussed in Section 2.4.1 (as well as 2.4.2 and 2.4.3). The performance of CMOS-1 at these two start rates provides a perfect illustration of the dangers involved in operating a fab beyond its limits.

If backing off on lot starts is not a viable option, i.e., production must meet the 1.43 lots/hr, then the only realistic, long-term solution is to buy more equipment. Generally, when a plant is undersized, it is usually at one or two key workstations, at least in the early days of operation. (Over time, process creep and new demands may seriously degrade capacity.)

Looking at the equipment utilization ratios (Table 5.9) for the 1 lot/0.7 hrs start rate shows that two workstations, #3 and #6, are bottlenecks for CMOS-1. Adding one machine at each workstation will debottleneck the fab (at any WIP level), as shown by Figs. 5.9(d)-(f). Furthermore, the capital investment is cheap if it is the only way to meet demand. But it is also cheap from a quantitative viewpoint as well; an additional 0.26 lots/hr for a capital investment of $3-4mm is money well spent. And that is not counting the other values gained in terms of stable operation, lower cycle times, lower WIP, etc.

But what if, by adding more equipment, we end up with excess capacity?

5.4 The Role of the Controlling Bottleneck

The possibility of buying more equipment and once again ending up with burst capacity brings us full circle back to where we started. Except that now we should have some answers to the questions:

- When is a bottleneck really a bottleneck?
- Is the existence of a bottleneck necessarily a bad thing?
- How should bottleneck(s) be handled?

A heavily-used workstation is truly a bottleneck only if it degrades fab capacity **below** the desired start rate. If a heavily-used workstation sets the limit of the fab, but the limit is at or above design, then the constraining behavior is not a 'bad thing', but rather a powerful mechanism by which the operation of the entire wafer fab can be controlled.

In a well-understood and well-controlled fab, one workstation is the planned 'bottleneck'. It throttles the flow of lots through the fab, never allowing too many or too few over a given period of time. In order to perform this function, the capacity-controlling workstation must have a utilization ratio close to 1.0, and it must be operating in a steady-state mode, in other words no unforeseen capacity losses.

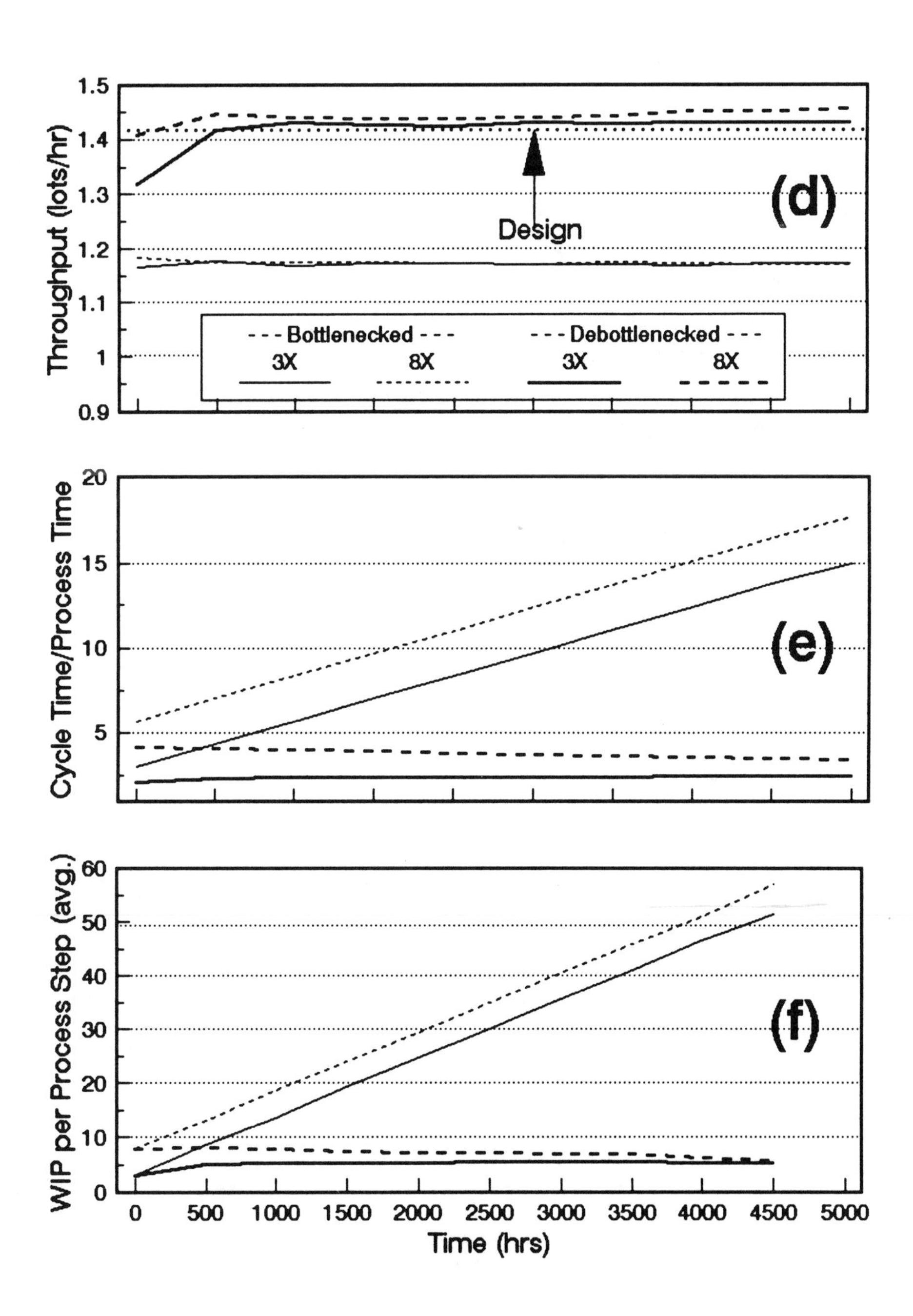

Figure 5.9(cont.): Debottlenecked CMOS-1 @ 1 lot/0.7 hrs

However, in the real world, there are always unforseen capacity losses. Since the behavior of the bottleneck or constraining workstation swings the behavior of the entire wafer fab, it is a serious occurrence when a machine fails unexpectedly. It will cause a transient response in fab performance. In fact, if a 'hold' is not put on lots upstream of the bottleneck, they will continue moving through the process sequence, and will build a bubble of inventory at the capacity-controlling workstation.

This type of transient behavior may not be critical if there is even a slight burst capacity at the workstation, or if the duration of the unplanned down is not excessive. In these circumstances, the bubble will clear when the machine comes back on-line.

If, however, there is no burst capacity at the controlling workstation, or if the unplanned down is excessive, then some form of lot-move control will be needed to avoid or eliminate inventory bubbles. The complexity of the control strategy may range from a fairly simple 'starvation-avoidance' technique [5.5] to a more sophisticated 'capacity-pacing' (lot-idling) scheme.

5.4.1 Lot-Move Control Options

The **starvation-avoidance technique** for lot-move control generally focuses on matching lot starts to the bottleneck's dynamic capacity. For instance, if equipment at the bottleneck workstation goes down, then the number of lots started through the line are reduced by the equivalent of that machine's capacity. The reduction in lot starts will continue until such time as the downed equipment comes back on-line. Then the lot-start rate is returned to normal.

The only problem with this type of control scheme is that it will generally make for an uneven distribution of inventory across the line. At the front end and back end, WIP will be sparse; at the bottleneck (and possibly at a few workstations upstream of it) inventory bubbles will build. Whenever an inventory bubble builds at a heavily-used workstation, it may take a long time to clear. If the workstation has a utilization ratio of

1.0, and the fab is running with large inventories, the bubble may never clear, but simply grow larger with each successive upset. Eventually, the fab may be required to 'purge' or scrap lots to clear a bubble.

A **capacity-pacing** or **lot-idling technique** (also called demand-pull by some) not only reduces the fab's lot-start rate to match the bottleneck's dynamic capacity, but it also implements selective idling of lots upstream of the bottleneck workstation. Perhaps a good analogy is an automobile race where a car has crashed and some portion of the track is no longer usable. The yellow flag is waved and the remaining cars merely pace the track, not moving out of relative position to the other cars, and not advancing the race, until such time as the track is cleared and the green flag is given for the race to recommences.

Similarly, when equipment goes down at the bottleneck workstation, the equivalent-lot capacity at each process step upstream is idled at its current step; i.e., it is not permitted to move on to the next step in the process sequence until the bottleneck equipment is back on-line.

Unlike the simpler starvation-avoidance technique, capacity pacing or lot idling does not allow inventory bubbles to build. Hence, when the lost capacity comes back on-line, the fab's return to normal, stable operation is immediate: cycle times will not have suffered and inventory will not have built. Unfortunately, with this control strategy, the throughput that is lost can never be recouped.

Such is not necessarily the case, however, if an inventory bubble has built at the bottleneck workstation, and if there is any burst capacity whatsoever at the bottleneck when the downed machine comes back on-line.

5.4.2 Impact of Lot-Move Control

To illustrate the impact of lot-move control, consider the scenario where CMOS-1 is operating at capacity (1 lot/0.80 hrs), and with an optimally-distributed WIP, 4.5 lots per process step. Under steady-state operation, the fab has a cycle time-to-process ratio of 2.367, a throughput of 1.25 lots/hr, and a burst capacity of 0.54% on the constraining workstation (Wkst #3). Over 5040 hours of operation, the fab produces 6310 lots.

Now, suppose one machine goes down at Workstation #3 for an unplanned 80 hours. Since Workstation #3 is the fab's constraining bottleneck, and since it has a total of seven machines, the fab has lost 1/7th of its capacity for almost half a week of production, a significant throughput hit and enough of an upset to seriously perturb operations.

Depending on the fab's bottleneck-control policy, the options at this point are: to do nothing, and let the fab recover naturally, or to instigate lot-move control.

The impact of these options are shown in Fig. 5.10. If no lot-move control is implemented, then lot starts will continue, uninterrupted, at the full-capacity rate throughout the 80 hours of unplanned outage. As a result, a bubble of inventory (an extra 32 lots) will build at Workstation #3. When the downed machine comes back on-line, it will take 15 weeks to work through the bubble.

During this time, the machines at Workstation #3 will be utilized virtually every minute they are available. In other words, over the 15-week recovery period, the workstation's 0.54% burst capacity will 'chew up' the inventory that has built, and is ready-and-waiting at the bottleneck. By virtue of having chewed up those bubble lots, the throughput that was lost as a result of the lost capacity is recouped. So, for this simplified fab, the policy of simply riding out the unexpected failure has no adverse effect on the quarterly production (Fig. 5.10(a)).

Not so, however, the impact on cycle time (and possibly the yield of the lots that have been waiting). If no control action is taken, the average cycle time for lots in the fab jumps by almost 25%, and does not return to its pre-upset value until well over 30 weeks of operation (Fig. 5.10(b)).

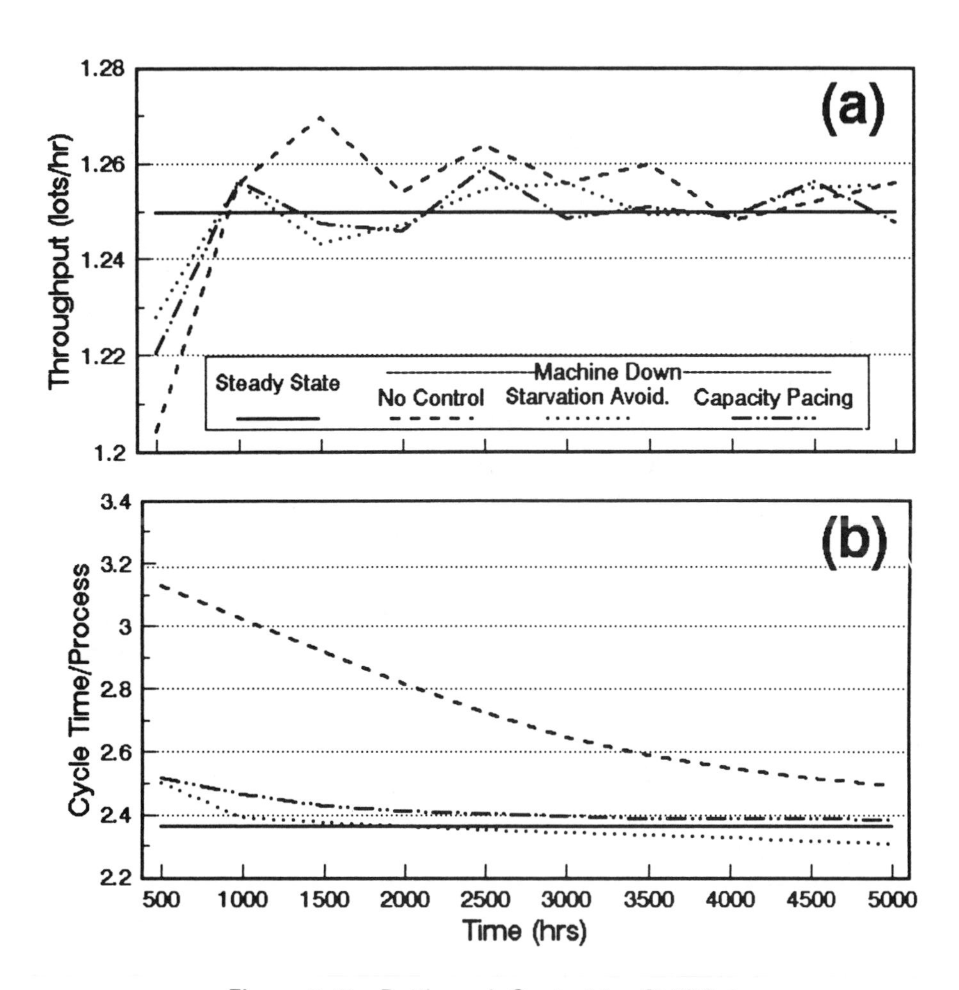

Figure 5.10: Bottleneck Control for CMOS-1

Such a cycle-time perturbation may not be particularly significant for a standard production line, since there will still be a steady flow of lots coming out. But this magnitude of delay could be disastrous for an ASIC line or where it involves engineering or development lots. Which brings up the other alternative: to implement some form of lot-move control once it is clear that the unexpected failure will be of significant duration.

Both starvation-avoidance and capacity-pacing techniques insure that average cycle time never rises more than a few percent, even in the early stages of the upset (Fig. 5.10(b)). The impact of either of these lot-move control options on throughput, however, does carry a penalty.

Because both techniques back off on lot starts for the duration of the outage, some throughput will be irretrievably lost (80 hours * 1/7(1.25 lots/hr) = 14.3 lots). Notwithstanding, both of these lot-move control techniques will bring the fab's throughput quickly back to near design (one week for capacity pacing, three weeks for starvation avoidance). But because a bubble of inventory still builds at the bottleneck for the starvation-avoidance procedure, the lot-completion profile for a quarter plus of operation will be much more erratic than for the much smoother capacity pacing.

Because fab cycle times are long, and it often takes weeks to notice the impact of an unexpected perturbation, much less any remedial action taken to offset it, developing an intelligible and consistent policy for lot control is quite difficult, and, for that reason, may even be counter productive. An effective policy and good implementation requires a clear understanding of:

- the current capacity-constraining workstation,
- whether the constraining workstation is a bottleneck,
- degree of constraint (before, during, and after transient),
- type and probable duration of unexpected perturbations,
- relative importance of the fab's performance parameters,

and, perhaps most importantly,

- impact of the remedial action, based on whether the constraining workstation is or is not a bottleneck.

In light of the fact that many wafer fabs do not know their constraining workstation, much less whether it is or is not a true bottleneck, little wonder the option of choice is to do nothing.

5.4.3 Mandatory Lot-Move Control

When an unexpected failure occurs at a **true** bottleneck, the 'do-nothing' options essentially disappears. Hence, the importance of knowing whether the constraining workstation is or is not a true bottleneck.

In the above example, Workstation #3 had a small burst capacity under normal operating condition, and thus was merely a constraining workstation. Now consider the same unexpected machine failure, but where CMOS-1 is running with a lot-start rate is 1 lot/0.79 hours (i.e., 1.2658 lots/hr). At this lot-start rate, Workstation #3 is a true bottleneck. It limits the steady-state capacity of the fab to a throughput production of 1.256 lots/hr, approximately 1% below design capacity. (Utilization ratio is 1.0 for Workstation #3 even without unexpected failures.)

If no control is applied when the machine fails, not only will the throughput never be recouped, but the number of lots lost will exceed the lost capacity (see Table 5.10).

Lost capacity = 80 hrs * 1/7(1.256 lots/hr) = 14.4 lots.
Lots lost, after 504 hrs and no control = 20
Lots lost, after 5040 hrs and no control = 30

TABLE 5.10: LOT-MOVE CONTROL STRATEGIES

Constraining Workstation vs "True" Bottleneck

Control Strategy:	Throughput (lots)		Cycle Time Ratio	
	Constraining Workstation	"True" Bottleneck	Constraining Workstation	"True" Bottleneck
Steady State	6,310	6,313	2.366	3.254
No Control	6,310	6,283	2.493	3.859
Starvation Avoidance	6,298	6,303	2.309	3.159
Capacity Pacing	6,292	6,293	2.385	3.276

Constraining Wkst Start Rate = 1 lot/0.80 hrs
"True" Bottleneck Start Rate = 1 lot/0.79 hrs
Performance after 5040 hours; machine down first 80 hours.

Thus, losing a machine at a true bottleneck makes the implementation of some form of lot-move control no longer merely optional, but highly desireable and, in some situations, even critical, as seen in Figs. 5.11(a)-(c).

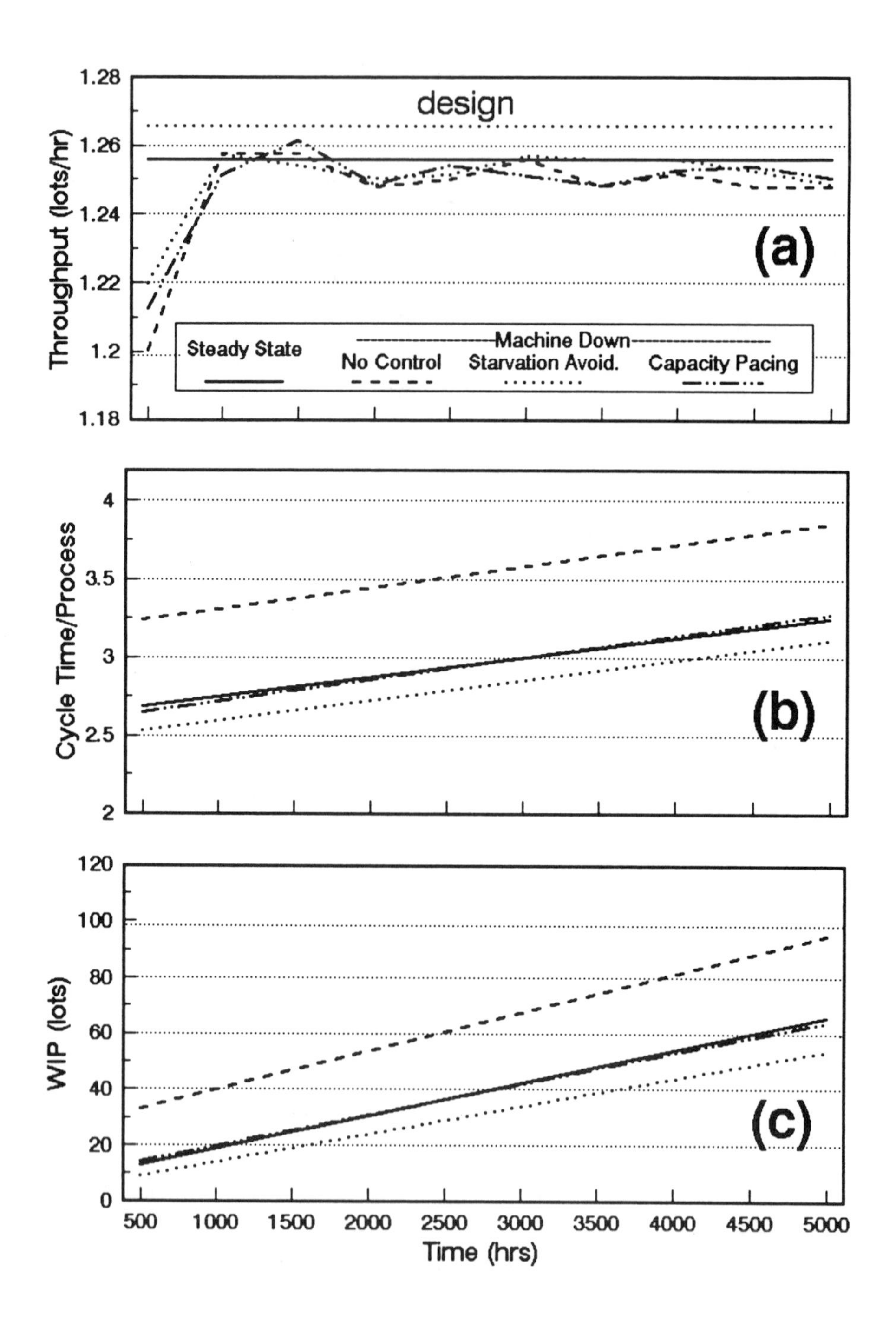

Figure 5.11: "True" Bottleneck Control, CMOS-1

The reason the 'no control' strategy is so detrimental when equipment fails at a true bottleneck is that, even before the failure, the fab is already in an inventory-building mode of operation (Fig. 5.11(c)). Lots started, over and above the plant's capacity to handle them, are added to inventory. When capacity decreases still further as a result of the downed machine, the too-high start rate will add even more inventory, compounding an already unstable mode of operation. Zero-burst capacity means that excess inventory can never be cleared. Every bottleneck upset exacerbates the situation.

The implementation of a lot-move control strategy does not solve the intrinsic problems associated with running a truly bottlenecked fab, but it will at least contain the damage resulting from an additional loss of capacity. As shown in Fig. 5.11(c), with either control scheme, the fab will still be in an inventory-building mode. But the absolute quantity of WIP will be less than in the steady-state case because of the reduction in lot starts during the unexpected outage.

Similarly, the throughput achieved is better with either lot control strategy (Fig. 5.11(a)) than with no control at all. With the capacity-pacing mode, the initial loss of lots is exactly what would be expected based on the lost capacity (14 lots after 504 hours). As the fab continues to operate, out to 5040 hours, there is no opportunity to recoup any of the loss. For this example, there is a modest additional drop in throughput (Table 5.10) due to the continued inventory build.

With the starvation-avoidance mode of control, the initial loss of throughput is somewhat less than the lost capacity (only 10 lots versus 14.4 after 504 hours). The reduction in lot starts, combined with continuing to process WIP already in the line, allows the fab to consume some existing inventory, recouping some of the loss; also, it makes for less WIP in the line. This smaller WIP, together with the back-off in lot starts, causes the inventory build rate (slope of line in Fig. 5.11(c)) to be less than in the steady state case. As a result, after 5040 hours of operation, there is no further drop in throughput (Table 5.10).

As is always the case when analyzing factory performance, cycle time is a much more sensitive parameter to perturbations in the fab than is the throughput. When a fab is truly bottlenecked, and day-to-day 'steady-state' operation is in an inventory-building mode, cycle times can quickly get out of hand. A fab that is building inventory will, by definition, have increasingly longer cycle times (Fig. 5.11(b)). Take this unstable operation and superimpose an unplanned disturbance, and the fab will not only have the regular build in inventory, but also an increase equal to the capacity lost because of the disturbance. In Fig. 5.11(b), not only is the cycle-time ratio much higher in the 'No Control' versus the 'Steady State' case, but the slope of the line is increasing 9% faster.

As with throughput performance, implementation of a lot-move control strategy does not suddenly solve the cycle-time problem, primarily because the intrinsic inventory-building situation still exists. But the implementation will at least curtail the harm resulting from an additional capacity loss. With either starvation-avoidance or capacity-pacing control, the fab will have increasing cycle-time ratios over time, but the magnitude of the ratio, as well as the rate of increase, will be at or below the steady-state case. Again, the containment is primarily due to the reduction in lot starts during the unexpected outage, thus curbing runaway inventory.

Such a sensitivity, as seen with cycle time to operational perturbations, would normally lead one to expect the parameter to be a good one to track, analyze, and use for factory control. Unfortunately, determining the cause of lengthening cycle time can be difficult, and analysis of the fab-operations tracking data is not performed routinely.

Finally, before moving on, it is relevant to point out that most wafer fabs **do** indeed have one or more real bottleneck. For these fabs, it is important, and sometimes even critical, to understand the cause-and-effect relationships of factory dynamics and bottleneck control. As a general rule, throughput that is lost as a result of lost capacity is lost forever. Only a fab with a state-of-the-art tracking system and a sophisticated, deadly-accurate factory model can hope to cheat the reaper by even a lot or two.

5.5 The Pros and Cons of Burst Capacity

Understanding how a truly-bottlenecked fab can quickly go south brings up an interesting dilemma: Is it better to operate with a 2-3% reserve of burst capacity at the bottleneck in order to be able to recoup throughput losses? Or is it better to just take the throughput hit whenever machines at the bottleneck fail?

Although economics should dictate the answer, in actuality, a fab's operating policy is generally decided on a much more enigmatic basis. Leaving 2-3% of a fab's capacity idle somehow goes against, if not the laws of nature, then at least against the laws of management. Whether it is a wise decision or not, will be seen with the help of CMOS-1.

Like most wafer fabs, CMOS-1 has an equipment set which is unlikely to change. The question now is, given that equipment set, and in light of the fact that unexpected equipment failures at the bottleneck will occur, does the fab keep some burst capacity as an insurance policy or not?

Table 5.11 shows the impact of slacking off on lot starts in order to provide burst capacity for coping with unexpected failures at the bottleneck.

TABLE 5.11: DESIGN vs ACTUAL CAPACITIES

	Burst Capacity (%)	Design Capacity (lots)	Lost Value ($mm)	Actual Thrupt (lots)	Lost Value ($mm)
1 lot/0.79 hrs	0.0	2,551	--	2,522	--
1 lot/0.80 hrs	0.5	2,520	4.3	2,512	1.4
1 lot/0.81 hrs	1.5	2,489	8.5	2,494	3.8
1 lot/0.82 hrs	2.5	2,458	12.8	2,464	8.0

Capacity & thrupt based on 12 wks steady-state operation.
Value based on $7000 wafers, 25-wafer lots.
Burst capacity is for constraining or bottleneck workstation.

Of course, when lot starts decrease, the 'design' capacity decreases; hence, throughput should decrease. As seen from the table, the value of the lost capacity makes for an expensive insurance policy. Over a nominal quarter of operation, 2.5% burst capacity reduces design capacity by 93 lots. If these are 25-wafer lots, with a value of $7000/wafer and a production cost $1500/wafer, the lost capacity is worth $12.8mm.

To complicate matter further, the existence of extra capacity for 'rainy-day use' necessitates implementing control policies to insure that it will only be used for such occasions. If control policies are not implemented, operators will load machines whenever the machines are available, and with whatever lot is available. The result will be depletion of inventory.

In Table 5.11, notice that the actual throughputs for the 1.5% and 2.5% burst-capacity cases are higher than expected based on their respective lot-start rates. The old adage that 'you can't get more out than you put in' makes for stable fab operation. Conversely, when throughput is higher than lot starts, it means that inventory is being consumed, and the fab is in a mode of operation which can not be sustained. At some point, lot starts will have to be ramped up in order to refurbish inventory. Unfortunately, when the start rate is eventually dropped back, inventory will again be depleted.

The end result of having burst capacity and no special control rules is that the fab will forever be in a 'binge/purge' mode of operation. While throughput can be somewhat managed despite the roller-coaster ride, cycle times on the other hand will be all over the map, making for erratic product development cycles and unpredictable and/or unstable schedules (especially in ASIC fabs).

Despite the apparent higher-than-expected throughputs with the larger burst capacities, and even overlooking the fact that WIP is being depleted in order to achieve them, throughput performance is still inferior to the cases where there is little or no excess capacity. Even for this simplified fab, the quarterly price tag for 1.5% of burst is $3.8mm, and $8.0mm for 2.5%. True, as seen in Table 5.12, the burst capacity does have some small cycle time advantages, but probably not enough to justify the throughput cost, as well as the added operational complexity.

TABLE 5.12: CYCLE TIME vs OPERATING POLICY

Burst Capacity (%)	CT/P @ 2016 hrs		CT/P @ 5040 hrs	
	Steady State	Machine Down	Steady State	Machine Down
0.0	2.87	2.73 *	3.25	3.11 *
0.5	2.41	2.82	2.36	2.49
1.5	2.28	2.49	2.23	2.32
2.5	2.20	2.36	2.16	2.21

*Starvation avoidance; others use no special control.

It should be noted that the 'lost values' shown in Table 5.11 for burst-capacity are for steady-state operation. The purpose, however, of providing for burst capacity is to handle unexpected equipment failures or other transients. The information in Table 5.13 is perhaps more informative in determining whether the burst insurance policy is worth the cost.

TABLE 5.13: COST OF OPERATING POLICIES

	Burst Capacity (%)	Actual Thrupt (lots)	Product Value ($mm)	Cost of Policy ($mm)
1 lot/0.79 hrs				
Steady State	0.0	2,522	346.78	--
Machine Down (SA)	0.0	2,511	345.26	1.52
1 lot/0.80 hrs				
Steady State	0.5	2,512	345.40	1.38
Machine Down (NC)	0.0	2,512	345.40	1.38
1 lot/0.81 hrs				
Steady State	1.5	2,494	342.92	3.86
Machine Down (NC)	1.0	2,491	342.51	4.27
1 lot/0.82 hrs				
Steady State	2.5	2,464	338.80	7.98
Machine Down (NC)	1.8	2,471	339.76	7.02

Values are for nominal quarter (2016 hrs).
'SA' indicates a starvation-avoidance policy.
'NC' indicates no particular control policy.

In the case where the fab is running slightly beyond capacity (1 lot/0.79 hrs), a bottleneck resource going down for 80 hours makes for a throughput loss worth $1.52mm (again assuming a wafer value of $7000 and a production cost of $1500). In the case where there is 0.5% burst capacity during steady-state operation (1 lot/0.80 hrs), the lost throughput, allocated for the burst, is worth $1.38mm, regardless of whether a bottleneck machine goes down or not.

The conclusion that may be drawn from these two cases is that perhaps the small half-percent of burst capacity (that effectively disappears during an upset) is not a bad place to operate. The fab is very stable with this small amount of burst, as evidenced by the fact that performance is basically the same with or without an upset (including a stable inventory profile); the cost of the insurance policy ($1.38mm) can easily be eaten up by one upset in the over-capacity case ($1.52mm); and, because the fab is not running beyond its capacity, cycle times are much shorter and less erratic. Furthermore, no special lot-move control schemes are needed as long as the failures are non-catastrophic.

Larger burst capacity at the bottleneck, however, may not be a wise operating policy for a wafer fab. The insurance policy becomes quite costly with each percent of excess capacity. Entire fab capacity is sitting idle, waiting for an upset to occur. Then, when and if an upset does occur, there still may be excess capacity on the bottleneck workstation; hence no inventory bubble will ever build there. If no inventory bubble builds, it eliminates the possibility of ever making up some or all of the lost throughput.

For the 1.5%-burst capacity case, not only does the fab lose throughput while waiting for something that may not occur, but even if it does occur, the fab does not recoup throughput when an outage comes back on-line. The contingent capacity is no real insurance policy; an outage is even more costly than when the fab is running at steady state ($4.27mm vs $3.86mm, respectively).

Operating with the large burst capacity (2.5%) may be the worst strategy of all. The insurance policy in this case cost $7.98mm in lost throughput for day-to-day operation. Even when the contingent capacity is needed and used, and throughput is recouped, the cost is still $7mm. Even then, the recouped throughput does not come from an inventory bubble in front of the bottleneck workstation__because none has built there. The excess capacity is simply too large to build a bubble during the 80-hour outage. Instead, the made-up throughput comes from consuming WIP. The excess capacity is 'sucking' WIP through the line, WIP which cannot be replenished by the lot-start rate.

All of which brings us back to the questions which started this section: Is it better to operate with a 2-3% reserve of burst capacity at the bottleneck to recoup throughput losses? Or is it better to just take the throughput hit whenever machines at the bottleneck fail? The answers to these questions for the simplified CMOS-1 fab also happen to be the same answers for any state-of-the-art wafer fab with today's level of lot-move control.

A 2-3% reserve of burst capacity at the bottleneck or constraining workstation is, in general, not a good operating policy. Burst capacity requires special control rules to govern: when the burst capacity will be used, when lots will move up and down the line, and how the appropriate WIP level will be maintained. Further, it is unlikely that this amount of burst capacity could ever recoup enough throughput to justify the dollar cost of the capacity set-aside.

On the other hand, operating with a small amount of burst capacity at the constraining workstation, on the order of 0.5%, may provide a comfortable and stable means for controlling a wafer fab. Under this policy, when the constraining workstation suffers unexpected losses, it will become a true bottleneck, but one which will clear, and, while it is clearing, will recoup throughput losses. The insurance-related expense for running this type of operation is low, and will, over time, be less costly than running at or slightly over capacity and simply suffering throughput losses whenever machines at the bottleneck fail.

5.6 Factory Performance Trade-offs

As has been seen repeatedly in this chapter, the performance achieved by a wafer fab is the result of the trade-offs that are made in its operation. Unfortunately, there is no single, accepted rule on how best to operate a fab. The driving forces for one fab may be entirely different from the driving forces of another. The management of one fab may be willing to live with long cycle times and high inventories in exchange for the immediate realization of higher throughputs and greater schedule stability. Another fab, which must respond rapidly to changing market demands, will forego throughput in exchange for fast cycle times. Yet a third fab may be operated with the primary objective of moving units and keeping equipment busy.

Rarely, if ever, is it possible to achieve one performance measure without sacrificing others. Fig. 5.12(a) is a reasonable representation of some of the major trade-offs involved.

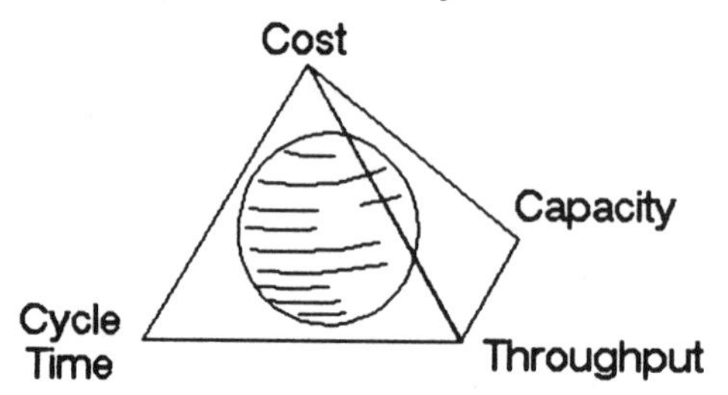

Figure 5.12(a): Performance Trade-offs

In light of the inexorably-bound relationships discussed earlier, Fig. 5.12(b) is simply another view.

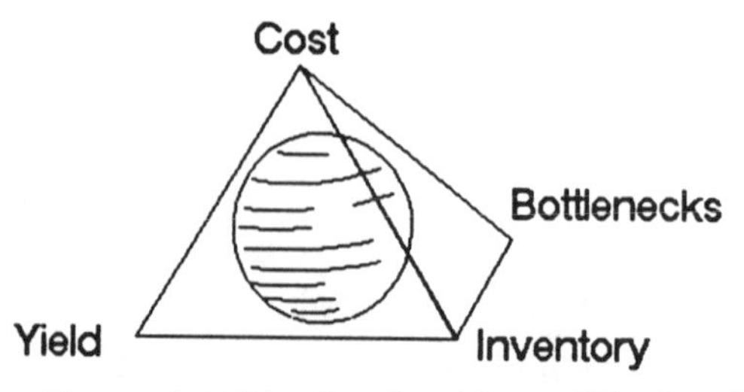

Figure 5.12(b): Another View of Trade-offs

Regardless of a fab's primary objectives, operation will never be at an extreme of the tetrahedron, nor should it be. For instance, if fast cycle time is the number one objective, then obviously the fastest cycle time will be achieved with a virtually empty fab. In the extreme case, a lot would not start until the preceding one had completed enough of its processing sequence to guarantee no wait. Needless to say, throughput would be atrocious.

This example illustrates the ramifications of tunnel vision in wafer fabrication. If one, or even several performance parameters are given too much weight, to the exclusion of other pivotal measures, performance will suffer. Products may be costly, delivery erratic and unpredictable, operations difficult and unstable.

If one considers the trade-off tetrahedra in Figs. 5-12(a) and (b), the most rational way to run a wafer fab is not at one of the geometric points; nor is it on one of the plane edges. While it could be somewhere towards the center of one of the triangular faces, even that type of operating policy is probably not wise in that it would still exclude important criteria. Exclusion of any key parameter is detrimental to fab performance.

The most viable trade-off compromises would take place somewhere within a sphere enclosed within the tetrahedra; the sphere's surface would be the outer limits of where a wafer fab could operate and still achieve good performance. Finding out where the surface of this sphere is for a given wafer fab, and using it effectively, involves:

- accurately knowing the capacity of the fab
- not exceeding the capacity of the fab
- paying close attention to process changes over time
- matching inventory to fab capacity
- understanding the impact of operating control rules
- understanding the give-and-take performance trade-offs
- using the give-and-take trade-offs for added fab control

Fig. 5.13(a)-(g) shows some of the trade-off relationships for various factory performance measures. Recognizing and dealing with these trade-off relationships becomes much more difficult as factory complexity increases. For specific cases, involving particular fabs and their stated performance issues and objectives, detailed analyses will usually require an accurate dynamic model.

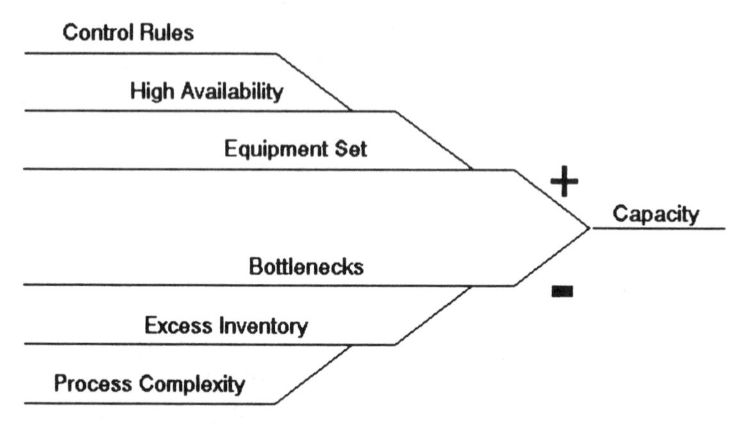

Figure 5.13(a): Capacity Trade-Off Relationships

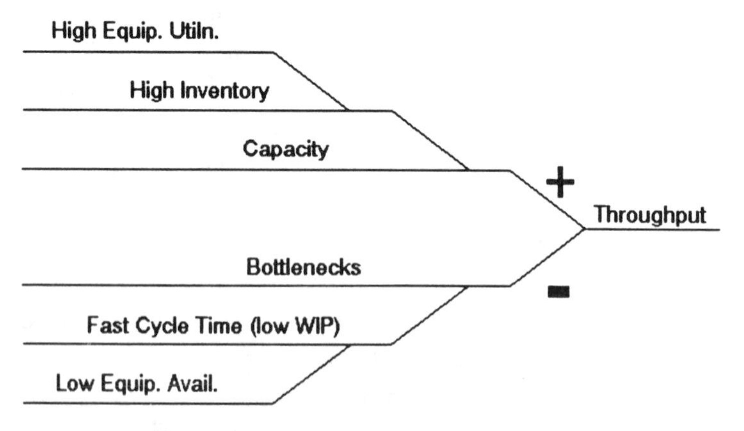

Figure 5.13(b): Throughput Trade-Off Relationships

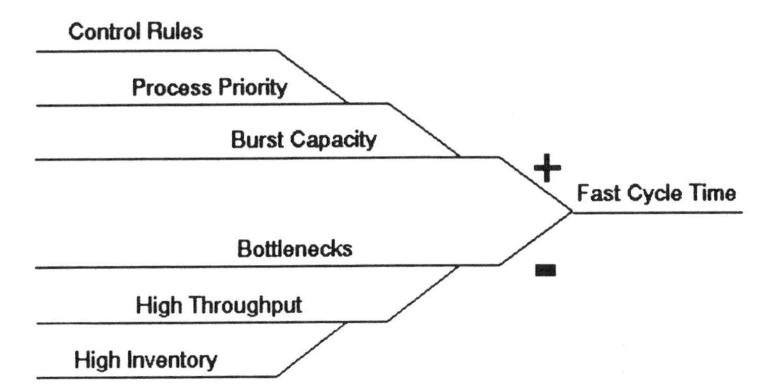

Figure 5.13(c): Fast Cycle-Time Trade-Off Relationships

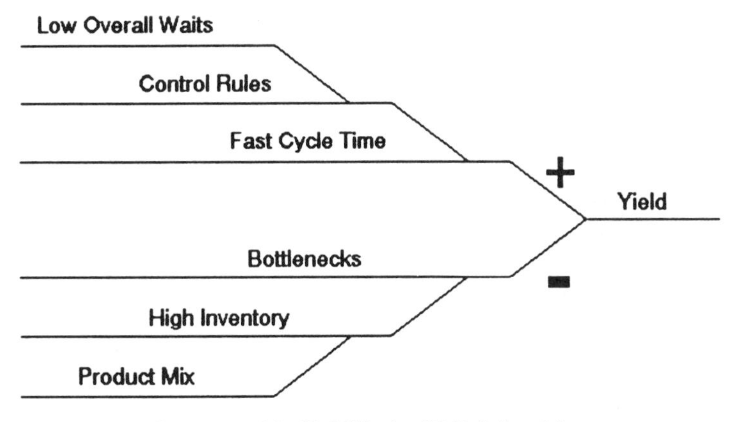

Figure 5.13(d): Yield Trade-Off Relationships

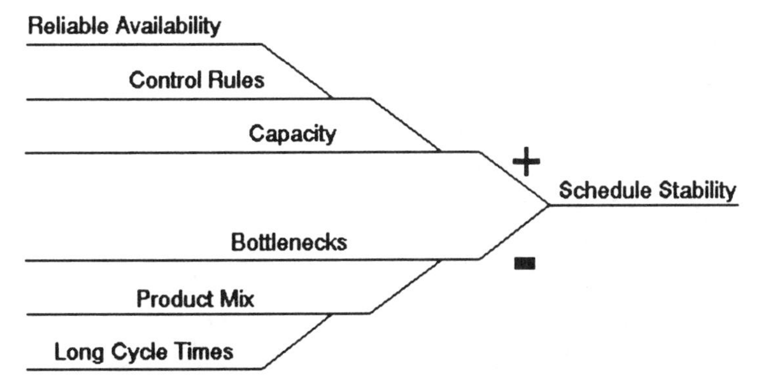

Figure 5.13(e): Schedule-Stability Trade-Off Relationships

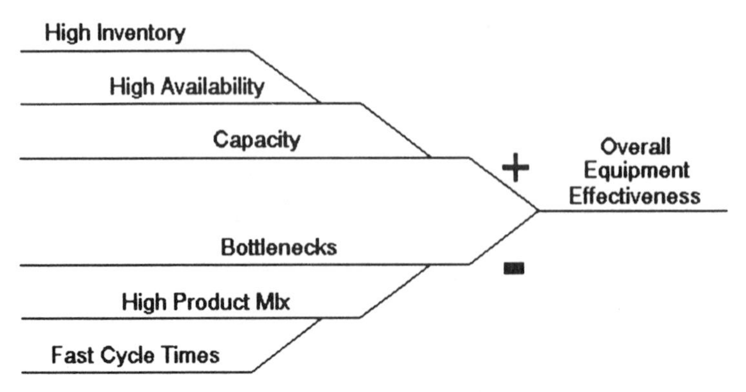

Figure 5.13(f): OEE Trade-Off Relationships

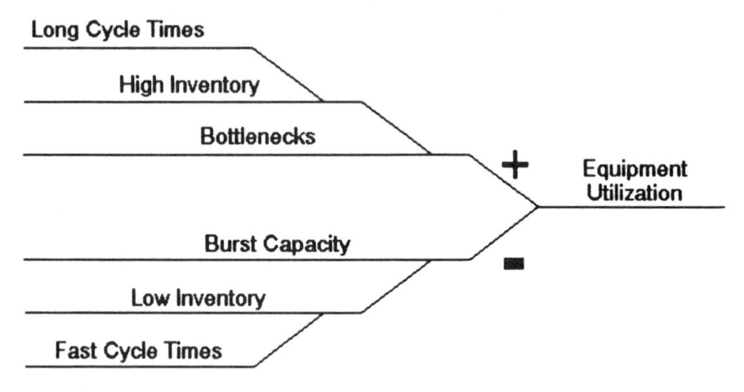

Figure 5.13(g): Equipment-Utilization Trade-Off Relationships

5.7 Summary

In this chapter, we have primarily used an idealized, highly-simplified version of a wafer fab called CMOS-1. While the results and understanding gained from the study of this toy factory are representative of fabs in general, we have yet to touch on topics that make wafer fabrication one of the world's most complex manufacturing processes, topics such as:

- multiple process (product) flows
- priority process flows and special control rules
- process creep and process migration
- equipment availability and catastrophic failures
- line imbalance and load mismatch
- process-step clustering and automation

But...

Half a league, half a league,
Half a league onward,
...
Into the jaws of death,
Into the mouth of hell
Rode the six hundred.

...these and other topics will be covered in Chapters 6 and 8.

Further Reading and Notes

1. Yield is treated exhaustively in Landzberg's treatise [5.3]. Yield models in particular are covered in [5.6].

2. Many issues of manufacturing operations for factories in general are treated by Vollman, et al [5.2] and Morton and Pentico [5.5].

References

[5.1] Atherton, R.W., L.F. Atherton, M.A. Pool, "Detailed Simulation for Semiconductor Manufacturing," **1990 Winter Simulation Conference Proceedings**, Society for Computer Simulation, December, 1990.

[5.2] Vollman, T.E., W. L. Berry, and D. C. Whybark, **Manufacturing Planning and Control Systems**, Dow-Jones Irwin, Homewood, Illinois, 1984.

[5.3] Landzberg, A. H. (ed.), **Microelectronics Manufacturing Diagnostics Handbook**, Van Nostrand Reinhold, New York, 1993.

[5.4] Weste, N. H. E., and Kamran Eshraghian, **Principles of CMOS VLSI Design, A System Perspective**, Addison-Wesley Publishing Company, Reading, Massachusetts, 1993.

[5.5] Morton, T. E., and D. W. Pentico, **Heuristic Scheduling Systems**, Wiley, New York, 1993.

[5.6] Bonk, P. J., M. R. Gupta, R. A. Hamilton, and A. Satya, "Manufacturing Yield,," in Landzberg, 1993.

Exercises

1. Figure 5.1 shows that increasing initial inventories in the high range causes cycle time to increase and throughput to decrease over the operating period studied. What happens to total WIP for the cases labeled 2X and 4X?

2. For several 500-hour periods, a widget fab was operated with no initial inventory, but with a range of lot-start rates. Several months later, the same start rates were run, but this time with an initial WIP of 77 lots.

Start Rate (lots/hr)	Throughput		Net Change in WIP	
	w/o inventory	with inventory	w/o inventory	with inventory
1	248	134	253	367
.67	251	178	83	156
.50	199	202	52	49
.40	157	222	44	-21
.33	133	209	34	-42

Plot 'Throughput' vs 'Start Rate' for the two cases (i.e., without and with WIP). Think about the reality of the two cases.

3. What is the dynamic capacity of the widget fab in Exercise #2? Why?

4. The production demand for a wafer fab is 45 wafers/hour. The measured throughputs of the top five constraining work centers are respectively 40, 42, 44, 46, and 48 wafers per hour. To which of these work centers should we add machines? Which are likely to be real bottlenecks?

5. In the situation described in Exercise 4, the uncertainty in availability is ten percent. Does this change your answer? If availability is 10% higher, where should equipment be added? If it is 10% lower, where? What if the uncertainty in availability is 20%?

6. In each of the scenarios described in Exercise 5, what measured-throughput value is sufficiently high that extra machines clearly are not needed to guarantee 45 wafers per hour under normal operating conditions?

7. The burst capacity of a wafer fab is 0.3 wafers per hour. How many weeks will be required to clear 1000 wafers of excess inventory?

8. The process flow and workstation assignments for Sammy T's state-of-the art PiZMOS fab are given below.

EXERCISE 5.8: SAMMY T'S PiZMOS FAB

STEP #	PROCESS STEP	TIME	WKST #	WORKSTATION
1	Al-SPUTTER	1.7	1	SPUTTER
2	PATTERN	3.2	2	LITHO
3	DEVELOP	1.4	3	DEVELOP
4	ETCH	2.9	4	P-ETCH
5	SPUTTER	1.7	1	SPUTTER
6	PATTERN	1.2	2	LITHO
7	DEVELOP	1.4	3	DEVELOP

times in hrs

(a) Draw the fab graph for PiZMOS-1.
(b) Determine the hub order for the workstations.
(c) How much throughput will each workstation require in order for the overall fab throughput to be 0.5 lots/hr?
(d) Assuming all equipment has an availability ratio of 0.9, and all machines have standard batch size (25 wafers), size the fab. (See Section 4.8, if needed.)
(e) Over time, the availability ratio of Workstation #2 begins to degrade, eventually dropping to 0.6. At what point does Wkst 2 become the fab bottleneck? How many new machines will remedy the problem?

Reality: What a concept!
-Robin Williams

Chapter 6
Advanced Performance Concepts

6.0 Introduction

In Chapter 5, we covered most of the basics of factory performance analysis as it applies to wafer fabrication. The relationships and trade-offs of the various performance parameters were explored for fabs running at capacity, under capacity, and/or over capacity. But other than several validation comparisons against real fab performance data, most of the discussions and case studies revolved around an idealized, highly-simplified version of a wafer fab known as CMOS-1. The reason for using such a vehicle is that it enables us to take a highly-complex type of manufacturing and make it tractable and comprehensible. The same strategy will be followed in this chapter as we explore the more advanced aspects of wafer fabrication.

It is rare and unusual to find a wafer fab that has only one process flow. In fact, if one correctly defines the engineering lot as a separate flow, then there are no single-flow fabs. The existence of multiple flows requires that the fab's equipment must run under diverse operating rules that may change often and radically. To complicate matters further, the imposition of a priority on one of the flows (which is often the case with development lots) may cause fab performance to crater, without management ever knowing the reason why.

Another complication in analyzing and/or predicting fab performance derives from a rather unusual source: the plethora of control rules that are imposed to deal with multiple processes in the factory. These control rules cover situations such as process priority, resource competition, setup charges, load mismatches and mixed batches, etc. Although the control rules are devised and implemented with the best of intentions, their true impact is often difficult, if not impossible, to ascertain. Because of the long cycle times involved, and the interrelationships of thousands of parameters, sleuthing out cause-and-effect of fab control rules is like trying to determine how the war was lost because of a nail. Fortunately, with a clear understanding of how and where to focus fab analysis, and which analytical tools to use, it is possible to select and refine suitable control rules to achieve the desired results.

Equipment availability is another source of problems in performance analysis. Availability, especially at the bottleneck, can be tantamount to fab capacity; it can also be a vast unknown if not tracked or handled wisely.

In the old days, quoted availabilities represented what gear was capable of in a pristine environment, run under regular and ideal conditions. Not too surprisingly, performance on the fab floor was usually no where near as auspicious. Good fab managers were able to confirm the actual availabilities, adjust their CIM standards accordingly (and their expectations), and move forward with the business at hand: making saleable product.

Unfortunately, along about 1990, the issue of overly-optimistic availability numbers moved from being a major, though understandable problem into a sideshow trip to the twilight zone. Because such an unwarranted hoopla was made about the *non sequitur*, cost-of-ownership, strange and counter productive games were played with availability and utilization numbers, making an already-serious problem even worse.

For those who bought into the cost-of-ownership doctrine, fab performance began being measured, not by something sensible such as cost per good die, but by how busy one could keep the equipment. Thus, wafer fabs insisted upon, and equipment vendors quoted, availability numbers that gave good cost-of-ownership values. It did not seem to matter that the quoted availabilities could only be realized when *the moon is in the seventh house and Jupiter aligns with Mars*. And sometimes not even then.

Good factory performance can be achieved despite poor equipment availability if the situation is recognized, and additional machines are bought. However, running a fab with phantom availability has the same repercussions as running one with phantom capacity. Some of the lots that are started will be beyond the fab's capability to process them, and, hence, will lead to: bottlenecks, long cycle times, low yields, high WIP, and possibly even lower throughputs. Naturally, high equipment availabilities would be nice, but not if they are fictitious.

Ironically, the business of demanding and generating falsely-optimistic availabilities in pursuit of good cost-of-ownership is senseless. Not only because the false numbers lead to bad capital and operating policies, but also because the driving force, cost-of-ownership, is not a good measure of fab performance in the first place. For a performance evaluation to have validity, it must have built into the evaluation criteria a measure of the fab's primary goal: production of salable product. But with cost-of-ownership, the focus is not on manufacturing productivity, but rather on equipment activity. With such a focus, the obvious way to run a fab is to turn the equipment on and keep it busy all the time, even if it is doing nothing more than processing dummy wafers.

Do not misconstrue the message here; availability is important, and striving to improve it is always a worthwhile activity, especially at the fab's bottleneck workstations. But immensely more important than seeking the holy grail of high availabilities is possessing true knowledge of the fab: its bottlenecks, the real availability at the bottlenecks, and what must be done in order to manufacture within the fab's means. Low availabilities are not crippling to fab performance; fictitious ones are.

Just as fab capacity is reduced by the realization of lower-than-expected availabilities, so too is it reduced, over time, by the natural aging or maturing of the factory. Reliability and availability of equipment decline, process times become longer, steps are added, process mixes change, and environments are less clean. Since fab performance is always limited, or capped, by factory capacity, the first step of performance analysis is to determine the fab's true capacity, as a function of everything that is going on inside the four walls.

Finally, one additional source of performance concern in a wafer fab is the intrinsic mismatch of equipment load size. One workstation may have single-wafer processing equipment and the next may have 200-wafer furnaces. Control rules are often implemented to smooth or balance the line, and in the process may throw away capacity since the end result will be better fab performance. But in order to devise such control rules, a clear understanding must be had of the causes of line imbalance, and the performance trade-offs involved in trying to ameliorate them.

6.1 Multiple Process Flows

In its standard operating mode, a wafer fab will usually fabricate products from several process flows simultaneously. Depending on the similarity or dissimilarity of the products, the process sequences can range from being almost identical to being radically different. The greater the difference, the more diverse the demands on the factory. Needless to say, determining the dynamic capacity of a multi-flow factory becomes a non-trivial problem because each flow, along with the interaction effects, are competing for factory resources. To study the complex and dynamic forces at work in wafer fabrication, we will follow a simplified factory, CMOS-2, through the life cycle of the plant.

From day one, we know that CMOS-2 will be a multi-flow fab, producing a variety of products. We also know that, over time, the product slate (and hence the processes) will change as technology evolves. But when the plant first comes on-line it will be running with the single 2-metal flow shown in Fig. 6.1 and described in Table 6.1.

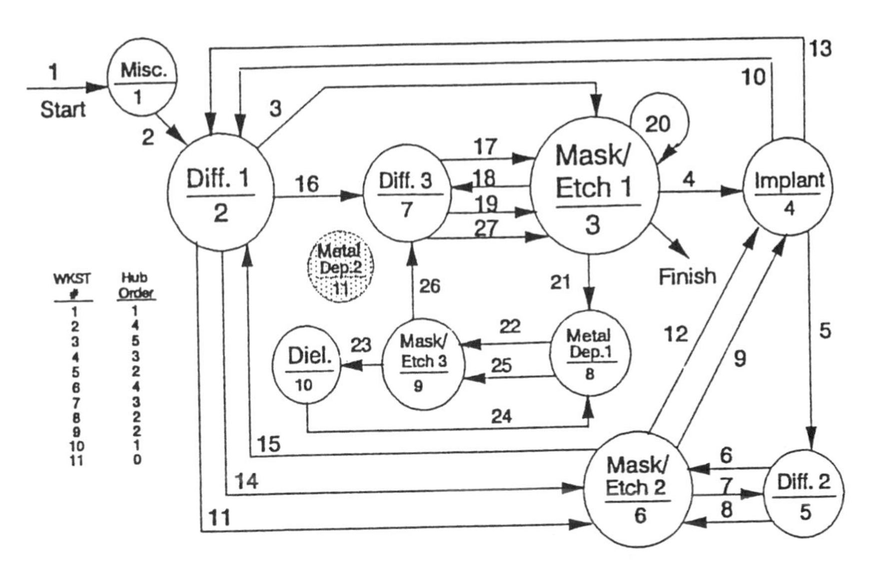

Figure 6.1: Fab Graph for Process #1, CMOS-2

TABLE 6.1: 2-METAL PROCESS #1, CMOS-2

STEP #	PROCESS STEP	TIME	WKST #	WORKSTATION
1	LASER MARK		1	MISC.
2	INITIAL OX		2	DIFF 1
3	P-TUB MASK		3	MASK/ETCH 1
4	P-TUB IM		4	IMPLANT
5	NITRIDE DEP		5	DIFF 2
6	FIELD MASK		6	MASK/ETCH 2
7	FIELD DIFF/OX		5	DIFF 2
8	SOURCE/DRAIN MSK 1		6	MASK/ETCH 2
9	S/D IM 1		4	IMPLANT
10	S/D OX 1		2	DIFF 1
11	S/D MASK 2		6	MASK/ETCH 2
12	S/D IM 2		4	IMPLANT
13	S/D OX 2		2	DIFF 1
14	NITRIDE STRIP		6	MASK/ETCH 2
15	GATE OX		2	DIFF 1
16	POLY DEP/DOPE 1		7	DIFF 3
17	POLY MASK 1		3	MASK/ETCH 1
18	POLY DEP/DOPE 2		7	DIFF 3
19	POLY MASK 2		3	MASK/ETCH 1
20	CONTACT MASK		3	MASK/ETCH 1
21	**METAL DEP 1**	**1.10**	**8**	**METAL SPUTTER 1**
22	**METAL MASK**		**9**	**MASK/ETCH 3**
23	**DIELECTRIC**		**10**	**DIEL-FAB**
24	**METAL DEP 2**	**1.10**	**8**	**METAL SPUTTER 1**
25	METAL MASK		9	MASK/ETCH 3
26	TOPSIDE DEP		7	DIFF 3
27	TOPSIDE MASK		3	MASK/ETCH 1

As is typically the case, engineering development for this flow, as well as the second flow, will have been done in an earlier plant (CMOS-1 in Chapter 5).

Running the single 2-metal Flow-1 in Table 6.1, the steady-state CMOS-2 has a lot-start rate of 1 lot every 0.9 hour, and achieves a throughput of 1.111 lots/hr. The cycle time of lots through the fab is 89.1 hours, and the cycle-time-to-process-time ratio (CT/P) is 2.332. Minimum processing time for the flow sequence is 35.5 hours.

Within months of plant startup and shakedown, it will be time to start phasing in Flow-2, shown in Fig. 6.2 and listed in Table 6.2. This second flow is more complex than Flow-1 in that it has three metal layers, longer metal process times, and additional associated interconnect steps. Thus, for the same number of finished wafers, Flow-2 will require more fab resources. If the plant were running solely with Flow-2, the capacity would be 1 lot/hr (0.111 lots/hr less than with Flow-1 alone), the production throughput would be 0.9988 lots/hr (0.1131 lots/hr less), the cycle time would be 123.4 hours (34.3 hours more), and CT/P would be 2.865 (vs 2.332). Minimum processing time for Flow-2 is 40.05 hours (4.55 hours more than Flow-1).

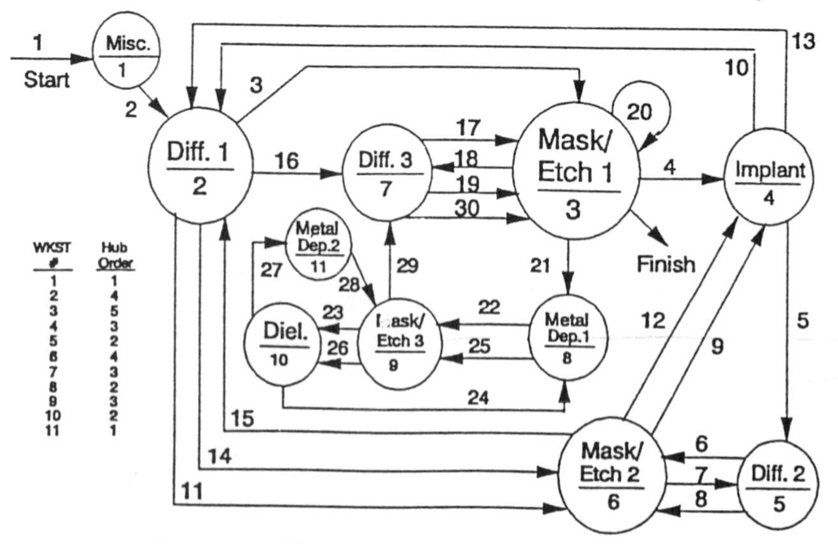

Figure 6.2: Fab Graph for Process #2, CMOS-2

TABLE 6.2: 3-METAL PROCESS #2, CMOS-2				
STEP #	PROCESS STEP	TIME	WKST #	WORKSTATION
1	LASER MARK		1	MISC.
2	INITIAL OX		2	DIFF 1
3	P-TUB MASK		3	MASK/ETCH 1
4	P-TUB IM		4	IMPLANT
5	NITRIDE DEP		5	DIFF 2
6	FIELD MASK		6	MASK/ETCH 2
7	FIELD DIFF/OX		5	DIFF 2
8	SOURCE/DRAIN MSK 1		6	MASK/ETCH 2
9	S/D IM 1		4	IMPLANT
10	S/D OX 1		2	DIFF 1
11	S/D MASK 2		6	MASK/ETCH 2
12	S/D IM 2		4	IMPLANT
13	S/D OX 2		2	DIFF 1
14	NITRIDE STRIP		6	MASK/ETCH 2
15	GATE OX		2	DIFF 1
16	POLY DEP/DOPE 1		7	DIFF 3
17	POLY MASK 1		3	MASK/ETCH 1
18	POLY DEP/DOPE 2		7	DIFF 3
19	POLY MASK 2		3	MASK/ETCH 1
20	CONTACT MASK		3	MASK/ETCH 1
21	**METAL DEP 1**	**1.20**	**8**	**METAL SPUTTER 1**
22	**METAL MASK**		**9**	**MASK/ETCH 3**
23	**DIELECTRIC**		**10**	**DIEL-FAB**
24	**METAL DEP 2**	**1.25**	**8**	**METAL SPUTTER 1**
25	**METAL MASK**		**9**	**MASK/ETCH 3**
26	**DIELECTRIC**		**10**	**DIEL-FAB**
27	**METAL DEP 3**	**1.30**	**11**	**METAL SPUTTER 2**
28	METAL MASK		9	MASK/ETCH 3
29	TOPSIDE DEP		7	DIFF 3
30	TOPSIDE MASK		3	MASK/ETCH 1

Determining the capacity of a plant for a single flow is a much simpler problem than determining it when there is a mixture of the two or more flows. The fundamental difficulty for a multi-flow factory lies in the large number of independent variables that affect capacity. Obviously, the capacitv of a multi-flow plant will lie somewhere between the highest- and lowest-capacity single-flow contributors.

Thus, for CMOS-2, the mixture's capacity will be somewhere between 1 lot every 54 minutes and 1 lot every 60 minutes, or 1 to 1.111 lots/hr. But exactly what the capacity is will depend on how much of each product is processed, how badly the fab is bottlenecked, whether the flows have the same bottleneck, and how much machine setup is needed since both products are using the same equipment. The situation where flows have different bottlenecks or where there is a bottleneck shift will be covered in Section 6.6. For the following discussion, all flows will have the same bottleneck.

6.1.1 Process Flow Mixing Rules

To provide a starting point for determining the capacity of a multi-flow plant, we will appropriate some concepts normally applied to chemical-mixing behavior in chemical engineering thermodynamics [6.1]. To run a complex, multi-flow plant, there are several key pieces of information that are either mandatory, or at least highly desireable: the lot-start rate for each flow, ls_i, the expected throughput for each flow, t_i, and the total expected throughput for the plant, T.

The lot-start rate of each flow will be set based on one of two criteria: the amount of fab resources, w_i, dedicated to each flow, **or** the desired production rate of the main flow, q_s. These criteria are usually not interchangeable.

Finally, for reasons covered later in the dispatching discussion, Section 6.4, lot starts will always be in terms of 1 lot per starting interval. This type of start rate will keep lots flowing into the factory as needed, not in bunches, thus helping to balance the line.

6.1.2 Mixing Rules Based on Fab-Resources Allocation

The resources in the plant, i.e., the machines, operators, etc., have a finite capacity. One method of setting the production of each flow is simply to allocate some fraction, w_i, of the fab's capacity to each, such that the allocation fractions all sum to '1'.

$$w_1 + w_2 + \ldots + w_n = 1$$

where n is the total number of flows.

Once the management decision is made on how resources will be allocated, it is then necessary to determine the lot-start rates, ls_i, which will give that apportionment. The logical and straightforward approach is for each flow to use some fraction of its single-flow lot-start rate, LS_i. The *single-flow lot-start rate* is the start rate that would be used if that flow were the only one using the plant. These rates are a key concept for decomposing the behavior of complex fabs.

The dimensions of both throughput and lot-start rate are in lots/hour. Thus, the *prorated* single-flow lot-start rates will give the hoped-for throughput, t_i, for each flow.

$$t_i = w_i * LS_i$$

But lots are discrete, and the start rate is actually the result of starting single lots at a specified interval, si_i. *Lot-start interval* is the natural control variable. In keeping with the criterion that start rates be in terms of 1 lot per starting interval, then for each flow, the lot-start rate, ls_i, is expressed in terms of the reciprocal of the starting interval.

$$ls_i = 1/si_i$$

Conversely, the lot-start interval, si_i, is the reciprocal of the throughput,

$$si_i = 1/t_i$$

To illustrate the fab-allocation mixing rule, consider the two process flows planned for CMOS-2. Under single-flow conditions, Flow-1 has a lot-start rate of 1 lot/0.9 hrs, and Flow-2 has a rate of 1 lot/hr. Six months into operation, the desired allocation of fab resources is 75% for Flow-1 and 25% for Flow-2. What then is the appropriate starting rates for the two flows at that time?

According to the mixing rule:

$w_1 = 0.75$
$LS_1 = 1 \text{ lot}/0.9 \text{ hrs}$
$t_1 = 0.75*(1/0.9) = 0.8333 \text{ lots/hr}$
$ls_1 = 1 \text{ lot}/1.2 \text{ hrs}$
$si_1 = 1.2 \text{ hours}$

$w_2 = 0.25$
$LS_2 = 1 \text{ lot}/1 \text{ hr}$
$t_2 = 0.25*(1/1) = 0.25 \text{ lots/hr}$
$ls_2 = 1 \text{ lot}/4.0 \text{ hrs}$
$si_2 = 4.0 \text{ hours}$

In a wafer fab, one flow will typically be displaced gradually by another, more advanced flow as market demands change. In the case of CMOS-2, during the early days of plant operation, the only process flow is the 2-metal Flow-1. As the plant ages, and as demands change, the 3-metal Flow-2 will take over the plant from Flow-1 in capacity increments of 25%. Table 6.3 gives the resource-allocation mixes that will be seen in CMOS-2 as it ages (75%/25%, 50%/50%, and 25%/75% for Flow-1/Flow-2).

TABLE 6.3: RESOURCE-ALLOCATION MIXING, CMOS-2

	Lot-Start 1 (lots/hr)	Thrupt 1 (lots/hr)	Lot-Start 2 (lots/hr)	Thrupt 2 (lots/hr)	Total Thrupt (lots)
100% Flow-1	1/0.9	1.1111	--	--	1.1111
75% / 25%	1/1.2	0.8333	1/4.0	0.2500	1.0833
50% / 50%	1/1.8	0.5556	1/2.0	0.5000	1.0556
25% / 75%	1/3.6	0.2778	1/1.33	0.7500	1.0278
100% Flow-2	--	--	1/1	1.0000	1.0000

Since the complexity of each flow in a multi-process fab may be markedly different, the fab's total throughput is likely to change as the mix of products changes. Again, this goes back to the fact that with a fixed resource allocation, a more complex flow will produce fewer lots. The fab's total throughput will be a sum of the throughputs for each flow.

$$T = t_1 + t_2 + \ldots + t_n$$

6.1.3 Mixing Rules Based on Production Allocation

Alternatively to setting the multi-flow lot-start rates based on fab-resource allocation, another method is to set them based on meeting the production demands of a standard flow. The lot-start rates for the other flows will be some percentage of the standard flow. In most cases, the flow that is of most importance, and hence the logical choice for the standard flow, is the one that is the most complex. (If another flow is selected as the standard, setting the lot-start rates may be more difficult.) The rank order of flow complexity is (in order of importance) based on:

- highest use of bottleneck
- number of process steps
- longest process time
- largest number of high-order hubs

For example with CMOS-2, Flow-1 and Flow-2 use the bottleneck workstation (Wkst 8) the same number of times, but because of the longer process times on steps #21 and 24, Flow-2 has the highest usage. If, however, the length of steps #21 and 24 were the same for both processes, then we would look to the second criteria, number of process steps, to determine the most complex process. Again, Flow-2 would be the winner with 30 versus 27 steps. In fact, Flow-2 ranks more complex than Flow-1 in every one of the above categories.

Once having determined the most complex process, it will be designated the standard flow, and its lot-start rate will be set based on desired production.

$$ls_S = q_S * LS_S$$

where q_S is the fraction of the single-flow rate to be produced. Note, that the production-start rate **must** be less than or equal to the single-flow lot-start rate for the standard flow, i.e.,

$$q_S \leq 1$$

The closer q_S is to 1, the less material will be started for the other flows. The sum of the single-flow fractions for all of the processes must be less than or equal to 1.

$$q_S + q_2 + q_3 + ... + q_N \leq 1$$

Once the standard flow's allocation has been accounted for, the remainder of the fab's production capability, q_R, may be allocated to the other flows:

$$q_R = 1 - q_S$$

and

$$q_2 + q_3 + ... + q_N \leq q_R$$

Thus, the lot-start rate for the second flow is selected such that,

$$ls_2 = q_2 * LS_S$$

where

$$q_2 \leq q_R$$

Before the next lot-start rate is selected, the remaining production fraction is updated:

$$q'_R = q_R - q_2$$

Each subsequent lot-start rate will be selected in a similar manner, all except for the last one. The last process basically will have access to the remaining fab capacity:

$$ls_N = q_N * LS_S + R$$

where

$$q_N = q_R$$

and R, the remaining capacity, is:

$$R \leq \sum_{i=1}^{N} (q_i * (LS_i - LS_s))$$

It is interesting to note that if q_s is large, i.e., most of the production is the standard flow, then the capacity remainder will be small. If q_S is small (most of the production is from other, simpler processes), then R may be sizable. Thus, R is inversely proportional to q_s.

$$R \propto 1/q_s$$

To illustrate the production-allocation rule with CMOS-2, we will seek the lot-start rates for various values of q_s. Since for this fab, Flow-2 is the more complex of the two production processes, it will be used as the standard flow for setting the lot-start rates. Thus,

$$LS_s = LS_2 = 1 \text{ lot/hr}$$

The lot-start rates for the first incremental take-over (nominal 75%/25% allocation for Flow-1/Flow-2, respectively) are:

$$q_S = 0.25$$

$$ls_S = 0.25 * 1 \text{ lot/hr} = 1 \text{ lot/4 hrs}$$

leaving

$$q_1 = q_R = 1 - q_S = 0.75$$

$$ls_1 = (q_R * LS_S) + R = (0.75 * 1 \text{ lot/hr}) + R$$

where

$$R \leq q_1(LS_1 - LS_s) \leq 0.75(1.1111 - 1.0)$$

Thus,

$$ls_1 \leq 0.75 + 0.75(0.1111) \leq 0.83333 \text{ lots/hr}$$

or in the form of 1 lot per interval,

$$ls_1 \leq 1 \text{ lot/1.2 hrs}$$

Table 6.4 shows the throughputs for the various production allocations over the life cycle of CMOS-2. At first glance, the results do not look markedly different from those derived using the resource-allocation mixing rule in Table 6.3. However, there is a subtle but important distinction between the two methods; it is one that may have a significant impact on fab operations.

The second method, production-allocation, sets inflexible start rates for all but the last flow (Processes **#1** through **n-1**). Furthermore, it bounds the start rate for the final flow, **n**. Thus, the throughputs for the first **n-1** flows are set, and for the **nth** flow is bounded.

TABLE 6.4: PRODUCTION-ALLOCATION MIXING, CMOS-2

	Std Flow (lots/hr)	2nd Flow (lots/hr)	Remainder (lots/hr)	Max 2nd Flow (lots/hr)	Total Thrupt (Min)	Total Thrupt (Max)
0% Standard	0.00	1.00	0.1110	1.1111	1.0001	1.1111
25% Standard	0.25	0.75	0.0833	0.8333	1.0000	1.0833
50% Standard	0.50	0.50	0.0556	0.5556	1.0000	1.0556
75% Standard	0.75	0.25	0.0278	0.2778	1.0000	1.0278
100% Standard	1.00	0.00	0.0000	0.0000	1.0000	1.0000

Since the lower bound 'Total Thrupt (Min)', derived with the production-allocation method, is tied to the fab's most complex process, it will generally be achievable. The higher bound 'Total Thrupt (Max)', derived with the resource-allocation method, is a different matter altogether.

In a real, multi-process fab, the throughputs expected, based on a resource-allocation mixing rule, are not achievable. Worse, if the calculated lot-start rates are indeed used, the fab will bottleneck and performance may become unstable and/or severely degraded. The root cause is that multi-process mixing rules, regardless of their basis, do not take into consideration the effects of process-process interactions. Such interactions may come in the guise of machine setups between inter-process usage, 'broken' batches at large furnaces, and/or hot, priority processes. (A *broken batch* is one which might have been filled in a single-process fab, but which will not be mingled between flows.)

So how far off might the mixing rules be?

It could be a little or it could be a lot, and it is extremely fab dependent. Contributing factors include, not only the relative complexities of the flows involved and the equipment set available to handle them, but also the fab's control rules on setup, batching, and dispatching.

6.2 Setup Rules

Setup is the machine preparation that must be performed when the next lot to be loaded is from a different process (or has different requirements) than the one that last used the machine. Each setup requires a certain amount of time to complete, and can take from minutes to hours, sometimes even overwhelming the lengths of process steps themselves.

During a setup, the machine involved is unavailable for processing. In addition to decreasing factory capacity, setups make for longer cycle times. Setup can be especially costly when the machine in question is:

- at a bottleneck workstation,
- at a high-order hub,
- at a workstation with many machines,
- at a workstation with small-batch machines, or
- an implanter and/or other setup-intensive gear.

In certain situations, such as when an especially long setup is involved (like implant), fabs may implement lot-dispatching rules at the workstation to minimize machine preparations. Rules such as dedicating a machine to a flow, holding up lots from one flow until a certain number from another flow have been processed, etc. are not uncommon. But even with special rules, setup in a multi-process fab is a difficult problem for spreadsheet modelers because it is one of the wafer fab's most dynamic variables.

The proper handling of setup requires knowing for certain exactly when, and under what circumstances, it will take place. Such knowledge requires a factory model that mimics the fab, down to the detail of moving every lot dynamically through the factory while observing every fab-specific variable and control rule.

To get a perspective of the setup dynamics at work, consider CMOS-2 with varying mixes of Flow-1 and Flow-2. Although in the early life of CMOS-2, the process mix will be a nominal 75% Flow-1 and 25% Flow-2, the other mixes, 50%/50%

and 25%/75%, will also be followed for the remainder of Section 6.2. In wafer fabrication, it is not the normal course of events for a simpler flow to displace a more complex flow, but by following all three mixes, we will cover all eventualities.

We know that the total hours spent in setup over a 12-week period for each of the single-flow cases are 3957.2 and 3660.1 for Flow-1 and Flow-2, respectively. If there were no setup penalty associated with process interaction, then the setup for a multi-flow case would simply be the sum of the appropriate percentages of the single-flow setups. For example, for a resource allocation of 75% Flow-1 and 25% Flow-2, the theoretical setup would be:

$$\text{Setup}_{EXP} = (0.75 * 3597.2) + (0.25 * 3660.1) = 3612.9 \text{ hrs}$$

This calculation, however, does not take into consideration the dynamics that take place when lots from different processes move through the factory, at different paces, in different orders, yet utilizing the same equipment set. Furthermore, whether or when a setup takes place depends on a lot's arrival at a certain machine, what lots are in front of it, and what loading rules may apply. The setup calculated above would only be representative if the lots from the two processes were kept segregated; i.e., all lots from Flow-1 are completed, followed by all lots from Flow-2, or vice versa.

But what happens when the flows are mixed?

6.2.1 Resource Allocation Mixing

If lots are started through the fab based on the Resource-Allocation Mixing Rule, the total time spent in setup is quite a bit different from the theoretical calculation above. For the 75%/25% case, the total setup that takes place during actual fab operations is 4110 hours, an additional 497.1 hours above ideal. This 13.8% setup penalty results because machines are flip-flopping back and forth between lots from the two flows.

Table 6.5 gives the setup penalties associated with the other mixes of lot-start rates. Not too surprisingly, the greater the mixing (50%/50% case), the greater the setup penalty (19.2%).

TABLE 6.5: SETUP PENALTY FOR MULTI-FLOW CMOS-2

	Single-Flow S/U (hrs)	Calc. S/U (hrs)	Actual S/U (hrs)	Inter-Process Penalty	Setup Penalty %
100% Flow-1	3597.2	--	--	--	--
75% / 25%	--	3612.9	4110.0	497.1	13.8
50% / 50%	--	3628.6	4326.0	697.4	19.2
25% / 75%	--	3644.4	4167.3	522.9	14.4
100% Flow-2	3660.1	--	--	--	--

Calculated S/U: sum of single-flow percentages.

On the surface, the setup penalties for running a multi-process fab appear significant. But are they really? Perhaps these are large percentages of very small numbers. Maybe setup does not really impact performance all that much.

If only that were the case.

For CMOS-2, Table 6.6 shows the performance penalties associated with the interaction of the two flows. At a glance, we can see that for either of the single-flow cases, performance is outstanding. Throughput is right where it should be; the cycle-time-to-process ratio is low; and inventory is not being built or depleted. The fab is operating in a very stable mode, and at capacity.

For all of the mixed-flow cases, however, the picture is not so rosy. From the inventory build alone, it is obvious that the fab has bottlenecked in every case. This observation is further verified by the higher cycle times that are occurring.

Since performance in the mixed-flow cases is even worse than for 100% of the most complicated flow (#2), the extra setup to handle process interactions must be the primary culprit. Actually, such a result is not only logical, but should have been expected. If Flow-1 by itself is running at fab capacity, and if

Flow-2 by itself is running at fab capacity, then **any** additional setup will push requirements beyond capacity.

TABLE 6.6: PROCESS INTERACTION PENALTIES

	Lot-Starts (lot/intvl)	Throughput (lots/hr)		CT/P		△ Inv. (lots)
		Expected	Actual	Expected	Actual	
100% Flow-1	1/0.90	1.1111	1.1119	--	2.332	-2
75% Flow-1 25% Flow-2	1/1.20 1/4.00	0.8333 0.2500	0.8180 0.2480	2.332 2.865	2.968 2.871	31 5
50% Flow-1 50% Flow-2	1/1.80 1/2.00	0.5556 0.5000	0.5397 0.4911	2.332 2.865	3.391 3.217	32 19
25% Flow-1 75% Flow-2	1/3.60 1/1.33	0.2778 0.7500	0.2703 0.7351	2.332 2.865	3.422 3.224	16 34
100% Flow-2	1/1.00	1.0000	0.9998	--	2.865	2

% of Flow by Resource-Allocation Mixing Rule.

Another way of looking at the situation is that additional setup shrinks capacity below the resources that are being allocated by the mixing rule. A necessary prerequisite for the mixing rule to work is that the single-flow lot-start rates, LS_i, must not cause the fab to bottleneck. But extra setup will do just that because the LS_i will be too large for the shrunk capacity.

Thus, for any real multi-process fab, the Resource-Allocation Mixing Rule should only be used to set the upper bounds on lot-start rates, bounds which can never realistically be reached. If lot-start rates are not backed off, and if production planning is done based on the unachievable rates, quarterly shortfalls will be sizeable__quarter, after quarter, after quarter (Table 6.7), and management will not be happy campers.

TABLE 6.7: QUARTERLY PRODUCTION SHORTFALL
Using Resource-Allocation Mixing Rule

	Throughput (lots/hr)			Quarterly Shortfall		
	Expected	Actual	Δ	lots	wafers	$mm
75% Flow-1	0.8333	0.8180	0.0153	31	775	6.975
25% Flow-2	0.2500	0.2480	0.0020	4	100	1.100
Total				35	875	8.075
50% Flow-1	0.5556	0.5397	0.0159	32	800	7.200
50% Flow-2	0.5000	0.4911	0.0089	18	450	4.950
Total				50	1,250	12.150
25% Flow-1	0.2778	0.2703	0.0075	15	375	3.375
75% Flow-2	0.7500	0.7351	0.0149	30	750	8.250
Total				42	1,125	11.625

% of Flow by Resource-Allocation Mixing Rule.
Flow-1 wafers $9000/ea; Flow-2 wafers $11000/ea.

6.2.2 Production Allocation Mixing

On the other hand, if the Production-Allocation Mixing Rule is used to set lot-start rates, the losses resulting from setup and other mixing effects are accounted for, in general, by using only a portion of the 'Remainder' capacity for the simplest, or least-important flow.

As before with CMOS-2, Flow-2 is designated the standard flow by virtue of being most complex. As such, **all** lot-start rates are set relative to its single-flow rate. Thus,

$$LS_S = LS_2 = 1 \text{ lot/hr}$$

If the production rate of interest is for 0.25 lots/hr to be from Flow-2, and the remainder from Flow-1, then:

$$ls_2 = q_2 * LS_S = 0.25 * 1 \text{ lot/hr}$$

and

$$Is_1 = (q_1 * LS_s) + R = (0.75 * 1 \text{ lot/hr}) + R$$

where

$$R \leq R_{MAX}$$

and

$$R_{MAX} = q_1(LS_1 - LS_s) = 0.75(1.1111 - 1.0) = 0.0833$$

But what, exactly, is the remaining capacity, R, after the mixing losses have kicked in?

Unfortunately, it will vary from fab to fab, depending on the process mix, equipment, bottlenecks, control rules, etc. But, as a general rule, R will be greater than zero, except under very bizarre condition such as humongous amounts of setup. On the other extreme, R will be less than the sum of the ratioed single-flow rates.

Perhaps a good rule of thumb on how to estimate R is to assume that it is proportional to the amount of mixing. Since setup is proportional to the amount of mixing, this is usually not a bad assumption, and it should be fairly close to the remaining fab capacity. Thus,

$$R \cong q_1 * R_{MAX} = 0.75 * 0.0833 = 0.0625$$

Therefore, we may estimate Is_1 as:

$$Is_1 = 0.75 + 0.0625 = 0.8125 \text{ lots/hr}$$

or, more specifically

$$Is_1 = 1 \text{ lot}/1.23 \text{ hrs}$$

Table 6.8 gives the anticipated lot-start rates for the other production allocations for CMOS-2. In it, the $ls_{1, MAX}$ and $ls_{1, CALC}$ are derived by:

$$ls_{1, MAX} = (q_1 * LS_S) + R_{MAX} \quad \text{and}$$

$$ls_{1, CALC} = (q_1 * LS_S) + (q_1 * R_{MAX}) = q_1(LS_S + R_{MAX})$$

TABLE 6.8: SETTING LOT-START RATES FOR CMOS-2
Using Production-Allocation Mixing Rule

ls_2	q_1	$q_1 * LS_2$	R_{max}	$ls_{1,max}$	$ls_{1,calc}$	$ls_{1,calc}$ (1 lot/Invl)
0.25	0.75	0.75	0.0833	0.8333	0.8125	1/1.23
0.50	0.50	0.50	0.0556	0.5556	0.5278	1/1.90
0.75	0.25	0.25	0.0278	0.2778	0.2570	1/3.89

Standard flow, based on complexity, is FLOW-2.
Single-flow lot-start rate for standard, LS_2 = 1 lot/hr
All rates in lots/hr.

Now, the key question: how do these mixing-rule-derived lot-start rates perform in the plant? Table 6.9 shows CMOS-2's mixed-process performance at the various production allocations. In general, the lot-start rates match factory capacity astonishing well in every case. Some further fine tuning may be necessary, but overall it may be said that the projected start rates from this rule (using the ratioed remainder):

- do not bottleneck the fab,
- give throughputs very close to planned,
- do not cause long cycle times
- do not excessively build or deplete inventory,
- have accommodated the burden of multi-processes.

At least, these conclusions may be drawn for CMOS-2, under the fab-specific conditions used in the analysis. But what happens if conditions change?

TABLE 6.9: START-RATE-to-CAPACITY MATCHING
Using Production Allocation Mixing Rule

	Lot-Starts (lot/Intvl)	Throughput (lots/hr)		CT/P	Total S/U (hrs)	△ Inv. (lots)
		Expected	Actual			
FLOW-1	1/1.23	0.8125	0.8140	2.430	4059.2	-1
FLOW-2	1/4.00	0.2500	0.2530	2.418		-5
FLOW-1	1/1.90	0.5278	0.5263	2.550	4257.8	1
FLOW-2	1/2.00	0.5000	0.5035	2.477		-6
FLOW-1	1/3.89	0.2570	0.2564	2.799	4076.6	0
FLOW-2	1/1.33	0.7500	0.7525	2.676		-1

So far, the multi-process analysis has concentrated on the burden of additional setup. But what about the other aspects of operation that may be impacted by mixing, such as equipment batching?

6.3 Batching Rules

The load size of a piece of semiconductor-processing equipment is known as a *batch*, and it may range from one wafer to hundreds of wafers. For example, there is a 8-wafer etcher on the market, and a known 16-wafer sputterer still in production in an number of older fabs. Either of these units will require multiple loads to process a standard lot. At the other extreme, oxidation and diffusion furnaces typically have a load size of 100 to 200 wafers, and thus may process several lots at once. A machine with a load size that differs from the fab's standard lot size will generally have *batching rules* that govern how lots are loaded into the machine.

Although lot size may vary at certain fabs, the industry standard for 8" or smaller wafers is 24 or 25 wafers per lot. The lot size for 12" wafers, though as yet undetermined, will be substantially smaller.

For the sake of convenience, we will get the topic of small-batch equipment out of the way first.

6.3.1 Small-Batch-Size Equipment

Small-batch equipment, which was not uncommon in the early 1970's, is gradually disappearing from the industry. Of course, there are some obsolete, antediluvian machines in operation in older fabs, and a few other small-batch models are still being sold. But most newer, state-of-the-art fabs do not have such equipment.

The introduction of 12" wafers, however, could change all that. The size and weight of these next-generation wafers will most likely accelerate the trend to single-wafer processing, and perhaps redefine the standard lot size. As a interim step, the concept of small-batch equipment may be resurrected for these 12" monsters.

Equipment with load sizes smaller than the standard lot size will have batching rules to determine how lots are divided up among multiple batches. (A *batch* is the actual wafers puts in a machine, regardless of rated load size or lot provenance.) If the number of wafers in a lot is evenly dividable by the load size, the answer is straightforward: the processing will simply be performed in 'x' batches, where x is the number of wafers in a lot divided by the load size.

Where the batching rules are needed, however, is when the lot size is not evenly dividable by the load size. In such a case, there will be a remainder number of wafers. Will the remainder then be processed with other wafers from another lot?

To insure lot integrity, the batching rule at most fabs is *no splitting of lots*, i.e., wafers from two split lots will not be in a machine at the same time. Lot integrity is necessary so that wafers have the same processing history. If lots are split between batches at one or more steps, it becomes difficult to analyze device characteristics and to correlate the device's performance to processing results.

The mixing of non-split lots in large-batch equipment is a different story.

6.3.2 Furnaces and Other Large-Batch Equipment

The type of processing performed in large-batch machines is primarily diffusion and oxidation, and it takes place in furnaces that may hold up to 200 wafers (8 lots). For such equipment, where the machine load size is larger than the standard lot size, the batching options are:

- wait for a full load
- wait for a specified number of lots (2, 3, etc.)
- load whatever is available and process (load & go)

Normally, these batching options only apply to lots from the same process **and** having the same process recipe (i.e., same gases, same process time, etc.) In a multi-process fab, especially in an ASIC, the possibility of mixed-process batches will arise. Such mixing may lead to performance improvements, if done with care, or yield crashes, if Murphy has his way. Mixing batches between processes is a very fab-specific decision.

Having a machine that can process numerous lots at a time sounds like the answer to a fab manager's wildest dream. However, the reality of having large-capacity equipment may lead to more headaches than wild dreams, since it is only human nature to want to utilize any piece of gear to its fullest extent. Hence, with a large furnace, the natural inclination is to wait for a full load, regardless of how many lots that may be, before processing begins. After all, why waste fab resources? It won't look good on the books.

For some process flows, though, it may be quite a wait before four, five, or more lots arrive at a work center, lots with exactly the same recipe.

Historically, the long waits that took place while waiting to fill furnaces did not go unnoticed, especially by operators. Over the years, and primarily at the initiative of operators, furnace-loading rules evolved which gave a reasonable compromise between high furnace utilizations and excessive waits. Concisely stated for a single-process fab, these empirical batching rules are:

If a fab is running close to its optimal inventory, furnace loads of two, or more likely three lots give better performance than waiting for full loads.

If a fab is basically a just-in-time (JIT) operation, the best furnace loading strategy is 'Load & Go'.

If the fab is glutted with WIP, full-furnace loading probably will not degrade performance any worse than it already is.

Table 6.10 shows how furnace-loading rules impact CMOS-2 while it is running a single process, Flow-1. The three diffusion work centers (Wksts 2, 5, 7) each have several 100-wafer furnaces. The batching options for each of the furnaces are: (1) to load and process whenever one or more same-recipe lots become available (Load & Go), (2) to hold loading until at least two same-recipe lots are available (50% Load), (3) to load only after three same-recipe lots are at the workstation (75% Load), or (4) to process only if a full load (4 lots) of same-recipe lots is available.

TABLE 6.10: FURNACE LOADING FOR FLOW-1 (CMOS-2)
100-Wafer Diffusion Furnaces

	Load & Go	50% Load	75% Load	Full Load
Cycle Time (hrs)	88.3	89.1	92.5	97.2
CT/P	2.312	2.332	2.421	2.543
Thrupt (lots/hr)	1.1171	1.1119	1.1131	1.1091
Add. Value ($mm)*	2.12	0.28	0.71	- 0.71

For CMOS-2, in going from 'Load & Go' to 'Full Load', fab performance generally declines. Cycle times gradually increase, from 88.3 to 97.2 hours; and throughput drops off, from 1.1171 lots/hr to 1.1091 lots/hr.

Even though the differences in the throughput look small, the wafer value over a quarter of operation (assuming a $9000 per wafers sale price and a $2000/wafer production cost) is worth $2.1mm more than the design value (1.1111 lots/hr) for the 'Load & Go' case, and worth $0.7mm less than design for the 'Full Load' case. Thus, there is almost a $3mm difference in value/cost ($12mm annually) attributable to nothing more than choosing the correct furnace-loading policy. And counter to intuition, the best policy is **not** to hold up lots waiting for full loads. In most current wafer fabs, the general rule of thumb is to run 3-lot furnace loads, regardless of furnace size.

The partial-furnace loading rule makes even more sense if one considers what happens in real wafer fabs, versus what is happening in CMOS-2. One must remember that the toy fab used in this chapter is a highly balanced factory, running with an optimal WIP of 4 lots per process step and optimally-spaced lot starts, perfectly matched to fab capacity. Thus, the results for CMOS-2 are not nearly as dramatic as they would be for a real fab running at a comfortable 3X inventory, yet having longer and unexpected equipment failures (covered in Section 6.9) that disrupt the smooth flow of lots to the large 200-wafer furnaces.

The more unbalanced the fab, the more performance will degrade by filling large-batch equipment. Fab imbalance may result for a variety of reasons: a too heavily- and/or erratically used bottleneck, disparate batch sizes, diverse granularity, grossly dissimilar availabilities, etc.

One way to magnify the impact of batch-size, while still using CMOS-2 for illustrative purposes, is to have the same fab capacity spread across multiple flows. Unless special batch-mixing rules are implemented, multiple processes exacerbate the fundamental mismatch between uniformly smooth-flowing lots and large-batch sizes.

Table 6.11 shows the cycle-time performance versus batch size for CMOS-2 with two flows. The only restrictions on lots going into a given furnace are that they be from the same flow and use exactly the same recipe.

TABLE 6.11: CYCLE-TIME PERFORMANCE vs BATCH SIZE
100-Wafer Diffusion Furnaces in CMOS-2

	Lot-Starts	Cycle Time (hrs)			
	(lot/intvl)	Load & Go	50% Load	75% Load	Full Load
FLOW-1	1/1.23	92.5	92.8	91.8	110.2
FLOW-2	1/4.00	100.7	104.1	104.9	183.3
FLOW-1	1/1.90	97.0	97.4	97.0	106.2
FLOW-2	1/2.00	106.8	106.6	108.2	120.7
FLOW-1	1/3.89	101.9	106.9	101.1	153.1
FLOW-2	1/1.33	112.2	115.2	111.2	116.6

As expected, the best cycle times are achieved when the furnaces are run in a 'Load & Go' fashion, i.e., loading whatever becomes available, whether it is one lot or four lots. Because 'Load & Go' operation does not require a group of matched lots, waiting is minimized. Since cycle time is predominately a measure of lots waiting, cycle time is thus minimized.

It should be pointed out that 'Load & Go' for large-batch equipment is not synonymous with single-lot loads. More often than not, with the amount of WIP being carried and the typical lot-start rates, furnace loads will have multiple lots while observing this rule. This observation helps to explain why both cycle time and throughput (Table 6.12) look similarly good for either 'Load & Go' or partial-loading.

TABLE 6.12: THROUGHPUT vs FURNACE BATCH SIZE

	Lot-Starts	Throughput (lots/hr)				
	(lot/intvl)	Expected	Load & Go	50% Load	75% Load	Full Load
FLOW-1	1/1.23	0.8125	0.8135	0.8140	0.8135	0.8070
FLOW-2	1/4.00	0.2500	0.2530	0.2530	0.2535	0.2426
FLOW-1	1/1.90	0.5278	0.5278	0.5263	0.5278	0.5228
FLOW-2	1/2.00	0.5000	0.5030	0.5035	0.5030	0.5010
FLOW-1	1/3.89	0.2570	0.2564	0.2564	0.2564	0.2505
FLOW-2	1/1.33	0.7500	0.7530	0.7525	0.7520	0.7495

Where fab performance craters, however, is when the 'Full Load' rule is imposed. Cycle times are long because of the waiting that transpires while trying to fill a load. Further, waits and cycle times are especially long for small flows. In these cases, the interval between lots arriving to fill a furnace may be excruciatingly long.

Throughputs also suffer with imposition of the 'Full Load' rule, again especially for small flows. The degraded performance for small flows is quite costly (Table 6.13) in terms of lost production. (The economics assume FLOW-1 produces wafers valued at $9000 and costing $2000 to make; FLOW-2 produces $11,000 wafers that costs $2500 to make.)

TABLE 6.13: ADDED THROUGHPUT VALUE vs BATCH SIZE

	Lot-Starts (lot/intvl)	Added Value/Loss ($mm)			
		Load & Go	50% Load	75% Load	Full Load
FLOW-1	1/1.23	0.454	0.680	0.454	-2.495
FLOW-2	1/4.00	1.663	1.663	1.940	-4.103
FLOW-1	1/1.90	0.000	-0.680	0.000	-2.268
FLOW-2	1/2.00	1.663	1.940	1.663	0.554
FLOW-1	1/3.89	-0.272	-0.272	-0.272	-2.948
FLOW-2	1/1.33	1.663	1.386	1.109	-0.277

On the surface, the degraded throughput seems to violate the fundamentals of fab performance. After all, throughput is the result of factory capacity; and it seems farfetched that a batch-loading rule could change capacity. Even if it could, it should go in the other direction since all of the furnace capacity is being forced into use. So, why is the factory losing capacity; or is the result merely an artifact of CMOS-2?

When in doubt, or totally dumbfounded, look to the factory bottleneck (or constraining workstation) for the first line of answers.

In CMOS-2, the constraining workstation is Wkst 8, 'Metal Dep. 1'. For all lot-start rates considered in Table 6.12, and for every furnace loading rule **except** 'Full Load', the utilization ratio of Wkst 8 is 1.0; in other words, the machines stay busy all the time they are not down for maintenance. Furthermore, for all cases other than 'Full Load', there is a good match between the combined throughput of the two flows and fab capacity. In other words, Metal Dep.1 is a constraining workstation, not a true bottleneck.

But when the 'Full Load' furnace-batching rule is in force, utilization of Wkst 8 ranges between 0.888 and 0.986. Capacity has been lost because the constraining workstation is no longer being fully utilized. (See Chapter 5 for more discussion on bottlenecks and constraining workstations.) In a sense, the fab's control mechanism has lost control.

6.3.3 Chaotic or 'Chugging' Behavior at Bottleneck

To understand how and why the 'Full Load' rule has such an adverse affect on performance, we will follow WIP through a quarter of fab operation. Initially, inventory will be spread out across the fab uniformly, with 4 lots per process step, divided up 75%/25% between the two flows. This WIP staging is shown on the fab graph in Fig. 6.3.

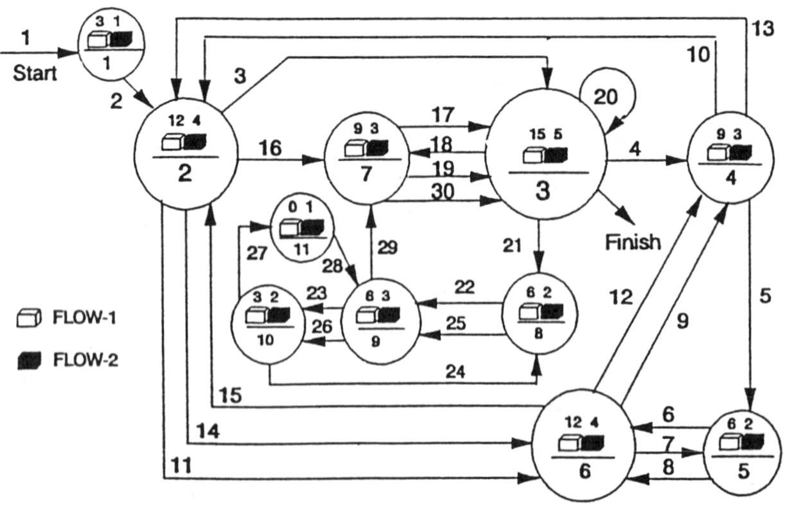

Figure 6.3: Initial Inventory for CMOS-2

As long as the 'Full Load' rule is not imposed, the constraining workstation keeps the fab running smoothly, and pretty much at capacity, by having roughly 8 lots of WIP to work on at any given time, 6 for Flow-1 and 2 for Flow-2.

The imposition of the 'Full Load' rule for the three diffusion furnaces (Wksts 2, 5, and 7) plays havoc with the WIP that Workstation #8 has, or does not have, to work with at any given time. As seen in Fig. 6.4, furnaces loads basically hold up lots, then release them all at once, causing WIP to move through the factory, and more importantly through the bottleneck, in a 'chugging' fashion. At times, the constraining workstation may have little or nothing to work on; at other times, a substantial portion of the fab's entire inventory may be sitting on its racks.

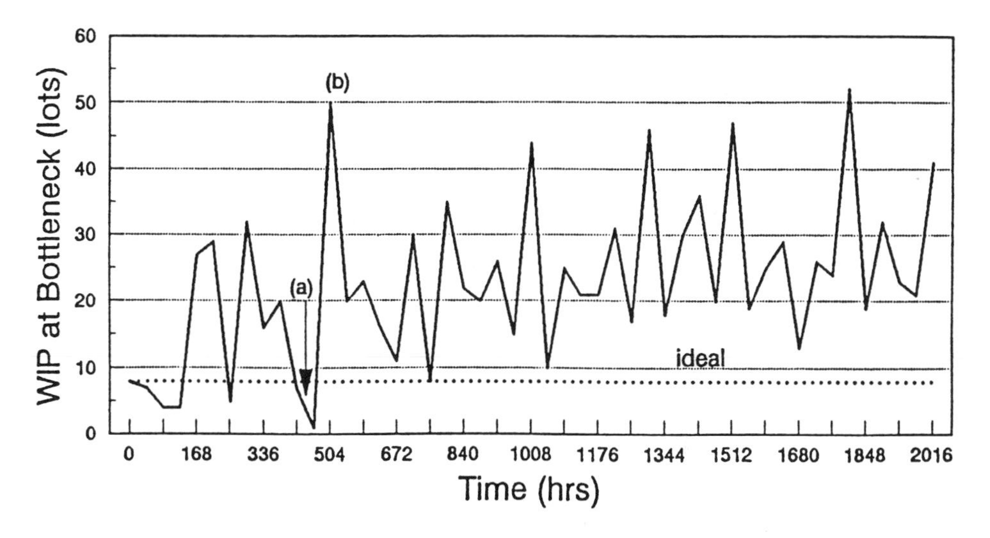

Figure 6.4: "Chugging" Behavior of WIP at Bottleneck

As cases in point: at 462 hours into the quarter (Fig. 6.5(a)), Workstation #8 has only one lot in inventory; at 504 hours (Fig. 6.5(b)), it has fifty lots awaiting processing.

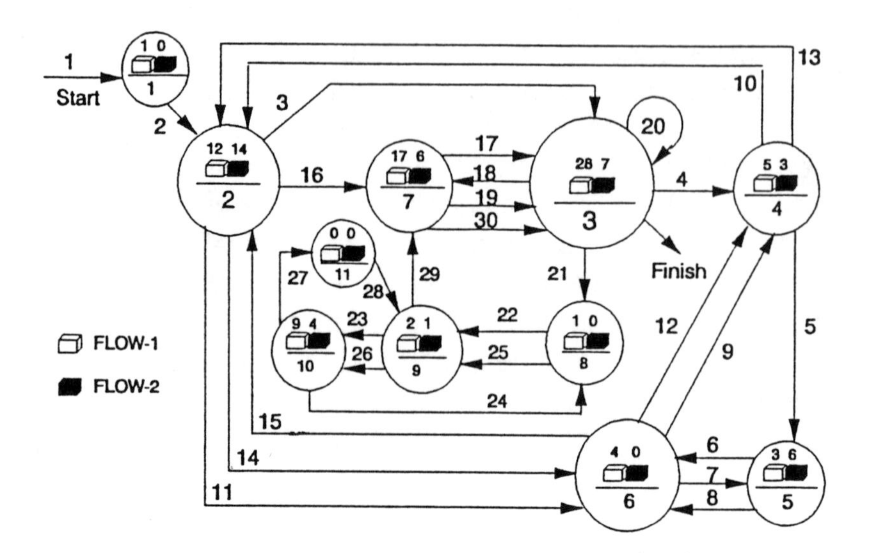

Figure 4.5(a): "Chugging" WIP at 462 hours

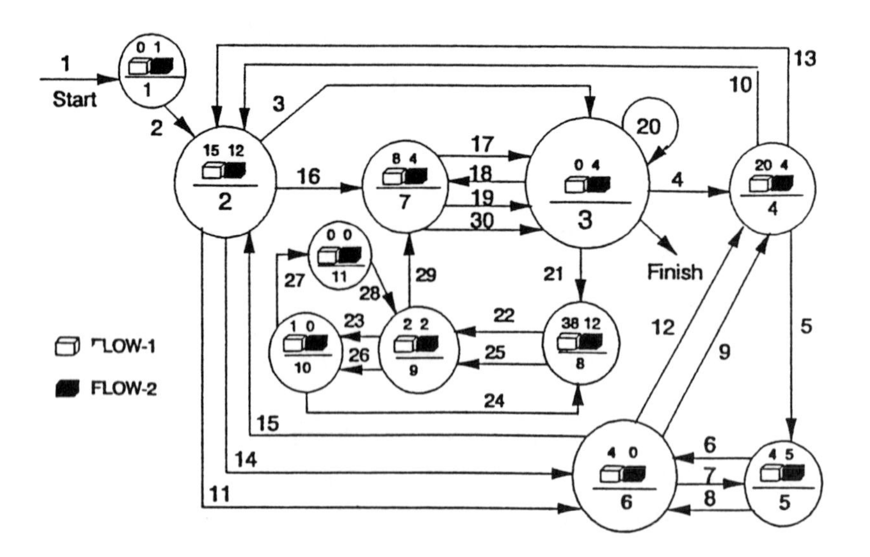

Figure 4.5(b): "Chugging" WIP at 504 hours

To see the WIP profile, upstream of the constraining workstation, look at the appropriate portion of the process sequences, given in Table 6.14. In the 462-hour case, there are 46 lots of WIP heading down the pike towards Wkst 8, and 20 of those lots are either in furnaces or waiting to be put in them. Meanwhile, Wkst 8 has only one lot of inventory to work on, no where near enough to keep the three machines at that work center busy.

At 504 hours, a big chug of the WIP, 35 lots, has made it through to Wkst 8 and is waiting to be processed by the three machines. Meanwhile, immediately upstream of Wkst 8, there is a total of 12 lots of WIP, all waiting at furnaces for a full load, but none being processed.

TABLE 6.14: LARGE-BATCH-INDUCED CHUGGING
Near CMOS-2's Constraining Workstation (#8)

STEP #	WKST #	WORKSTATION	WIP @ 462		WIP @ 504	
			FLOW-1	FLOW-2	FLOW-1	FLOW-2
15	2	DIFF 1	2	2	2	2
16	7	DIFF 3	7	1	3	1
17	3	MASK/ETCH 1	8	4	0	0
18	7	DIFF 3	6	2	2	2
19	3	MASK/ETCH 1	4	0	0	0
20	3	MASK/ETCH 1	8	2	0	0
21	8	METAL SPUT-1	1	0	27	8

While the type of chaotic behavior [6.2] shown in Fig. 6.4 may not be detrimental at other workstations, it is totally unacceptable at the bottleneck or constraining workstation. In many physical systems, when the controlling element goes chaotic, the entire system goes chaotic. It is a mark of the robustness of wafer fabrication that, with the bottleneck in a state of chaos, the only fab-wide impact on performance is a loss of a few percent of capacity. It goes without saying, however, that this is not a good way to run the factory, as witnessed by the gradual, but definitive trend in degradation seen in Fig. 6.4.

The overriding conclusion that falls out of every economic analysis, not only in this book but also in the real world with real fabs, is that money spent on equipment, if it buys throughput, is well spent. As far as semiconductor-processing equipment goes, furnaces are cheap. Why lose throughput, and why operate an unstable, roller-coaster fab when, for less than a million dollars a furnace, the problems will be fixed? **And** the payback period will be measured in days, or at most weeks. Such a simple concept, so easy to prove, and yet beyond the grasp of all but a select few world-class manufacturers (Table 6.15). For the others, throughput is regularly sacrificed for want of a piece of gear or two.

TABLE 6.15: CATEGORIES OF MANUFACTURERS

Wafer Fabrication

	World-Class	Wanna-Be's	No Delusions
Products	State-of-Art	1 Generation Back	Jelly Beans
Volume	Very High	High	Whatever you want
Market Penetration	First	Second	When others leave
Yield Policy	At any price	At some price	Bad ones come back
Capital Policy	Spend for excellence	Cost conscious	Categorically none
Operating Policy	Spend what it takes	Expect yield crashes	Run to extinction

Like a dog with a bone, financial managers tenaciously hold on to the belief that more throughput can be had from any existing equipment set, if only the personnel will work harder. It is standard practice to categorically refuse requests for capital expenditures unless the fab is in truly dire straits, and sometimes not even then. So, when 'working harder' simply doesn't work, the next step is the implementation of some fairly complicated batching and dispatching rules in the hope that they will lead to capacity, and hence throughput.

6.3.4 Mixing Batches to Recoup Furnace Capacity

In large-batch equipment, mixing batches is generally not a bad operating policy, with the proviso that the inter- or intra-process steps being mixed require exactly the same recipe. With a good CIM or tracking system, the policy is not even overly difficult to implement. But how much does such a policy improve performance? The answer may actually be rather surprising.

Going back to the three standard multi-process scenarios, Table 6.16 compares the performance of CMOS-2 with and without inter-process batch mixing, and using a Load & Go policy. As may be seen, the multi-process furnace mixing for the most part does not help, and may in fact slightly hurt performance for the two cases where one flow is considerably larger than the other. In these cases, the dynamics involved in the large-flow/small-flow interaction swamp possible mixing advantages.

TABLE 6.16: INTER-PROCESS BATCH FURNACE MIXING

	Lot-Starts	Throughput (lots/hr)			Cycle Time (hrs)	
	(lot/Intvl)	Expected	No Mixing	Mixing	No Mixing	Mixing
FLOW-1	1/1.23	0.8125	0.8135	0.8145	92.5	92.8
FLOW-2	1/4.00	0.2500	0.2530	0.2520	100.7	109.3
FLOW-1	1/1.90	0.5278	0.5278	0.5293	97.0	93.8
FLOW-2	1/2.00	0.5000	0.5030	0.5025	106.8	107.5
FLOW-1	1/3.89	0.2570	0.2564	0.2564	101.9	102.4
FLOW-2	1/1.33	0.7500	0.7530	0.7490	112.2	114.1

No waiting policy, i.e., 'Load & Go'

However, in the case where the flows are basically equal, the mixing does provide an small overall improvement. With flows that are essentially even, enough of the equivalent-recipe lots arrive at the furnaces regularly to provide some inter-process mixed batches. Unfortunately, the performance boost is not exactly what one might have hoped for.

At this point, the operations researchers in the audience are probably saying that we have overlooked the sure winner: requiring the furnaces to run with full loads, but allowing the batches to be mixed between processes. So let's put that dead horse to rest once and for all.

Full-furnace loading will never perform better than the case where there is only one flow in the fab. For obvious reasons, this is the situation where the maximum number of recipe-equivalent lots arrive at the furnaces for processing in the quickest order. The performance for the two bounding cases, 100% Flow-1 and 100% Flow-2, are presented in Tables 6.10 and 6.17, respectively.

TABLE 6.17: FURNACE LOADING FOR FLOW-2 (CMOS-2)

	Load & Go	50% Load	75% Load	Full Load
Cycle Time (hrs)	119.0	123.4	120.8	126.4
CT/P	2.765	2.865	2.805	2.936
Thrupt (lots/hr)	0.9975	0.9988	0.9970	0.9940

Expected Thrupt = 1.0 lots/hr.

Requiring full loading of large-batch equipment almost always degrades fab performance. The exceptions arise when the fab is already being operated in a mode that guarantees poor performance: gross bottlenecks, excessive WIP, binge-purge lot starts, etc. In situations where the fab is already running about as poorly as it can, completely filling large batch equipment may not further hurt performance, and even if it does, it will not be traceable.

As a rule, the entire fab should not be put on a hiccup ride in pursuit of furnace-loading__loading that will not meet minimum performance requirements anyway. Even complicated batching rules do not fool the system into reimbursing capacity lost in the effort to fill large-batch equipment. The logical and cost-effective approach is to buy a sufficient quantity of furnaces, irrespective of load size, and run them in a manner that will permit the smooth flow of lots through the line.

Now we have exhausted batching, are there other, less deleterious control rules that may facilitate multi-flow fab operation, by recouping some of the performance lost as a result of process mixing? Dispatching rules, perhaps?

6.4 Dispatching Rules

In general, *dispatching rules* determine which lot(s) will be worked on next at a given workstation. In the broadest definition of dispatch, the rules may also extend to lot-move control and process priority.

The decision on which lots will be loaded into equipment, and when, will generally be scheduled by the fab's tracking or CIM system. The basis for the decision, i.e., the dispatching rule, differs from fab to fab, and is usually decreed by the operations manager. Tomes have been written on dispatching rules [6.3, 6.4], and in some rare instances, the work has actually been applied to wafer fabrication [6.5].

The dispatching rules that have been used or considered for wafer fabrication cover a wide spectrum, ranging all the way from the simplest, FIFO, to one of the more popular, critical ratio, to one of the most ill-conceived, random choice. There are in excess of fifty dispatching schemes that have, at one time or another, been bandied about the industry. Only a few of those will be described here.

FIFO (First In, First Out also known as First Come, First Serve): This rule prescribes that lots are processed in the order in which they arrive at a workstation. With centuries of real and synthetic experience using this rule in wafer fabrication, it may be categorically stated that it works exceptionally well and is easily implemented.

Critical Ratio: This rule selects the next lot based on a ratio of the time remaining till shipment to the work still needed to be performed. The critical ratio indicates whether a lot is on time, behind schedule, or ahead of schedule. The lot with the smallest critical ratio is processed first. This rule is widely used and very popular in wafer fabrication because, unlike FIFO, it ties scheduling to due dates. Further, it is a dynamic scheduler in the sense that, if due dates change, dispatch priorities change.

Slack Time: This rule provides an alternative to critical ratio, while still tying scheduling to due dates. The slack time is the time remaining till the due date minus all remaining setup and processing times. Lots are dispatched based on the smallest amount of slack time.

Process Priority (Hot Process): One or more processes are rank-ordered over other processes to establish lot priority. All lots from the hottest process will be dispatched before lots from the next hottest, before lots from the next hottest, and so on. Lots in the queue from within a hot process will be dispatched FIFO. Likewise, lots from processes which are not rank-ordered will de dispatched FIFO relative to other unranked lots. This dispatching scheme is often applied to engineering or development lots, giving them priority over production lots.

Least Setup: By this rule, the next lot selected will minimize the amount of setup needed on a machine before processing. While this rule may not be desireable fab-wide, it could be useful at work centers with machines that require long setups (such as Implant). Furthermore, of all the dispatching rules, this is the only one that directly affects factory capacity, and hence throughput performance. Smart fabs implement this rule at selected workstations, regardless of other fab-wide dispatching rules.

Earliest Lot Due Date: As implied, this rule prioritizes lots in a queue based on earliest due date. The rule is generally not as good as critical ratio in that it takes a myopic view of lots at a work center, as opposed to how the lots are progressing through their entire process sequence. Even the nearsighted workstation view is somewhat myopic since no distinction is made between hubs and non hubs, and whether the lot is awaiting processing for step 'x' or step 'x+n'.

Least Lots Next Queue: For small, boutique fabs, this rule makes considerable sense, though not for the reason usually given. The rule bases lot selection on which lot is going to the workstation with the smallest queue. The theory behind the rule is to maximize machine utilization. But as we know from Chapter 5, maximizing machine utilization is not necessarily a worthy goal. The reality of using the rule, especially in a fab with few pieces of gear, is that it keeps lots moving smoothly through the fab, much as they would in a highly-granular or JIT operation. But even in a small fab, this is a difficult rule to implement because it requires dynamic look-aheads to workstations downstream in the fab's various process sequences.

Shortest Next Process Step: Based on this rule, the next lot selected is the one that requires the least processing time at the workstation. This is not a good dispatching scheme for a wafer fab where a significant fraction of the workstations are hubs, some of order twenty or more.

Least Work Remaining: This scheme looks at the lots in the queue, finds the one with the least work remaining, and hustles it through. This is not a good rule for wafer fabrication. Some lots could stay in the fab forevvvvvvver!

Random: This dispatching scheme allows the operator to pick any lot from the queue, randomly, to work on next. Any wafer fab even remotely considering this option should not be in the business.

One way to pigeonhole the various dispatching schemes is given in Table 6.18.

TABLE 6.18: PRIORITY DISPATCHING SCHEMES

Line Dispatch	Workstation Dispatch
Critical Ratio	FIFO
Slack Time	Process Priority
Short Next Queue	Least Setup
Least Work Remaining	Due Date
	Short Next Step
	Random

'Line' dispatching schemes are those which make decisions based on what is happening across the line or the factory. More specifically, information from the downstream portion of the process sequence for particular lots is factored into the decision process. The 'Workstation' schemes are those which make dispatching decisions based solely on the lots and information at a specific workstation.

This categorizing brings up an interesting question. Will the global type of dispatching provided by the Line schemes inherently give better factory performance than the myopic dispatching of the Workstation schemes? If we eliminate random dispatching from the list as being too silly to even consider, which of the other rules should be used in wafer fabrication? Which gives the best factory performance? The answer may surprise you.

They are all about the same.

As unbelievable as this statement seems, it is nonetheless true. Throughput, which is generally accepted as the archetypal measure of factory performance, is directly determined by factory capacity. With the exception of setup minimization, no other dispatching scheme impacts capacity, making either more or less of it available. Thus, dispatching schemes have no direct impact on the throughput performance. In this regard, equipment batching has much more of an impact since full-furnace loading does decrease fab capacity.

However, if the definition of factory performance is widened to include other measures such as fast cycle times, high yields, schedule stability, operating stability, etc., then the 'best' dispatching scheme will be a function of the fab's true, overall objectives. (See Section 5.6.) We can say that, in general, Line dispatching does have more of a tie to schedule manipulation than does Workstation dispatching.

6.4.1 Bunched or Clumped Lot Starts

Before we look at some of the more popular dispatching schemes, we need to take a quick look at a trap that many wafer fabs, at one time or another, have fallen into: the policy of starting lots in bunches or bursts. Under this policy, the total number of lots that should be started over a given period, will be started into the fab all at once, usually at the beginning of the shift, the day, or the week.

The policy of burst starts originated as a matter of convenience and also as a labor-saving strategy. Like paying the monthly bills, burst starts took care of the physical operation, as well as the CIM bookkeeping, all in one fell swoop. Because the fab was already loaded with WIP, logic said that loading it up with a few more excess lots would not matter. The lots would eventually spread out and work their way through the system.

However, everything we have seen in wafer fabs indicates that whenever lots move through the plant in 'chugs' or 'bunches', poor performance generally results. The bunching of WIP does not provide a smooth flow of material to the bottleneck or constraining workstation; thus it leads to less-than-full utilization. Further, although the bursting of starts may be more convenient initially, it tends to make the staging, management, and dispatching of WIP more difficult to perform adequately across the line.

The trend away from burst starts has been changing steadily over the last few years, with more logic going into lot-starting policies. The more automated a fab (i.e., bar coded WIP, automated CIM entry, etc.), the more likely the policy of single-lot starts, where the start interval is as long or short as is needed to keep the line supplied with WIP, but **at** the capacity of the fab. Thus, the start rate in these automated fabs is tied directly to the capacity of the bottleneck or constraining workstation.

To explore the policy of burst starts, consider four different lot-starting policies for CMOS-2: (1) 1 lot at an interval as required by the line, (2) all the lots that will be needed over a shift, (3) all lots that will be needed over a day (3 shifts), and (4) all lots that will be needed over a week.

As may be seen in Table 6.19, the fab's performance for the weekly starts is about as abysmal as expected. Throughputs have taken a major hit, and cycle times are very long. The once-a-week burst of material into the fab causes the same type of chugging behavior at the bottleneck that is observed when the full-furnace-loading rule is enforced.

TABLE 6.19: LOT-START INTERVAL FOR CMOS-2

Convenience vs Performance

	Throughput (lots/hr)				Cycle Time/Process Time (CT/P)			
	1 lot/req'd	once/shift	once/day	once/week	1 lot/req'd	once/shift	once/day	once/week
75% FLOW-1	0.8140	0.8150	0.8189	0.7922	2.430	2.539	2.455	5.433
25% FLOW-2	0.2530	0.2530	0.2560	0.2614	2.418	2.442	2.536	5.347
50% FLOW-1	0.5263	0.5298	0.5332	0.5253	2.550	2.595	2.492	5.376
50% FLOW-2	0.5035	0.4995	0.5074	0.4896	2.477	2.613	2.499	5.839
25% FLOW-1	0.2564	0.2579	0.2599	0.2634	2.799	2.658	2.697	4.300
75% FLOW-2	0.7525	0.7500	0.7584	0.7361	2.676	2.699	2.626	5.520

Surprisingly, the other three lot-start policies do not look markedly different, nor particularly bad. These results are not what we would expect after the recent furnace-loading fiasco. But before we get too excited and decide once-a shift or once-a-day starts are just as good as continuous loading of single lots, we must keep in mind two things. First, these results are for CMOS-2, a perfectly-sized, properly WIP-loaded, optimally-constrained wafer fab. In the real world, CMOS-2 would be in a new category beyond 'world-class' manufacturer.

The second point to be made about this example is that the lot-start intervals have been massaged for Flow-1 in order to keep the start rates within the fab's capacity. Thus, for the nominal 75%/25% case when it is loading up for a shift, Flow-1 has a start rate of 6 lots every 7.38 hours instead of 6 (or 7) lots every 8 hours. Similarly, all of the other Flow-1 intervals have been shaved of some time.

In a real fab, when it comes time to start a burst of lots for the shift, day, or week, then 'x' lots will be started. These will be whole lots, and there will be no fudging on the time at which they are started. The probability is small that the burst starts (remember whole numbers, no fractions) will match the fab's capacity. Furthermore, human nature being what it is, the number of lots started will probably be rounded up, not down. Table 6.20 shows the massaged starts versus both rounded-up and rounded-down scenarios just to be thorough.

TABLE 6.20: CAPACITY MATCHING, SHIFT-LEVEL STARTS

	Lot-Starts	Thrupt (lots/hr)	CT/P	ΔInv. (lots)	Btlnk Utln Ratio
FLOW-1 FLOW-2	6 lots/8 hrs 2 lots/8 hrs	0.7624 0.2555	1.922 1.955	- 19 - 9	0.9625
FLOW-1 FLOW-2	6 lots/7.38 hrs 2 lots/8 hrs	0.8150 0.2530	2.539 2.442	1 - 4	1.0000
FLOW-1 FLOW-2	7 lots/8 hrs 2 lots/8 hrs	0.8180 0.2361	4.175 3.986	122 30	1.0000

Metal Dep is Bottleneck (Constraining) Wkst #8.

In the case where fewer lots are started than the fab can handle, 6 lots for 8 hrs, the fab chews up WIP looking for material to work on. With the fab being drained, cycle times are very fast, and throughput for Flow-1 is down 6% from where it should be. No one will run a fab this way. The other, rounded-up, scenario is much more likely.

When one more lot a shift is started (7 lots for 8 hrs), the fab bottlenecks like a clogged artery. Because of the extra lot, WIP is building at a horrendous rate, cycle times have more than doubled, and although the fab is meeting production on Flow-1, the production of Flow-2 is seriously deficient (5.6%).

Similar trends are seen for the other two process mixes (50%/50% and 25%/75%) with the once-a-shift starts. However, the rounding up or down on Flow-1 has less impact as Flow-2 becomes more dominate in the process mix. (Production shortfall for Flow-1 are lowest for the 25%/75% case.)

Starting lots once-a-day, as opposed to once-a-shift, makes it somewhat easier to start the correct number of lots. (Rounding up or down has less impact over the larger number of lots and the longer interval.) Production shortfalls for this bursting policy are in the 1-2% range, as opposed to the 5-6% range seen with once-a-shift starts.

Starting lots once-a-week makes it easiest yet to zero in on the correct number of lots to start. Round-up or -down is small. Thus, the results in Table 6.19 are very close to those with unmassaged start intervals. Production shortfalls for starting lots once-a-week are in the 2-3% range, using either massaged or unmassaged intervals.

The defining characteristic in each instance where production falls below design is inefficient utilization of the bottleneck or constraining workstation. When the start interval is massaged to match starts with capacity, the bottleneck can be effectively used all the time for the once-a-shift and once-a-day burst starts. With the once-a-week burst starts, however, the bottleneck is not effectively utilized because the chugs of WIP going through it are too dramatic. Like the results shown in Fig. 6.4, there will be times when the constraining workstation has nothing to work on and other times when it will have 50+ lots waiting in queue.

Although these results are specific to CMOS-2, and the shortfall percentages will vary for real fabs, the conclusions are nonetheless valid. The bursting of lot starts will rarely (if ever) match the performance achieved by continuously feeding the line. Performance will be especially degraded in plants that use large bursts to supply long intervals.

6.4.2 Selected Minimization of Setup

As discussed earlier, dispatching rules do not typically impact factory capacity or overall factory throughput. The one exception, however, is a dispatching scheme that minimizes setup at certain workstations, specifically the bottleneck and other heavily-used workstations that may become bottlenecks during the course of operation. At other workstations, minimizing setup may be an academic exercise.

There are several ways in a wafer fab to minimize setup, including;

- dedicating specific machines to one recipe
- selecting lot and searching for an already setup machine
- searching through the queue to find same-recipe lots
- waiting for *N* same-recipe lots before switching

In most fabs, when setup minimization is used, it is used in conjunction with other, broader dispatching schemes such as critical ratio, process priority, etc. The broader scheme is the dominant dispatcher, and setup rules are generally used as a subset. For instance, if critical ratio is the dominant dispatcher then only the first or second of the above rules could be used.

6.4.3 First-In/First-Out (FIFO) versus Process Priority

In general, dispatching rules: (1) determine how capacity is apportioned between flow, (2) juggle the priority of lots within and between flows, (3) provide an extra measure of WIP control, and (4) lengthen or shorten cycle times (which provides an added degree of freedom for schedule manipulation).

FIFO, also known as first-come/first serve, is often dubbed the 'fair' dispatching rule. With this rule, lots are processed in the order in which they arrive at a workstation, irrespective of the process flow to which they belong. While this rule does not facilitate schedule 'tweaking', as do some of the line-dispatching schemes, it is nonetheless an easy rule to implement and, perhaps more importantly, does not leave any unexpected and/or unpleasant surprises in its wake.

Process-priority or hot-process dispatching is at the other end of the spectrum from FIFO. When a process is given priority over others, then its lots will 'bump' lots from lower-priority processes, even though they arrived at a work center first. As a result of the bumping, this dispatching scheme may, to some degree, override the lot-start profiles in determining throughputs. When a priority lot bumps another lot in a queue, it is effectively appropriating more than its fair share of the factory's capacity in its race for a faster finish. But since the fast finish is the overriding objective, the dynamic shift in throughput is something wafer fabs are willing to tolerate. It does, however, make for some uncertainty in the production slate.

Table 6.21 shows the performance differences for CMOS-2 between FIFO and priority dispatching for both Flow-1 and Flow-2. The three standard process mixes (nominal 75%/25%, 50%/50%, and 25%/75%) are given.

TABLE 6.21: PROCESS PRIORITY vs FIFO

Common Dispatching Rules in CMOS-2

	Throughput (lots/hr)			CT/P		
Priority:	FIFO	FLOW-1	FLOW-2	FIFO	FLOW-1	FLOW-2
75% FLOW-1	0.8140	0.8279	0.8219	2.430	1.489	2.253
25% FLOW-2	0.2530	0.2455	0.2574	2.418	3.787	1.438
50% FLOW-1	0.5263	0.5382	0.5303	2.550	1.381	2.728
50% FLOW-2	0.5035	0.5040	0.5149	2.477	2.644	1.446
25% FLOW-1	0.2564	0.2624	0.2545	2.799	1.372	3.941
75% FLOW-2	0.7525	0.7574	0.7698	2.676	2.550	1.548

As expected, once a process is given priority, its cycle time drops dramatically. Also as expected, the cycle time for the lower-priority, competing process climbs.

Because the priority process takes more than its fair share of the factory's capacity, its throughput is always higher than with FIFO. Surprisingly, though, more often than not, the non-priority flow also has a higher throughput. The reason goes back to setup; once a hot process has been put into the plant, there is usually a significant reduction in setup__unless the priority flow is quite small (see Section 6.2).

With priority dispatch, all priority lots that are available will be processed first; only then will lots from the second flow be processed. Less switching of equipment back and forth between the two processes means less setup. Thus, priority- or hot-process dispatch has, as a side benefit, self-imposed setup minimization. (See Table 6.22.)

TABLE 6.22: PERFORMANCE CONTRIBUTORS

Process Priority vs FIFO Dispatching

	Total Setup (hrs)			Bottleneck Utilization		
Priority:	FIFO	FLOW-1	FLOW-2	FIFO	FLOW-1	FLOW-2
75% FLOW-1 25% FLOW-2	4059.2	3618.1	3629.8	1.0000	0.9940	1.0000
50% FLOW-1 50% FLOW-2	4257.8	3946.1	3949.7	1.0000	1.0000	0.9993
25% FLOW-1 75% FLOW-2	4076.6	3741.9	3661.9	1.0000	0.9990	0.9961

Furnaces @ 50% loading.
No inter-process mixing in furnaces.

The only situation where the throughput declines for the non-priority flow is when that flow is only a small portion (25%) of the process mix. The factory, by stacking up large numbers of priority lots, followed by small numbers of non-priority lots, manages to get into a chugging mode once again. At times the constraining workstation has adequate inventory to work on, at other time little or none. As always, the chugging behavior is reflected in the less effective utilization of the bottleneck or constraining workstation.

6.4.4 Bottom Line on Dispatching Schemes

So, with all the options available, and their relative pluses and minuses, what type of dispatching is currently being used in most wafer fabs? Our experience in the U.S. and Pacific rim indicate it is predominantly a hybrid scheme that includes: hot processes and selective minimization of setup, in conjunction with either critical ratio or FIFO.

6.5 Engineering or Development Lots

An *engineering or development lot* is a lot of wafers produced solely for the purpose of engineering evaluation, not production. In many fabs, the specialized flow or flows that fabricate these lots usually have priority over production flows.

Knowing what we now do about priority flows, whenever a hot engineering lot is in the fab, it will get more than its fair share of the fab's capacity or resources. As a result, production will generally decline disproportionately to the number of these hot lots made. Further, since the quantity of engineering lots is likely to be small, the reduction in setup that is often seen with priority flows may not materialize. Depending on how different the masking is for the development lots, and the frequency of the lots, setup may be substantially higher.

Though the existence of these engineering lots is a well-recognized fact in wafer fabrication, they are, nonetheless, seldom taken into account during the design and sizing of a plant. As a result, after the fab comes on-line, and engineering appropriates 10 to 50% of the production capacity, the only possible response is the gnashing of teeth, the pointing of fingers, and the rolling of heads. The proverbial honeymoon is over, history has repeated, and no one can understand how the fab was so grossly undersized.

The answer is simple: just as there is no such thing as 'a free lunch' neither can engineering lots be 'snuck through the fab'; it is a physical impossibility. Fabs have a finite capacity;

production lots are usually started at a rate to fully utilize that capacity; and even a few additional lot can degrade performance.

As a case in point, consider CMOS-2 which is about midway into its life cycle. Flow-2 is responsible for about 50% of the plant's production, Flow-1 accounts for the other 50%, and the engineers are working feverishly to develop the next generation of product. Part of the feverish activity is a more or less constant stream of engineering lots (Table 6.23 and Fig 6.6).

TABLE 6.23: 4-METAL ENGINEERING FLOW

STEP #	PROCESS STEP	TIME	WKST #	WORKSTATION
1	LASER MARK		1	MISC.
2	INITIAL OX		2	DIFF 1
3	P-TUB MASK		3	MASK/ETCH 1
4	P-TUB IM		4	IMPLANT
5	NITRIDE DEP		5	DIFF 2
6	FIELD MASK		6	MASK/ETCH 2
7	FIELD DIFF/OX		5	DIFF 2
8	SOURCE/DRAIN MSK 1		6	MASK/ETCH 2
9	S/D IM 1		4	IMPLANT
10	S/D OX 1		2	DIFF 1
11	S/D MASK 2		6	MASK/ETCH 2
12	S/D IM 2		4	IMPLANT
13	S/D OX 2		2	DIFF 1
14	NITRIDE STRIP		6	MASK/ETCH 2
15	GATE OX		2	DIFF 1
16	POLY DEP/DOPE 1		7	DIFF 3
17	POLY MASK 1		3	MASK/ETCH 1
18	POLY DEP/DOPE 2		7	DIFF 3
19	POLY MASK 2		3	MASK/ETCH 1
20	CONTACT MASK		3	MASK/ETCH 1
21	METAL DEP 1	1.20	8	METAL SPUTTER 1
22	METAL MASK		9	MASK/ETCH 3
23	DIELECTRIC		10	DIEL-FAB
24	METAL DEP 2	1.25	8	METAL SPUTTER 1
25	METAL MASK		9	MASK/ETCH 3
26	DIELECTRIC		10	DIEL-FAB
27	METAL DEP 3	1.30	11	METAL SPUTTER 2
28	METAL MASK		9	MASK/ETCH 3
29	DIELECTRIC		10	DIEL-FAB
30	METAL DEP 4	1.30	11	METAL SPUTTER 2
31	METAL MASK		9	MASK/ETCH 3
32	TOPSIDE DEP		7	DIFF 3
33	TOPSIDE MASK		3	MASK/ETCH 1

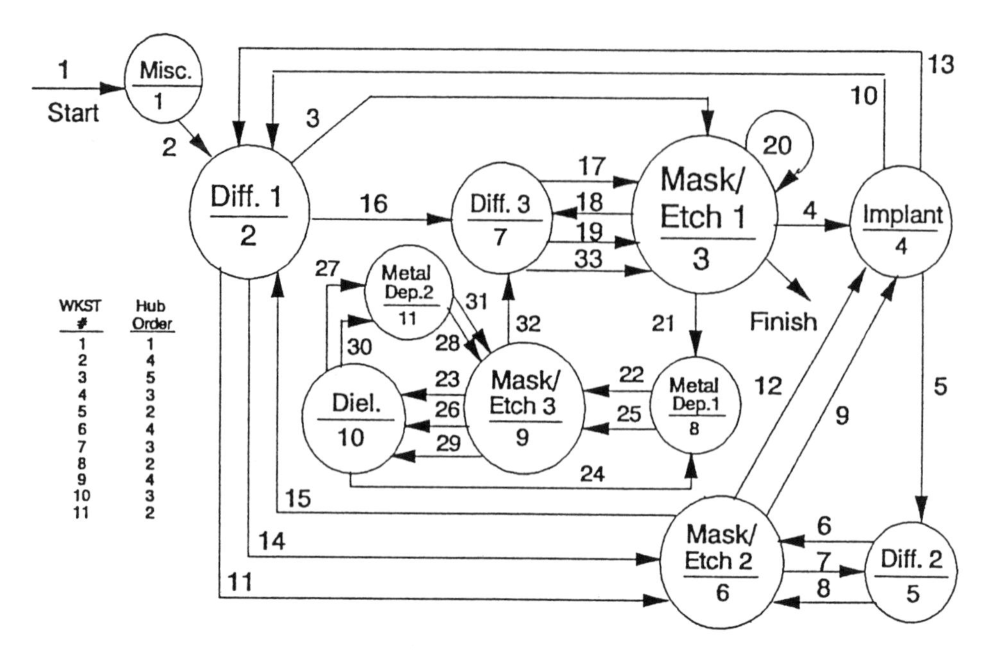

Figure 6.6: Fab Graph for Engineering Flow, CMOS-2

In the early stages of development, about one engineering lot a week is being shot through the line as a 'hot' lot. Later the frequency will increase to one lot a day, and later still to one lot a shift. Table 6.24 shows the fab's performance as the frequency of engineering lots increases.

TABLE 6.24: IMPACT OF ENGINEERING LOTS

Engg Rate	Flow	Cycle Time (hrs)	CT/P	Thrupt (lots)	△Inv. (lots)
None	FLOW-1 FLOW-2 ENGG	97.4 106.6 --	2.550 2.477 --	1,061 1,015 --	1 (6) --
1 lot/week	FLOW-1 FLOW-2 ENGG	106.7 115.6 148.9	2.794 2.686 3.125	1,053 1,010 12	9 (1) 1
1 lot/day	FLOW-1 FLOW-2 ENGG	152.6 162.6 76.3	3.994 3.776 1.602	1,005 956 82	57 53 3
1 lot/shift	FLOW-1 FLOW-2 ENGG	266.5 276.5 68.9	6.976 6.424 1.447	889 849 244	173 160 9

Well, obviously the engineers were right; it is possible to sneak engineering lots through the fab. But the rate at which it can be done without impacting performance is very low. In the case of CMOS-2, it is one lot a week. Over a 12-week period, at the start rate of 1 lot/week, the 12 engineering lots will cost production a total of 13 lots. Almost a wash. But at any of the higher rates, production really starts to suffer.

At an engineering rate of 2 lots/week, over the same 12-week period, 24 engineering lots are made and production falls by 28 lots. At 1 lot/day, 82 engineering lots result in a production decrease of 115 lots. At 1 lot/shift, the 244 engineering lots cause production to fall by 338 lots. (Fig. 6.7 shows the production costs for the three engineering start rates of 1 lot/week, 1 lot/day, and 1 lot/shift.)

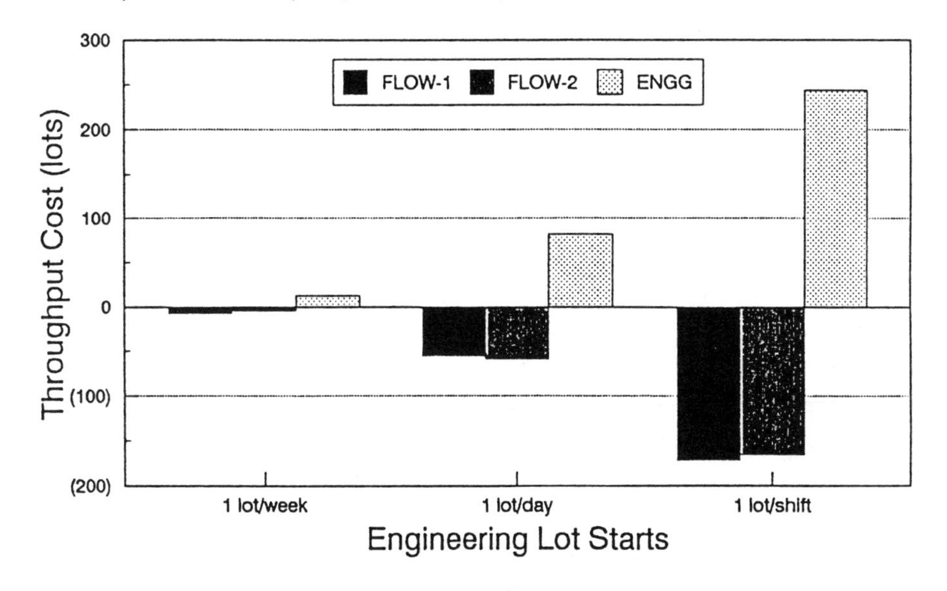

Figure 6.7: Production Costs of Engineering Lots

Note, that even at the highest rate of 1 lot/shift, the engineering flow is theoretically only using 12.5% of the fab's capacity. However, instead of the 2076 lots that were being produced back in the old days before the engineering flow was introduced, a total of only 1982 lots are produced. The fab has lost 4.5% of its capacity. True, the engineering flow is only consuming 12.5% of the capacity, but there is a lot less capacity to go around. Besides which, the fab did not have the spare capacity in the first place.

For stable fab operation, once the rate of engineering lots begins to degrade throughput, i.e., at 1 lot/day for CMOS-2, the lot-start rates of the production flows should be reduced. Lots that are started beyond the fab's capacity to handle them are simply building WIP. If the fab can not back off on production, there is really only one alternative: buy more equipment for the bottleneck or constraining workstation(s). Although this solution is usually unpalatable, it is nonetheless a cheap way for keeping production up. Furthermore, it may be the only feasible solution. The future of the business depends on engineering, and they will have their lots.

With the introduction of an engineering flow, a fab loses capacity for two reasons. First, it is usually a much more complicated flow, having more steps, longer process times, higher order hubs in the interconnect area, etc. Secondly, at some lot-start rate, an engineering flow may cause a bottleneck shift. Once the bottleneck shifts, all estimates previously made about capacity and resource (or production) allocation are no longer valid.

6.6 Bottleneck and Capacity Estimator

Sizing a fab and estimating equipment needed at a given workstation, based on wants and desires, is one thing (Section 4.8), determining what is feasible with a given equipment set is something quite different. Fab sizing is usually done once or twice during the lifetime of a plant; bottleneck and capacity reevaluations should be performed every time something significant in the fab changes.

Significant changes may include:

- addition of a flow (engineering or production)
- process creep (lengthening of process steps)
- process migration (addition of process steps)
- addition of equipment
- changes in equipment availability
- altered maintenance schedules
- etc.

The tool presented in this section, a method for bottleneck and capacity estimation, provides a static determination of what is possible, though not necessarily achievable, in a fab. The beauty of the method is that it does not require actual performance data, only a complete and accurate description of the process flows and the workstations, such as those given in Tables 6.23 and 6.25 for CMOS-2.

TABLE 6.25: FACTORY INFORMATION, CMOS-2
Two Production Flows and One Engineering Flow

Workstation		# Mach.	Load (waf)	Availability (up/down)	Process Time @ Wkst (hrs)		
					FLOW-1	FLOW-2	ENGG
1	MISC.	2	25	0.99/0.01	0.81	0.80	0.80
2	DIFFUSION 1	5	100	0.95/0.05	8.00	8.00	8.00
3	MASK/ETCH 1	7	25	0.97/0.03	5.00	5.00	5.00
4	IMPLANT	3	25	0.85/0.15	1.50	1.50	1.50
5	DIFFUSION 2	3	100	0.95/0.05	4.00	4.00	4.00
6	MASK/ETCH 2	6	25	0.97/0.03	4.00	4.00	4.00
7	DIFFUSION 3	4	100	0.95/0.05	6.00	6.00	6.00
8	METAL DEP 1	3	25	0.85/0.15	2.20	2.45	2.45
9	MASK/ETCH 3	4	25	0.97/0.03	2.00	3.00	4.00
10	DIEL-FAB	5	25	0.97/0.07	2.00	4.00	6.00
11	METAL DEP 2	2	25	0.85/0.15	0.00	1.30	2.60

The first step in looking for the fab's most probable bottleneck is to determine the total process time, P_{ij}, taking place at each workstation, j, for each flow, i. Next, we need to know the number of machines, n_j, at each workstation. (These numbers are given in Table 6.25 for CMOS-2.)

We then calculate the throughput, w_{ij}, for each flow at each workstation. (If a flow does not use a workstation, w_{ij} should be set to zero, i.e., do not divide by a zero process time.)

$$w_{ij} = n_j / P_{ij}$$

Since we will need a reference throughput, w_{sj}, for normalization purposes, select the lowest one (of the various flows) at each workstation.

$$w_{sj} = \min_{i=1,N} (w_{ij})$$

We then make an estimate of the degree of difficulty, deg_{ij}, of each flow on each workstation.

$$deg_{ij} = w_{ij} / w_{sj}$$

The actual demand of each flow, x_{ij}, on each workstation is given by:

$$x_{ij} = ls_i / deg_{ij}$$

where ls_i is the lot-start rate of flow i. Thus, the total demand, D_j, of all flows on the workstation is:

$$D_j = \sum_{i=1}^{N} (x_{ij})$$

The normalized bottleneck indicator, BI_j, of each workstation is:

$$BI_j = (D_j / a_j) / w_{sj}$$

where a_j is the workstation availability ratio.

Thus, BI_j is a static indicator of how much demand is being put on a workstation by a given lot-start profile; it takes into account the availability of the workstation's resources to meet that demand.

A value of '1' or more is almost a sure sign that the workstation is a bottleneck under the conditions specified. In addition, workstations with BI_j near '1' should not be overlooked as real or potential bottlenecks, because the actual factory dynamics may put demands greater than workstation resources. Even if these high-value workstations are not currently bottlenecks, they may become so with only slight changes in demand (process creep, process migration, more engineering lots, etc.).

6.6.1 Predicting Impact of Engineering Lots

To see how the bottleneck indicator works with CMOS-2, we will explore how well it predicts the impact of adding increasingly more engineering lots. Since only a portion of the process sequence is different between the three flows (Flow-1, Flow-2, and ENGG), we will concentrate only on the workstations involved in those portions. The steps that change from flow to flow are: steps #21-24 in Table 6.1, step #21-27 in Table 6.2, and steps #21-30 in Table 6.23. These steps represent the interconnect portion of the processes where more metal layers are being added with each successive generation of chip. The workstations performing these steps are Workstations #8-11. Table 6.26(a) compares the bottleneck indicator to the actual workstation utilization ratio Workstations #8-11 (over 2016 hours of fab operation).

As may be seen from the table, the impact of increasing the engineering load, from none, to 1 lot/week, to 1 lot/day, to 1 lot/shift, causes Workstations #9, 10, and 11 to become busier, but by no means do they approach bottleneck status. Only Workstation #8 suffers that fate.

TABLE 6.26(a): BOTTLENECK AND CAPACITY INDICATOR
Anticipating Impact of Engineering Flow

Start Intvls F1/F2/ENG	Workstation 8		Workstation 9		Workstation 10		Workstation 11	
	BI	Utln.	BI	Utln.	BI	Utln	BI	Utln
1.9/2.0/none	0.93	1.00	0.68	0.73	0.63	0.70	0.57	0.49
1.9/2.0/168	0.94	1.00	0.69	0.73	0.64	0.71	0.39	0.49
1.9/2.0/24	0.97	1.00	0.73	0.74	0.68	0.73	0.45	0.52
1.9/2.0/8	1.05	1.00	0.82	0.77	0.78	0.77	0.57	0.58

Start Intervals for 1 lot: FLOW-1:FLOW-2:ENGG
No ENGG WIP.

Since the constraining workstation (#8) is already at maximum capacity (Utln = 1.0) with no engineering lots, it is impossible to see the effect of the additional lots through the utilization ratio. However, a perusal of the change in WIP at the workstation gives an indication of their impact. WIP is growing even more rapidly than the extra number of lots started. This type of fab operation can not be sustained; the lot-start rate for Flow-1 or Flow-2, or both, will have to be reduced, and the sooner the better.

The steady deterioration in performance is anticipated by the bottleneck indicator, which climbs from 0.93, when the workstation is merely constraining, to 1.05 when the fab is seriously bottlenecked. When the indicator jumps past '1' (at 1 lot/shift), the prognosis for fab performance is grim at best, and is more adequately described as *down the dumper.*

Rather than blithely starting lots for which there is no capacity, and/or making this kind of determination from real-time experimentation on the factory floor, a few quick calculations using the bottleneck indicator will tell the operations manager that production can not be sustained.

6.6.2 Predicting Bottleneck Shifts

By similar logic, the same calculations may be used as a capacity indicator when imposing new processes or process mixes on an existing set of equipment. As the fab ages, the process mix **will** change. In a sense, adding engineering lots is a change in product mix. More specifically, with the development of each new generation of chip, its production will use increasingly more of the fab's capacity, and the older chips decreasingly less.

CMOS-2, for example, will at some point in its future make predominantly 4-metal chips. However, production of the 2-metal and 3-metal chips will not be totally abandoned since there will still be a market. Rather, they will be gradually phased out.

Table 6.26(b) shows how the capacity indicator may be used to search for feasible start rates for new product mixes. (Note that the third flow is still labeled ENGG even though it is now a full production flow.)

TABLE 6.26(b): BOTTLENECK AND CAPACITY INDICATOR
Anticipating Impact of 4-Metal Flow

Start Intvls F1/F2/ENG	Workstation 8		Workstation 9		Workstation 10		Workstation 11	
	BI	Utln.	BI	Utln.	BI	Utln	BI	Utln
8.0/8.0/1.33	0.95	0.98	0.97	0.88	1.08	0.99	1.24	0.97
8.0/8.0/1.5	0.87	0.93	0.88	0.86	0.98	0.99	1.12	0.97
8.0/8.0/1.8	0.76	0.86	0.76	0.85	0.84	0.92	0.95	0.97
6.0/6.0/1.5	0.94	0.99	0.94	0.94	1.03	0.99	1.15	0.97

Start Intervals for 1 lot: FLOW-1:FLOW-2:ENGG
ENGG WIP representative of fab reqmts.

The anticipated product mix for CMOS-2 will be in the ranges: 12-20% Flow-1, 12-20% Flow-2, and 60-75% Flow-3 (ENGG). Thus, the first four start-rate profiles we will run through our capacity indicator are:

1 lot/8 hrs, 1 lot/8hrs, 1 lot/1.33 hrs,
1 lot/8 hrs, 1 lot/8hrs, 1 lot/1.5 hrs,
1 lot/8 hrs, 1 lot/8hrs, 1 lot/1.8 hrs, and
1 lot/6 hrs, 1 lot/6hrs, 1 lot/1.5 hrs.

We know even before putting pencil to paper that there are likely to be problems and possibly even a bottleneck shift with the new product mix. The reason we know this is that the most complex flow, Flow-3 (ENGG), now claims the lion's share of the factory's resources, especially in the interconnect area.

Going strictly by the calculated BI_j, we can immediately verify that there has indeed been a bottleneck shift, away from Workstation #8, to either Workstation #10 or #11. But will the new bottleneck(s) be so bad that good factory performance is not possible?

Searching the possible lot-start profiles for a case where the BI_j are less than '1', the only feasible candidate is: 1/8, 1/8, 1/1.8. For this lot-start profile, the constraining workstation appears to be #11, Metal Dep-2, but it does not look to be a true bottleneck. This conclusion is verified by the actual utilization ratios (as well as no significant changes in WIP).

The other lot-start profiles, on the other hand, should not be imposed on the fab. In each case, either Workstation #10 or #11 is a true bottleneck and the lot-start profile will lead to unstable performance. (This conclusion is verified by increases in WIP.)

If the feasible lot-start profiles identified by the capacity indicator are not acceptable, and more production is a must, the solution, as usual, is to buy more equipment.

It should be noted before leaving this case that care must be exercised when interpreting actual utilization data for workstations not used by all flows, such as Wkst 11. Because Flow-1 is still in the fab, there will be a small increment of idle time at Wkst 11 due to interference from Flow-1 lots. Flow-1 lots will, at various times, actually impede lots from other flows as they make their way to Wkst 11. In such a situation, even with some idle time, a workstation may still be a true bottleneck.

6.6.3 Predicting Impact of Adding Equipment

It is interesting that CMOS-2, which can get 1.11 lots/hr throughput for Flow-1 alone, or 1 lot/hr for Flow-2 alone, can only get an overall throughput of 0.76 lots/hr for the three process flows when Flow-3 is the major product. This throughput is the result of using the sole lot-start profile which will not bottleneck the fab: 1/8, 1/8, 1/1.8.

However, with the bottleneck indicator we can quickly ascertain that the addition of one machine at Workstation #10 and one machine at Workstation #11, will permit a lot-start profile of 1/8, 1/8, 1/1.33. Actual fab performance for this start profile is shown in Table 6.27, and is compared with the no-added-equipment case, 1/8, 1/8, 1/1.8.

TABLE 6.27: EQUIPMENT FOR FLOW DISPLACEMENT
Adding Two Machines to Accomodate FLOW-3

	Start Rate (lot/hrs)		Throughput (lots)		Product Value ($mm)		Value Gain ($mm)
	w/ equip	w/o equip	w/ equip	w/o equip	w/ equip	w/o equip	
FLOW-1	1/8	1/8	261	249	45.68	43.58	2.10
FLOW-2	1/8	1/8	262	224	55.68	47.60	8.08
FLOW-3	1/1.33	1/1.8	1,438	1,067	341.52	253.41	88.11

Total Gain = $98.29 mm for 84 days operation = $1.17 mm/day

Payback = $4.5 mm/($1.17mm/day) = 3.85 days

FLOW-1: $9000 wafer w/ prodn cost of $2000
FLOW-2: $11,000 wafer w/ prodn cost of $2500
FLOW-3; $13,000 wafer w/ prodn cost of $3500

With the added equipment, an overall throughput of 0.97 lots/hr is achievable. A capital investment of $4.5mm ($2mm for dielectric and $2.5mm for sputtering) will have a payback period of less than four days.

6.6.4 Pros and Cons of Static-Analysis Tool

There are several points which should be made about the bottleneck/capacity indicator before we leave the topic. First, the method can only be an indicator because it is a static model. It does not take into account dynamic events such as setup. But in most instances, dynamic events will be more likely to cause rather than remove bottlenecks. Therefore, if the indicator says there **is** a bottleneck, it deserves serious attention.

Secondly, in one sense, this static model may be more useful than actual fab operating data in that it will identify all potential bottlenecks. With fab operating data, it is not uncommon for an upstream bottleneck to mask primary or secondary downstream bottlenecks. Therefore, even if there is a plethora of fab operating data available, and even if there is an accurate, validated factory-specific dynamic model available, the bottleneck/capacity indicator is, or should be, the starting point for analyzing how demands may impact the factory.

One final caution, since the bottleneck indicator does not have a priority rule, the calculation assumes the fab is being run FIFO.

6.7 Process Creep

As process flows mature, process steps will sometimes lengthen, usually in pursuit of improved yield and/or process performance. This lengthening of steps is known as *process creep*, and over the development and production cycle of a flow may be substantial.

Of course, longer process steps tie up machines longer. Hence, if the lengthening steps take place at the bottleneck, constraining workstation, or other heavily-used workstation, it can reduce the capacity of the entire fab. Under the worst-case scenario, it may even cause a bottleneck shift, the consequences of which may be severe, especially if unexpected.

As a case in point, consider CMOS-2's Workstations #10 and #11 which became heavily bottlenecked as Flow-3 (formerly the engineering flow) took over most of the production capacity of the fab. Gear was finally added at these two workstations because it became impossible to meet production demands otherwise. With the addition of the two machines (one each at Wksts 10 and 11), the bottlenecks went away, and Workstation #8 once again became the constraining workstation of the fab.

Of course, purchase of the machines should have been a 'no brainer'. After all, no previous adjustments had been made to the equipment set, despite the fact that Flow-3 is a more complex flow, and is producing much more valuable wafers. For that reason alone, the gear was justified. However, capital purchases are often a hard sell to financial managers, even when there is a ridiculously-short equipment-payback period.

With CMOS-2, the final, convincing argument for adding the two machines was that Flow-3's process times at Workstation #11 (steps #27 and 30) would continue to lengthen as the process matured, from 1.3 hours for each step up to possibly 2.0 hours. Presumably, the new equipment would not only debottleneck the current fab, but would be able to handle any future creep.

CMOS-2's savvy operations manager, however, no longer makes such presumptions without implementing his newly-acquired skill, the bottleneck indicator. Using the tool, the operations manager quickly learns that a process creep to 1.5 hours for each step can be tolerated with no adverse effects. But at a creep of 1.6 hours for each step, Workstation #11 again bottlenecks.

But, per usual, it takes proof on the factory floor before the pocketbooks loosen for more gear. As seen in Fig. 6.8, the bottleneck indicator's predictions are right on.

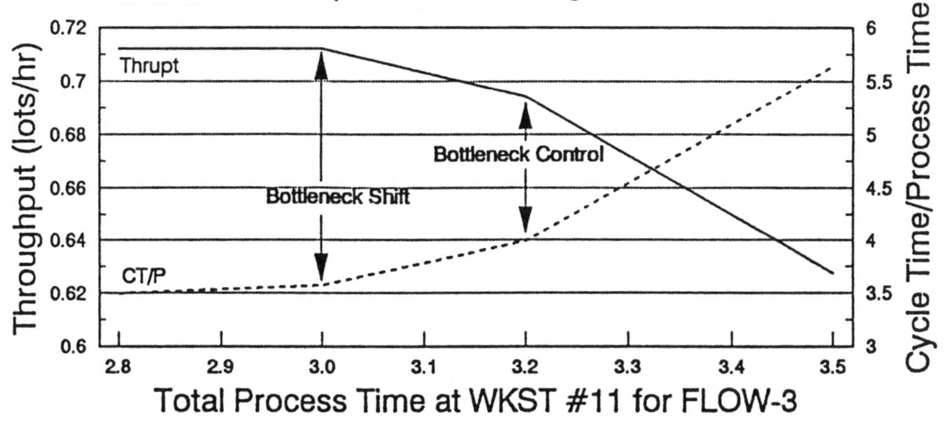

Figure 6.8: Impact of Process Creep

Shortly after Flow-3's process times for steps #27 and #30 creep past 1.5 hours each, the bottleneck shifts and Workstation #11 is the constraining workstation. At 1.6 hours each step, the performance of Flow-3 has cratered; throughput and cycle time are unacceptable and WIP is building at a fast rate. As the process times creep out to 1.75 hours each, the fab is crippled and management finally agrees to more equipment.

Meanwhile, guess how much production was lost in a capacity-short marketplace.

6.8 Process Migration

Another similar problem occurs over the life cycle of a process flow: extra process steps may be added to improve yield, device characteristics, or reliability. Fine-tuning the process or structures may come in the guise of an additional metal layer for a redundant conductor (or shunt) or a double implant for higher yield. These additional steps are known as *process migration* because the process sequence has migrated away from the original flow.

Process migration, like process creep, puts an additional burden on the fab. When an extra step reuses a workstation, the hub order of the work center increases, and the demand increases accordingly. If the additional demand causes the workstation to bottleneck (i.e., insufficient capacity for demand), then throughput will suffer.

To illustrate, consider CMOS-2's 4-metal flow given in Table 6.23. Despite the fact that Flow-3 is now the major process in the fab (making up more than 75% of production), fine-tuning will continue for as long as the devices are being manufactured.

With incessant tinkering, it does not take engineering long to realize that a double implant at step #12 (Fig. 6.9) dramatically improves yield. But perhaps even better, doing the extra step does not negatively impact either throughput or cycle-time performance. The utilization of the Implant workstation prior to adding the step is so low that the extra step helps to make the line more evenly balanced. As a result, there is actually a slight improvement in performance by doing the double implant.

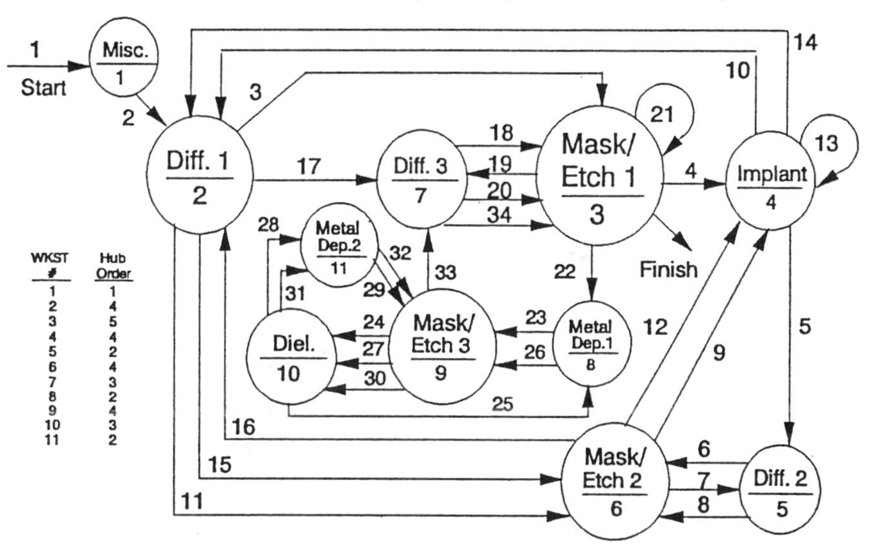

Figure 6.9: Fab Graph for MIgration of Flow-3

Naturally, if the double-up works once, why not for all three Implant step? (That is, steps #4, #9, and #12 in Table 6.23.)

By now, we know the answer: the workstation has a finite capacity, and at some point it will be exceeded. As seen in Table 6.29, the three machines at Workstation #4 can handle two double implants, but on the third the factory's bottleneck shifts from Metal Sputter-1, Wkst 8, to Implant, Wkst 4.

TABLE 6.28: IMPACT OF FLOW-3 MIGRATION
Doubling of Implant Steps

	FLOW-3 Performance		Workstation #4 Performance		
Implant:	Thrupt	CT/P	Δ Inv	Utln	BI
0 doubles	0.7133	3.499	base	0.7341	0.5894
1 double	0.7138	3.477	3	0.8748	0.7368
2 doubles	0.7108	3.487	72	1.0000	0.8842
3 doubles	0.5575	7.170	490	1.0000	1.0316

If the yield improvement in going from two double implants to three double implants is worth pursuing, then another implanter must be bought. The fab simply can not maintain performance otherwise. But, of course, we do not have to put in that final double implant and watch performance nosedive in order to realize it will require another machine. The Bottleneck Indicator, BI_4, has already told the story.

6.9 Equipment Availability

It is time now to put CMOS-2 to rest. We have seen the factory through its life cycle, from its early days of running a simple process, to a mixture of processes, through an engineering development, and through flow displacement by the constantly-evolving next generation chip. In taking up the final advanced topic, we will revert to a simpler era and a simpler fab, CMOS-1, so as not to muddy the waters.

Perhaps it is fitting that this discussion on fab performance ends on one of the 'grayest' subjects in all of wafer fabrication: *equipment availability*. In real factories, equipment may be unavailable for large segments of time, and for a variety of reasons. The reasons may include: planned maintenance, unexpected sporadic equipment failure, and other, non-productive machine uses, such as calibration steps, waiting for a full or near-full batch, etc. Whatever the reason, when a machine is not processing, it is not part of the fab's capacity, and hence is not contributing to throughput.

Although the end result is the same in each of the above cases, i.e., unusable equipment, the cause of the unavailability is completely different in each instance. Planned maintenance is, as the term implies, planned. It is, therefore, the least offensive type of unavailability, and the easiest to factor into factory operation. With the aid of equipment vendors, most wafers fabs have this type of unavailability under control. Machines are taken out-of-service on a regular, planned basis; tuned up or repaired for a known length of time; and then returned to service. There are few surprises in the frequency, duration, and impact of planned maintenance on a wafer fab.

The second type of unavailability, sporadic and unexpected equipment failure can be a fab's worst nightmare, especially if it happens at a bottleneck or a heavily-used workstation. The nature of these unexpected outages is that they are often catastrophic, i.e., of long duration, possibly 24 hours or more. When such an outage occurs at a controlling workstation, fab performance can go south quickly, and may even be unrecoverable unless appropriate remedial action is taken harshly and swiftly. Such action will usually mean cutting back on lot starts, even though the fab will not be able to meet production demands and customers will be disappointed.

The third type of equipment unavailability, non-productive machine usage, can also be a nightmare, but one which may easily be avoided. The initial fab sizing should take into account how much extra capacity will be needed at the heavily-used workstations in order to offset this down time. For example, equipment vendors probably have a good idea how much time needs be spent on a machine's calibration, etc. As far as the other non-production uses, such as batching losses, etc., a good factory-specific, dynamic model will predict how much slack must be built into the sizing to compensate for them.

Still, there is a huge gray area when it comes to equipment availability, and it will probably remain the case for the foreseeable future. Overly-optimistic availability quotes, naive expectations, and a poor understanding of fab dynamics, all lead to an on-going war between equipment vendors and fab houses. While the war rages, poor operating and maintenance policies are implemented. Those policies end up costing hundreds of millions of dollars every year in lost revenue.

6.9.1 The Regular Maintenance Cycle

First things first: **availability ratio and MTTF (mean time to failure) are not equal.** An availability ratio is the fraction of time, on average, that a machine is up and available for processing, as opposed to being down and unavailable. Thus, an availability ratio of 0.80 indicates that a machine will be available 80% of the time. But the ratio gives no indication whether the 80% availability comes in the form of 80 hours up/20 hours down or 8 hours up/2 hours down, etc.

MTTF, on the other hand, expressly states the expected length of time, on average, until a machine is expected to fail. An MTTF of 48 hours, and a MTTR (mean time to repair) of 12 hours, will have an availability ratio of 0.80, but the specific up time (on average) will be 48 hours. But does this distinction between availability ratio and MTTF have any relevance?

Most emphatically, if the equipment in question is either at the bottleneck, a potential bottleneck, or the constraining workstation. Every issue associated with machine availability, maintenance, failure, and/or other equipment-related subjects takes on supreme importance with these workstations.

Conversely, if the equipment in question is not at one of the fab-controlling workstations, most efforts aimed at improving availability will be futile from a fab perspective. Performance will still be dominated by the bottleneck, and, hence, will not change, no matter how much time and effort are spent on non-critical equipment. This is a hard lesson to learn, and one that is not immediately obvious even at some of the best-run fabs.

Improvement at Non-Critical Workstation

To illustrate the impact of improving availability on a non-critical work center, consider the situation at the old CMOS-1 fab. This elderly fab has gone through most of its life cycle since we last visited it in Chapter 5. It is still a simple fab, running only one flow, a single-metal, 24-step process sequence. With time and experience, the fab has settled into an at-capacity, optimally-run operation. Under steady-state conditions, the fab has no real bottleneck, merely a tightly-constraining work center at Metal Dep, Workstation #8. The process time for this step has crept to a lofty 1.6 hours in pursuit of higher yield and a better performing device.

Despite this rosy picture, age in the old fab is starting to tell; performance is not what it used to be. But since CMOS-1 is looking at the end of its days, buying equipment is not exactly a desireable recourse. Therefore, all eyes turn to the next most obvious solution, getting more out of the equipment that is already there, i.e., improving availability.

An analysis of the recent operating history of the fab's equipment, rank orders the availability at each workstation and finds that the etchers at Workstation #9 have, by far, some of the worst track records. Now that a target has been sighted, the push is on.

Under normal, benign conditions, Workstation #9 has a MTTF/MTTR of 80-hrs up/20-hrs down. With this availability, the workstation has a utilization ratio 0.54, the fab's throughput is 0.99 lots/hr, and the flow's cycle-time-to-process ratio is 2.5.

After a lengthy and expensive effort, the MTTF/MTTR for Workstation #9 is improved to 97-hrs up/3-hrs down.

The overall impact on fab performance: zero. The fab's throughput is still 0.99 lots/hr, and the flow's cycle-time ratio is still 2.5. But, for whatever it's worth, the workstation's utilization ratio has climbed to 0.71.

A good decision, improving availability, led to an expensive and ineffectual outcome. The reason was that the decision was not applied intelligently. The only availability improvement likely to impact fab performance is that of the constraining work center, Workstation #8. Not that availability at other work centers is unimportant; striving to improve equipment performance is always a worthwhile activity. But it is unrealistic to expect it to have fab-wide impact. After all, the availability at Workstation #9 had very little impact when it was poor, so why should it have more impact when it is improved?

Improvement at a Constraining Workstation

Now that we have seen how time and effort may be wasted on a non-critical workstation, we will pay attention where attention is due: the bottleneck or constraining workstation. In the case of CMOS-1, that is Workstation #8, Metal Dep.

Under normal, unimproved conditions, Workstation #8 also has a MTTF/MTTR of 80-hrs up/20-hrs down. With this availability, the workstation has a utilization ratio 0.99. Again, the fab's throughput is 0.99 lots/hr, and the flow's cycle time-to-process ratio is 2.5.

After a lengthy and expensive effort, the MTTF/MTTR for Workstation #8 is improved to 90-hrs up/10-hrs down, an absolute miracle. The fab's throughput takes a modest jump to 1.0 lot/hr, and the flow's cycle-time ratio drops to 2.1.

However, after just a few short weeks of the new, improved performance, availability at Metal Dep again starts falling off. It seems to be inherent with the Wkst-8 machines that they simply must have an hour of maintenance for every four hours of operation. Thus, the availability ratio of 0.8 looks like it can not be circumvented. But perhaps there is another way to bring about improved fab performance: by altering the regular maintenance cycle on the bottleneck or constraining workstation.

Fig. 6.10 shows the fab performance that results by vary MTTF/MTTR while keeping the availability ratio constant at 0.80.

At long up times and long down times, fab performance is poor. Likewise, at short up times and short down times, performance is again poor. The reason for the poor performance is different for the two extremes.

At the 80/20 repair cycle, the **down** time is too long for the constraining or bottleneck workstation; the fab gets into a chugging mode. WIP builds during the long down times, then gets rapidly chewed up when equipment comes back on-line. The chugging that results because of long-downed equipment is a form of line imbalance, and leads to less than optimal performance. (A balanced line, i.e., constant throughput per workstation, keeps WIP moving through the fab like water through a fire-bucket brigade.)

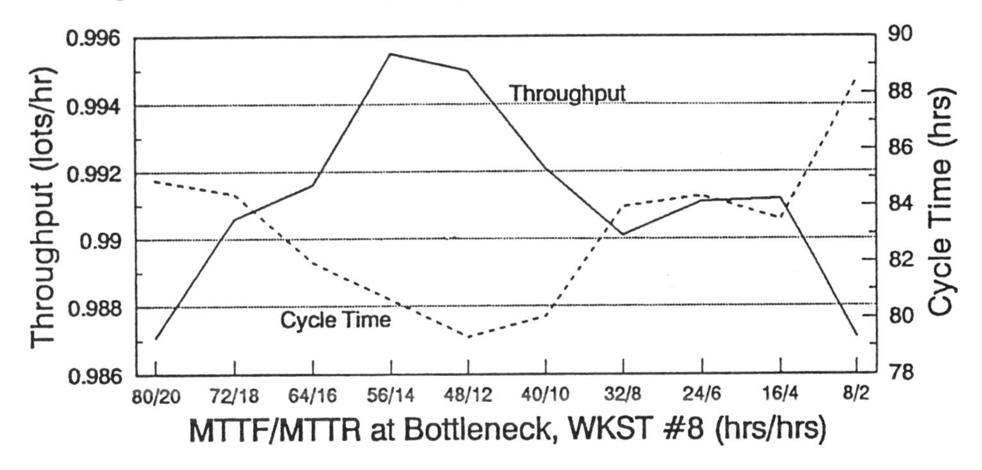

Figure 6.10: Maintenance Cycle at Bottleneck

The poor performance with the 8/2 repair cycle is because the up time is simply too short. As a result equipment at the work center spends less time processing and more time in setup than either the long cycle or an intermediate cycle.

As might be expected, the ideal maintenance cycle is an intermediate one (56/14, 48/12, etc.) where the up times are long enough to maximize processing, and the down times are not so long that they cause chugging or line imbalance.

Pause, you say! What about this big push in the industry to make MTTF as long as feasibly possible, and to hell with MTTR?

Well, the response to that bit of homespun logic is shown in Fig. 6.11. Setting MTTF for Workstation #8 constant at 80 hours, and varying MTTR, we note that as we shrink the down times (equivalent to lengthening the up times), there is really no performance improvement beyond a certain point. The fab is a total system, not a single set of equipment. Improving availability at one or several workstation, beyond the capacity of the fab, will not make the system's capacity greater, it will simply make for lower utilization of the higher-availability equipment.

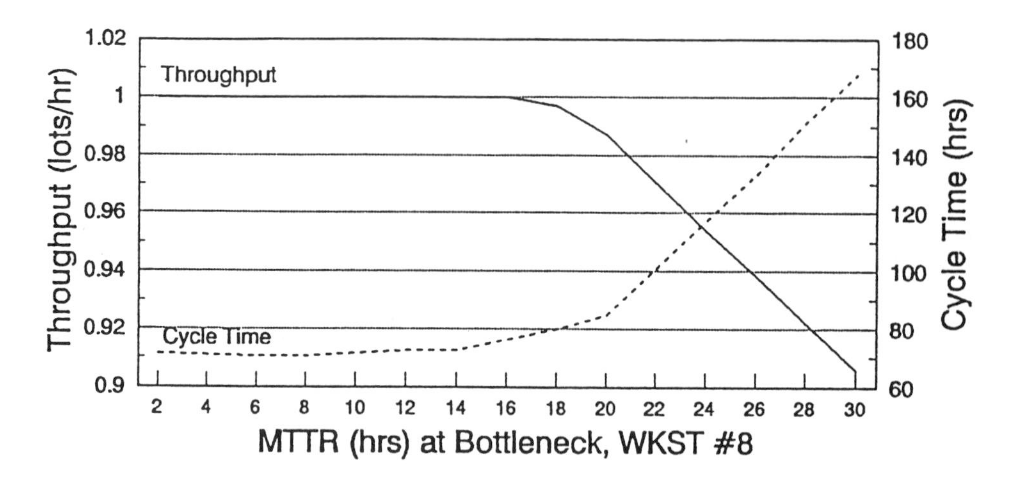

Figure 6.11: Ups & Downs of Maintenance Cycle

As a case in point, consider the earlier scenario where we improved the MTTF/MTTR on Wkst 8 from 80/20 to 90/10. The impact on the fab was to increase throughput from 0.99 lots/hr to 1 lot/hr, and reduce the cycle-time ratio from 2.5 to 2.1. Now suppose we 'improve' the maintenance cycle even further, to an MTTF/MTTR of 95/5. We will get further improvement in fab performance, right?

Wrong. While the cycle-time ratio holds constant, at 2.1, throughput actually goes down just a tad, from 1.0045 lots/hr in the 90/10 case to 1.0025 lots/hr in the 95/5 case. The utilization ratio, however, goes from 0.99 for 80/20, to 0.90 for 90/10, to 0.85 for 95/5.

So, are we spinning our wheels or gaining momentum by working so hard to make MTTF as long as possible while ignoring MTTR? The answer is crystal clear from Fig. 6.11.

At some length of MTTR, we will crater the fab. Not a wise thing to do. In fact, quite insane considering that the extra-long up time will, most likely, not have bought anything in terms of throughput or cycle-time performance.

Ah, but those killer downs can be catastrophic. Delaying or ignoring maintenance on key equipment, in pursuit of phantom performance, is definitely **not** recommended.

6.9.2 Catastrophic Equipment Failure

In addition to the regular maintenance cycle, the other availability issue facing wafer fabs is the one of unexpected failures. If the unexpected failure is of short duration, and happens regularly, then, by rights, the event should be factored into the fab's availability numbers. But even in the case where the event does not happen regularly, as long as it is of short duration, fab performance should not suffer markedly.

Of course, the exception is when the fab is trying to run beyond its capacity and is bottlenecked. Then, all unexpected events could have serious repercussions. (If the fab is severely bottlenecked, then an unexpected failure is just one more drop in the bucket.)

The remainder of this discussion will concentrate on well-run fabs which under normal operations have nothing more than a constraining workstation or very slight bottleneck. With such a fab, a short, unexpected outage will be barely noticeable. However, when a failed piece of equipment remains out-of-service for an extended period, even the well-run fab will feel the effects on performance. These long-duration failures are called *catastrophic* when the effect on performance is severe, and possibly even unrecoverable without remedial action (such as backing off on lot starts).

Whether a failure is catastrophic or not will be fab and workstation dependent. For example, if CMOS-1 is running its constraining workstation with a regular maintenance cycle of 16/4 MTTF/MTTR, a failure of 16 hours or more would be considered catastrophic (Fig. 6.12(a)). If the fab is running a regular maintenance cycle of 56/14, catastrophic failure occurs when the outage exceeds 24 hours (Fig. 6.12(b)). In either case, once the failure has reached a catastrophic length, WIP builds past the point of natural recovery, throughput falls, and cycle time climbs.

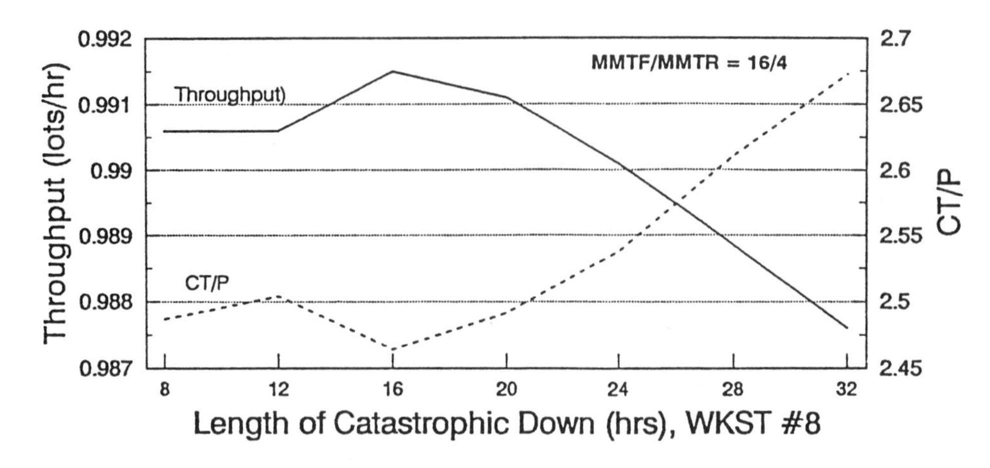

Figure 6.12(a): Catastrophic Downs at Bottleneck, 16/4 Maintenance Cycle

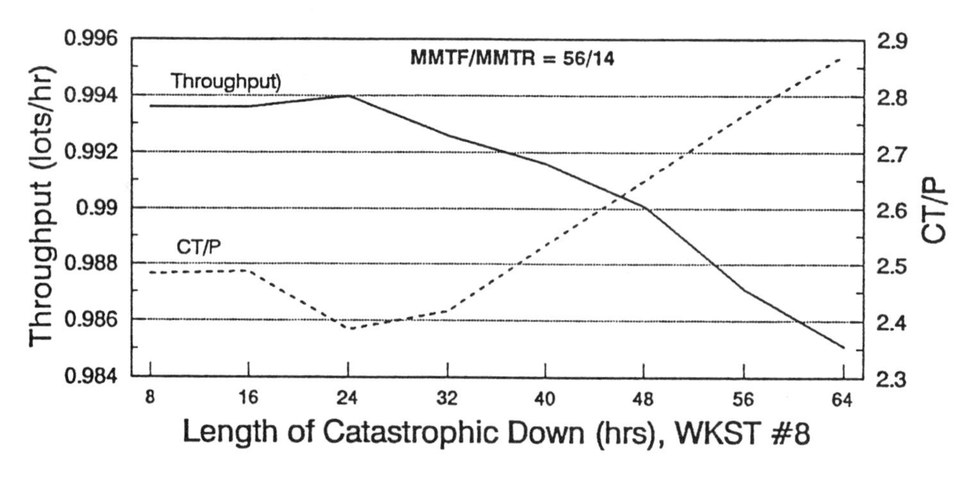

Figure 6.12(b): Catastrophic Downs at Bottleneck, 56/14 Maintenance Cycle

If possible, an even worst scenario occurs if the fab suffers one or more of these catastrophic failures a quarter. Each successive failure builds on the cumulative mess created by its predecessors (Fig. 6.13). Because of the element of uncertainty about when and if the next big failure is coming, it is a wise policy to clear excess inventory and stabilize the fab after each unexpected upset of long duration.

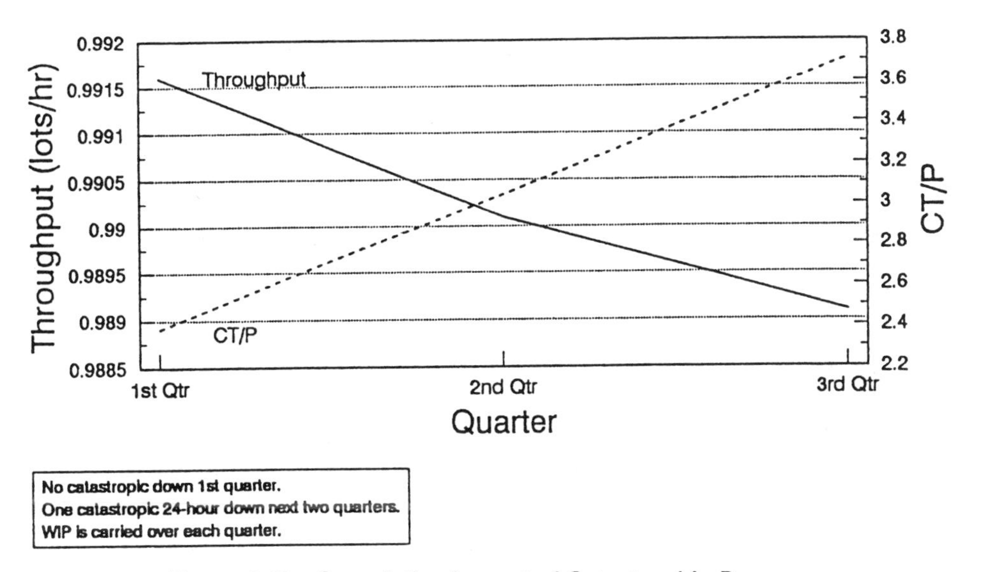

Figure 6.13: Cumulative Impact of Catastrophic Downs

6.10 Achieving Good Factory Performance

After the long and tortuous journey through the life cycles of two wafer fabs, certain hard and fast rules should now be ingrained on how to achieve good factory performance. Hopefully, in those areas where the policy is not so clear and direct, we have developed enough insight to understand the cause and effect relationship of the operating policies under consideration. We should, at a very minimum, be able to put a dollar value on the outcome of a given policy so that we may make an informed decision on the trade-offs involved.

The following rules are the best guides to good factory performance we can offer.

Rule #1: Know the factory bottleneck!

This is the single, biggest club available to control the beast. If it is not used, the beast will control every aspect of fab performance, usually to the detriment of everyone.

Corollary 1.1: Know under what conditions the bottleneck will shift, and where it will shift to.

Rule #2: Do not exceed dynamic capacity (i.e., the capacity of the bottleneck)!

Efforts to violate this rule will degrade performance, not improve it. Unless the factory has been overdesigned, there is no excess capacity, not even one lots worth.

Corollary 2.1: If you must have the extra lots, buy the extra equipment.

Corollary 2.2: When capacity decreases, back off on lot starts for how ever long the change is in effect.

Corollary 2.3: Beware of engineering lots, process creep, process migration, or anything else that will take a bite out of capacity. Expect less production and run the factory accordingly.

Rule #3: Keep WIP within the fab's comfort zone; have it staged along the line.

WIP is not a security blanket. It is work-in-progress, and it should treated as such, with *progress* being the key word. Excessive WIP will certainly stabilize performance, but at level that can be embarrassingly pathetic.

Rule #4: Avoid operating policies that will cause 'chugging'.

The best factory performance will be achieved by a smooth, steady movement of lots through the line, i.e., a balanced line. 'Chugging' will have WIP moving through bottlenecks or constraining workstations in clumps, and will lead to inefficient utilization of these critical work centers. Chugging policies include:

- waiting for full batches at large furnaces,
- bursting lot starts,
- having large priority flows,
- having long MTTR (regardless of MTTF) on bottlenecks,
- etc.

Rule #5: Do not stint on equipment at key workstations!

This may actually be the second most important rule whenever more production is desired, or whenever the fab's performance is not what it should be. The payback for equipment can be measured in days or, at most, weeks. But, the caveat here is that the equipment will only make a difference if it is added at a controlling workstation, i.e., the bottleneck or constraining workstation. Go back and review Rule #1.

Rule #6: Have good availability estimates.

Quoted availabilities will always be optimistic because they are for equipment operating in an ideal environment. When dropped into the middle of a monster fab, the dynamics and discrete nature of the beast can make actual availabilities 30%+ lower than expected. The reality of sizing a fab without a dynamic model means that, from day one, the fab may never achieve design (nor even comes close). If the capacity is not there, the options are to:

- back off on lot starts,
- find the true capacity,
- estimate the actual availabilities,
- and/or buy more equipment.

Rule #6: Don't stint on maintenance, especially on key equipment.

Poor maintenance policies on key equipment cost capacity. Capacity cost throughput. And throughput is why we are in the business. Besides, compared to the value of a single lot, maintenance is a peanut.

Now that we have the rules, summarized in big, bold letters in Table 6.30, let's go out and run those fabs like real men!

TABLE 6.29: ACHIEVING GOOD FAB PERFORMANCE

• Know the bottleneck.
  – Know under what conditions it will shift, and where it will shift to.
• Do not exceed fab's dynamic capacity.
  – If more throughput is needed, buy more equipment.
  – When capacity changes, back off on lot starts.
  – Derate capacity for engineering lots, process creep, process migration, etc.
• Keep WIP moderate (2-4 lots/step), and staged along the line.
• Avoid operating policies that cause "chugging".
• Buy sufficient equipment, especially at key workstations.
• Develop reasonably accurate availability estimates.
• Have good maintenance policies, especially for key equipment.

Further Reading and Notes

1. Mixing rules in chemical thermodynamics are covered in [6.6].

References

[6.1] Atherton, R. W., "Dynamic Capacity Planning Using Simulation Models", **Proceedings of the Fourth Symposium on Automated Integrated Circuit Manufacturing**, Electrochemical Society, Princeton, New Jersey, 1988.

[6.2] Moon, Francis C., **Chaotic Vibrations: An Introduction for Applied Scientists and Engineers**, Wiley, New York, 1987.

[6.3] Vollman, T. E., W. L. Berry, and D. C. Whybark, **Manufacturing Planning and Control Systems**, Dow-Jones Irwin, Homewood, Illinois, 1984.

[6.4] Morton, T. E., and D. W. Pentico, **Heuristic Scheduling Systems**, John Wiley & Sons, New York, 1993.

[6.5] O'Neil, P., "Performance Evaluation of Lot Dispatching and Scheduling Algorithms through Discrete Event Simulation," International Semiconductor Manufacturing Science Symposium, Semicon West, May 21-23, 1991.

[6.6] Smith, B., **Design of Equilibrium Stage Processes**, McGraw-Hill, New York, 1963.

Exercises

We will be following the life-cycle of Sammy T's PiZMOS factory that was sized in Exercise 5.8. The factory information is:

EXERCISE 6.0: FACTORY INFORMATION, PIZMOS-1

Workstation		# Mach.	Process times (hrs)		
			FLOW-1	FLOW-2	FLOW-1
1	SPUTTER	3	3.4	2.4	2.4
2	PATTERN	3	4.4	4.4	4.4
3	DEVELOP	2	2.8	2.8	2.8
4	ETCH	2	2.9	5.8	5.8
5	PVD	1	0.0	0.0	3.2
6	Strip (opt)	1	0.5	0.5	0.5

1. Summarize the life-cycle of a wafer fab as presented in this chapter.

2. The process recipe for FLOW-1 is presented in Exercise 5.8. Steps #1 and 5, at Wkst 1, are 1.7 hours each. Over the next year of operation, these times are expected to creep up to 2 hours each. At what point do these creeping times cause the workstation to bottleneck (i.e., total process time exceeds capacity)? (Reminder: all availabilities = 0.9; no setup; lot-start rate = 0.5 lots/hr)

3. Although we are not expecting a migration of FLOW-1, *per se*, we do anticipate considerable rework. Rework will have a similar impact as adding process steps. (Rework entails the stripping-off of a patterned film if it does not meet spec.) Compare the case of no rework (recipe in Exercise 5.8) to the following three rework recipes:

1 rework: sputter, pattern, develop, etch, sputter, pattern, strip, pattern, develop

1 rework: sputter, pattern, strip, pattern, develop, etch, sputter, pattern, develop

2 reworks: sputter, pattern, strip, pattern, develop, etch, sputter, pattern, strip, pattern, develop

Resize PiZMOS-1 to accommodate the three rework cases.

4. With the lot-start rate for FLOW-1 at 0.5 lots/hr (no rework), resize the fab for engineering lot-start rates of: 1 lot/week, 1 lot/day, and 1 lot/shift, using FLOW-2 as the engineering flow.

5. Add one machine at Wkst 4 and use the bottleneck indicator to evaluate the lot-start rates of 0.25/0.25/0.05 lots/hr for FLOW-1/FLOW-2/FLOW-3, respectively. Repeat with the availability on Wkst 4 at 0.80, then at 0.70.

The purpose of computing is insight, not numbers.
- Richard Hamming

Chapter 7
Factory Modeling

7.0 Introduction

Karlin [7.1] describes a model as "a suitable abstraction of reality preserving the essential structure of the problem in such a way that its analysis affords insight into both the original concrete situation and other situations which have the same formal structure." In previous chapters, we have covered many aspects of fab design, factory dynamics, and performance analysis through a wide spectrum of case studies. Clearly, we have been doing a great deal of factory modeling. In particular, many of the performance analyses have been aided by detailed models of specific factories.

In this chapter, we will discuss factory modeling in general, as well as those critical modeling aspects that are essential for predicting and/or analyzing performance of wafer fabs. The foregoing chapters should make it clear that our subject is industrial-strength modeling. Perhaps, the biggest surprise in modeling real factories is how complicated and difficult it is. Unlike classical physics, it is not a matter of a few parameters and a few equations. The minimum model description typically requires thousands of parameters.

Building an industrial-strength, dynamic model capable of treating tens of thousands of input parameters is an enormous project. As a measure of the size of the undertaking, a data base system is required just in order to manage the input and output parameters. Afterwards, expert systems are often needed to aid in the interpretation of results.

Clearly, great care and forethought must be given to the features of such a model and the modeling methodologies used.

Wafer fabs are highly complex, man-made (synthetic) systems. As novel artificial systems, in the sense of Simon [7.2], it should not be surprising that novel modeling methods are required. Furthermore, for a fab performance model to be of industrial strength, it must be dynamic in nature, and it must be validated by data and experience with actual factories. (Fab graphs, for example, have a proven track record as a sound basis for wafer fab modeling.)

Just as *the proof of the pudding is in the eating*, the value of a model is in its successful use. The modeling methodology presented in this chapter works, and works very effectively. The detailed, calibration-validation procedure described below simply would not work, if the model's structure were not sound.

A accurate factory-specific model is developed through a procedure that involves initialization, calibration, and validation. Model initialization involves completely describing a specific factory in terms of process flows, equipment set, WIP, fab-specific control rules, and any pertinent initial conditions (such as downed equipment, etc.). More particularly, the data-driven description must specify: types and numbers of machines, process flow sequences and process times, possible setup times, machine load sizes, maintenance schedules, characterizations of reliability, parts' characteristics, staffing, etc.

As previously remarked, these data may in fact amount to 10,000 parameters or more. Fortunately, data tables of parameters provide a very natural description of a factory. They are immediately familiar to operations and process personnel.

The second part of the modeling procedure involves calibrating the factory-description data that have been assembled. With thousands of parameters, even those coming from the fab's own tracking or CIM system, are bound to contain errors. In calibration, the values of the parameters describing the factory are carefully evaluated for accuracy.

The final part of the modeling procedure, validation, involves establishing that the model accurately describes the factory's behavior. In practice, a set of simulation cases are developed to predict well-characterized factory operating conditions. The detailed-model's simulated predictions are then compared to measurements of actual factory performance for cycle-time, throughput, equipment utilization, and bottlenecks. Only when a good match is obtained is the factory adequately modeled. A *good* match should be at least 90% accurate, though 95%+ is expected for a well-run fab.

In developing the mathematical structure of a model, great care must be given to fitting the model to real situations. Once a model is formulated, many man-years of effort may be spent analyzing the results from the model and developing synthetic experience. If a poor model structure has been chosen, then much of the effort will have value only as puzzle solving.

A validated mathematical model, however, can provide the basis for accumulating synthetic experience. The virtue of synthetic experience is that it can be as accurate as real experience, but acquired at a fraction of the time and expense. If a model is an accurate representation of reality, it is a valuable tool for interpreting and correlating data and for extrapolating into unknown situations.

7.1 Fab Graphs

The fab graph is the fundamental tool of performance analysis for wafer fabs. The defining feature of a wafer fab is its process flow(s). However, a process flow can not be executed without equipment. Consideration of how a process flow maps onto a set of equipment leads naturally to the fab graph.

In addition to providing a useful pictorial representation of the process flow, the fab graph is a legitimate mathematical entity. It is a directed multigraph [7.3]. Thus, it provides a basis for modeling and performance analysis. In addition, the fab graph can be encoded in a data structure and used for computing and simulation. Early performance analyses using fab graphs are described in [7.4] through [7.7]. More recent work is covered in [7.8] through [7.11].

We will consider two alternative representations for the fab graph. The first follows directly from its origin as a process flow. The nodes of the graph are defined by sets of processing equipment, or workstations. A sequence of integers defines the order in which the process steps are performed on the nodes.

$$p(n) = w_n$$

where n=1,2,.....L, and L is the number of steps in the process flow. w_n is an integer taken from the set of labels for the nodes.

As a concrete example, consider the CMOS-1 process flow of Table 4.1. The fab graph sequence (order in which the workstations are used) is given by the fourth column. It is as follows:

$$p(n) = \{1,2,3,4,5,6,5,6,4,2,6,4,2,6,2,7,3,7,3,3,8,9,7,3\}$$

This fab graph has been well studied in Chapter 4.

The second fab-graph representation is in terms of the sets of process steps performed at each node. If there are W workstations, then there are W such sets. These sets were previously introduced as hubsets, and the number of steps in each set is the hub number or hub order. The hubsets are defined as:

$$H(m) = [s_j] \qquad j=1,..,\text{hub\_order}(m)$$

where s_j is the step number of a process performed at workstation m. Continuing the example of CMOS-1, the hubsets are as below.

$$\begin{aligned}
&H(1) = [1] && \text{hub_order}(1) = 1\\
&H(2) = [2,10,13,15] && \text{hub_order}(2) = 4\\
&H(3) = [3,17,19,20,24] && \text{hub_order}(3) = 5\\
&H(4) = [4,9,12] && \text{hub_order}(4) = 3\\
&H(5) = [5,7] && \text{hub_order}(5) = 2
\end{aligned}$$

H(6) = [6,8,11,14]	hub_order(6) = 4
H(7) = [16,18,23]	hub_order(7) = 3
H(8) = [21]	hub_order(8) = 1
H(9) = [22]	hub_order(9) = 1

The hubsets and hub_orders above provide a hub analysis of the flow of CMOS-1. The hub analysis may be developed in three ways: by parsing the process sequence p(n), by direct inspection of the fab graph, or as output from the fab analysis software described below.

The sets of integers above may seem to be abstract and have little real-world utility. However, in earlier chapters we have discussed the total processing requirement at a workstation P_{ij} and its importance in establishing capacity. The total processing requirement is easily calculated once the hubsets are known. As shown in earlier chapters, the fab graph and hub analysis provide a great deal of structure for analyzing the performance of wafer fabs.

Fabs with multiple process flows provide additional dimensions to the hub analysis. Each process flow requires its own hub analysis. The total set of process steps at each work center, from all flows, is the union of the hubsets for that equipment type. Hence we may consider the total recipe returns at each workstation. The total *recipe returns* for any equipment type is the sum of the recipe returns over all processes. This 'total return' value indicates the number of different process steps performed at a workstation.

In addition to indicating the amount of processing performed *at* a work center, the hub analysis also shows the traffic routes of material moving to the work center. The hub_order indicates the total number of routes to the equipment set, the hubset names the routes. Thus, the hub analysis may also be considered a traffic analysis.

To illustrate the impact of multiple process flows, we consider the traffic analysis given in Tables 7.1, 7.2, and 7.3. (These tables are from fab performance analysis software.)

Table 7.1 provides the hub number at each workstation for three simplistic process flows that are sharing the same factory. The described flows actually represent the processes at the level of description of a technology flow. Note that each flow does not necessarily use all the equipment in the factory. The 'Total Returns', given in column four, is the sum over all three flows.

Table 7.1

********** WORKSTATION TRAFFIC ANALYSIS **********

WORKSTATION #	NAME	---RETURNS FOR PROCESS--- PR1-DMP1	PR2-DMP2	PR3-DMP3	TOTAL RETURNS
1	SUBSTRATE	1	1	1	3
2	WELL FORMATION	1	1	1	3
3	ISOLATION	1	1	2	4
4	ELECTRON EJECT	0	0	2	2
5	TUNNEL OXIDE	0	0	2	2
6	GATE OXIDE	1	1	2	4
7	CHANNEL DOPING	1	1	2	4
8	GATE ELECTRODE	1	1	2	4
9	INTERPOLY OXID	0	1	0	1
10	DUAL LEVEL GAT	0	1	0	1
11	N+/P+ BURIED	1	1	2	4
12	P+ BURIED LAY	1	0	0	1
13	POLY LOAD	1	0	0	1
14	POLY WIRING	1	0	0	1
15	CONTACT	1	1	2	4
16	REFLOW	1	1	2	4
17	1ST METAL	1	1	2	4
18	2ND METAL	1	1	3	5
19	PASSIVATION	1	1	1	3
20	TEST	1	1	0	2

Table 7.2 provides the traffic analysis for a single realistic fab flow. As illustrated, the total number of returns can be quite large. Two masking workstations are shown with a total hub number of 27. In a real fab, multiple flows of this complexity are present and the total number of hub returns is four to ten times the numbers shown. Also, note the large variety and number of cleans and inspects.

Table 7.2

********** WORKSTATION TRAFFIC ANALYSIS **********

WORKSTATION #	NAME	-----RETURNS FOR PROCESS----- CHAP7__P1
1	Met Etcher	4
2	Ox-N Si Etcher	1
3	Field Ox Etcher	2
4	Diel Ox Etcher	1
5	Poly Ox Etcher	5
6	Pass Ox Etcher	1
7	Poly 1 Etcher	1
8	Poly 2 Etcher	1
9	HF Sink SCP	1
10	Fuse Etch SCP	1
11	Selec Etch SCP	1
12	Phos Sink SCP	1
13	Solvent SinkSCP	11
14	Sulf Sink SCP	15
15	FSI Merc Solv	14
16	FSI Merc ACid	15
17	Plasma 1 Resist	14
18	Plasma 2 Resist	13
19	Plasma Ecr Dep	4
20	Teos SCVD	1
21	Teos Dep	1
22	CMP I	1
23	CMP II	4
24	Pass I Dep	1
25	Pass 2 Dep	1
26	Sel W	4
27	W Etch	4
28	Rapid Therm 1	2
29	Rapid Therm 2	1
30	Ti Metal Sput	4
31	Al Alloy Sput	4
32	Sputter Cluster	1
33	HF Sink SCP	4
34	Vert Furn Well	3
35	Vert Furn Space	2
36	RTP Poly	1
37	HF Vap Poly	1
38	Vert Furn Trenc	3
39	Vert Furn Poly	1
40	Clust Tube Gate	2
41	Vert Furn Nitrd	1
42	Vert Furn Field	3
43	HF Vap Poly	1
44	HF Sink SCP	9
45	Implant, Hi En	1
46	Implant, Hi Cur	4
47	Implant, Med Cr	8
48	Duv Step Trac 1	15
49	Duv Step Trac 2	12
50	MRC Develop	10
51	SOG Track	1
52	Silyation TRk	10
53	Wafer Scribe	1
54	Initial Ox Sink	1
55	Semitool	44
56	Met, Prometric	21
57	Photo Step	44
58	Overlay Check	25
59	CD Measurement	59
60	Optical Inspect	3
61	Dose Unif Check	13
62	Met, Mapping	3
63	E-Probe-A	3

Table 7.3 gives a Workstation-Use Analysis. By process, all steps performed at each workstation are shown. This table provides the hubset. It shows the mapping of each process flow onto the equipment. The total process time at each process sector, e.g., the process requirement, is given in column three.

Table 7.3

********** WORKSTATION-USE ANALYSIS **********
***** PROCESS # 1: CMOS-BP1 *****

WORKSTATION #	NAME	PROCESS STEP #	NAME	TOTAL PROCESS TIME (hrs)
1	MISC	1	LASER MASK	.80
2	DIFFUSION 1	2	INITIAL OX	
		10	S/D OX 1	
		13	S/D OX 2	
		15	GATE OX	8.00
3	MASK/ETCH 1	3	P-TUB MASK	
		17	POLY MASK 1	
		19	POLY MASK 2	
		20	CONTACT MASK	
		27	TOPSIDE MASK	5.00
4	IMPLANT	4	P-TUB IM	
		9	S/D IM 1	
		12	S/D IM 2	1.50
5	DIFFUSION 2	5	NITRIDE DEP	
		7	FIELD DIFF/O	4.00
6	MASK/ETCH 2	6	FIELD MASK	
		8	SRC/DRAIN MS	
		11	S/D MASK 2	
		14	NITRIDE STRI	4.00
7	DIFFUSION 3	16	POLY DEP/DOP	
		18	POLY DEP/DOP	
		26	TOPSIDE DEP	6.00
9	MASK/ETCH 3	25	METAL MASK	1.00
10	MET-SPUT	21	TI METAL DEP	2.20
11	MET-CVD	22	TUNS-CVD	.90
12	PLAS-ETCH	23	TUNS-ETCH	.50
13	MET-SPUT-A	24	AL METAL DEP	1.10

The underlying structure of the fab, and its process flows, provide the necessary fundamental information for performance analysis. Understanding this structure is especially important for fabs with multiple process flows. Fortunately, much of this structure can be understood in terms of the fab graph.

7.2 Process/Equipment/Lot Interactions

The next level of detail in modeling is to move wafers, grouped as lots, through the process flows, from workstation to workstation. One of the fundamental features of a dynamic model is that it can deal with the details of lot movement and lot-machine interactions.

At each workstation, machines operate on lots. The rules describing these operations provide for the rich set of dynamics described in previous chapters.

The dynamic interaction between lots and machines is, in part, described in terms of fundamental events such as start_a_lot, move_a_lot, load_a_machine, etc. The events are implemented in the model, just as they are in the fab, using if-then rules, equipment and lot tracking, control rules, and other such relationships. The comprehensive, yet fab-specific rules provide a detailed model of how machines process lots.

7.2.1 Fundamental Events

In wafer fabs, materials movement and machine operations are discrete events. A lot (or part) moves from 'A' to 'B'; the lot waits; the lot is processed; the lot moves on to the next processing step. Each discrete event generates future events.

Thus, an effective wafer fab model must be based on formulas, rules, logic, and decision tables that describe these actual discrete events. The model uses realistic abstractions of what takes place routinely on the factory floor.

A list of typical fundamental events for a wafer fab is provided in Table 7.4. Each fundamental event is defined in terms of changes to the data tables for lots, machines, workstations, etc.

Table 7.4: Fundamental Events
Start_a_lot
Move_a_lot
Complete_a_lot
Load_a_machine
Unload_a_machine
Machine_fails
Machine_is_repaired

The first three events are concerned with movement of lots along a process flow, or simply following the process flow. A very rudimentary model results from the rule that, after a process time has elapsed, move_a_lot to the next workstation. Such a model will result in a cycle time equal to the sum of the process times, which is, of course, grossly in error. Such a crude model also assumes unlimited capacity at the workstation, and thus ignores the rich dynamics of machines processing lots.

The next two events in Table 7.4, Load_a_machine and Unload_a_machine, address the interaction of lots with machines. As described below, these operations can be highly complex. These fundamental rules are at the heart of establishing factory capacity.

The last two events address the issue that machines are not always available for processing. Hence, machines fail and are repaired. A large variety of models are possible for describing machine availability (Section 7.2.4).

7.2.2 Data Structures

The fundamental entities of the model are, of course, process flows, machines, workstations, and lots. In order to make a model specific and concrete, data tables and structures must be defined for each fundamental entity.

Computationally, data tables are overlaid with multiple linked lists. Deo [7.12] discusses aspects of computing with graphs, and more general data structures are described in [7.13].

The seven events listed in Table 7.4 must be defined in terms of the fundamental entities' data tables. For example, if a lot is loaded into a machine, then the machine table must reflect that the individual machine is loaded and that the particular lot is loaded into the machine.

Similarly, the lot table must reflect the number of wafers in the lot, the process flow to which the lot belongs, the current process step, and the machine processing the lot_if it is currently loaded.

In addition to the fundamental physical events, there are also data tables for queues and other events. Multiple linked lists are especially useful for establishing the order of events; in particular, the next event to occur.

7.2.3 The Workstations

The first fundamental entity in the wafer fab (some might even say it *is* the fab) is the equipment set. The equipment set, however, is naturally structured into groupings of like pieces of equipment for process-step assignments. Each like-equipment grouping is called a *workstation*, and has at least one machine available to perform the requisite processing. The workstation thus also assumes the role of a fundamental entity in describing the fab and establishing the model.

As a fundamental entity, workstations are represented in the model by a data table containing the physical and operating specification of each workstation or process sector. A key element of this data structure is the number of machines at each workstation.

Table 7.5 gives an example of such a data table, where each workstation in the fab is described by machine type and machine characteristics such as load size and availability parameters.

Table 7.5

********** WORKSTATION TABLE **********

WORKSTATION #	NAME	RESOURCE QUANTITY	LOAD SIZE	BATCH WAIT	AVERAGE UP	AVERAGE DOWN	RANDOM UP	RANDOM DOWN
1	MISC	2	25	-1	99.00	1.00	.00	.00
2	DIFFUSION 1	5	100	2	95.00	5.00	.00	.00
3	MASK/ETCH 1	7	25	-1	97.00	3.00	.00	.00
4	IMPLANT	3	25	-1	85.00	15.00	.00	.00
5	DIFFUSION 2	3	100	2	95.00	5.00	.00	.00
6	MASK/ETCH 2	6	25	-1	97.00	3.00	.00	.00
7	DIFFUSION 3	4	100	2	95.00	5.00	.00	.00
8	METAL DEP 1	3	16	-1	85.00	15.00	.00	.00
9	MASK/ETCH 3	4	25	-1	97.00	3.00	.00	.00
10	DIEL-FAB	5	25	-1	97.00	7.00	.00	.00
11	METAL DEP 2	2	25	-1	85.00	15.00	.00	.00

As discussed in Chapter 2, silicon wafers in cassette carriers travel through the fab in lots. Lot size may vary from six to fifty wafers and require one or two cassettes. As the next step in our modeling effort, we must determine how lots interact with machines, specifically how long a lot will be at a machine, i.e., its residence time.

Some additional concepts are required before we can make this determination; the first being the distinction between load size and batch size. The *load size* is the maximum number of wafers that the equipment can process at one time. The *batch size* is the number of wafers actually processed in a batch, and as such is a dynamic variable that is less than or equal to the load size. It is determined by the interaction of loading rules, lot size, and load size. Thus, batch size is the dynamic load size associated with each type of equipment or resource.

Another important concept is needed for the case where multiple lots may be loaded into a machine: *batch wait*. This parameter indicates how close to a full load a machine must be before it performs its function. To demonstrate, consider a furnace that can load 150 wafers or 6 lots. But depending on factory dynamics, from 1 to 6 lots may be available for loading at any given time. If only 2 lots are available, do you load the machine or not? The answer depends on the fab's furnace-loading rules, and is indicated by the batch-wait parameter.

A positive or a zero number indicates that there is a minimum batch size required before a machine can load. The machine will wait until sufficient lots are available before beginning processing; hence the term batch-wait.

Batch Wait = Full Load - Number of Lots Needed to go

or, more appropriately,

Number of Lots Needed to go = Full Load - Batch Wait

Thus, if the 6-lot furnace must wait for a full load or 6 lots before loading, the Batch Wait is 0 (6-0=6). If two-thirds of a load (4 lots) must be available before loading, the batch wait is 2 (6-4=2). Other batch waiting values are all positive numbers between 0 and the full batch capacity (in lots) of the machine.

A batch wait of '-1' has a special meaning: *load and go*. In other words, the multi-batch piece of equipment will not wait for any specific load size; instead it will load all available lots (up to the maximum load) and then proceed to process them.

7.2.4 Equipment Availability Models

The remaining key data needed for the physical description of the workstation is information on equipment availability. Just as the number of machines at a workstation is critical for setting dynamic capacity, so too is the availability of those machines critical. The types of availability parameters shown in Table 7.5 are familiar to most operations managers. The first two, Average Up and Average Down, are especially well-known.

Average Up and Average Down times correspond to mean-time-to-failure (MTTF) and mean-time-to-repair (MTTR), respectively. A machine or resource is 'Up' if it is capable of performing its task (processing for a given time). It is 'Down', if it is under repair or not capable of performing its function. The operating history of equipment and operators in a fab generally provides reasonable values for these availability averages.

The other availabilities, Random Up and Random Down, allow the introduction of uncertainty into an operating schedule [7.14, 7.15, 7.16]. The length of time that equipment (or operators) is 'Up' is based on a random sampling from an exponential distribution whose mean is established by the Random Up time. Likewise, the 'Down' time is set by sampling a second distribution whose mean is established by the Random Down time. Additional two-parameter distributions such as normal, log-normal, or Weibull may also be specified [7.16].

Thus, at a minimum, a dynamic model using information in Table 7.5 will allow specification of four different models for equipment availability (or reliability). These are: (1) no equipment failure, (2) fixed or deterministic equipment failure, (3) random equipment failure, and (4) a superposition of random and fixed equipment failure. The application of each to wafer fabrication is discussed below.

Case 1: NO EQUIPMENT FAILURE

The inference of this type of availability model will mean that equipment is always available for processing. This mode establishes a best case for equipment availability and allows the study of queuing losses, setup effects, and batching effects without the added complications of equipment failure. (The factory is modeled as a deterministic dynamic system.)

Case 2: DETERMINISTIC EQUIPMENT FAILURE

Equipment at a workstation will fail with a fixed pattern governed by the average-up and average-down parameters in the workstation table. For example, if these two numbers are 90 and 10, then a piece of equipment will be available for processing for 90 hours and then unavailable for 10 hours. The pattern will then repeat. A small perturbation is introduced, however, by the rule that the equipment will not fail during processing. A machine-up period will continue until the batch is completed.

This failure model is very popular in wafer fabrication because it is realistic and easy to interpret. (The factory is again modeled as a deterministic dynamic system.)

Case 3: RANDOM EQUIPMENT FAILURE

With this model, equipment at a workstation will fail with a random pattern established by sampling from an appropriate statistical distribution [7.14, 7.15, 7.16]. Similarly, the repair time is established by sampling from a second distribution.

A convenient one-parameter distribution is the exponential distribution. It is a non-symmetric, one-parameter distribution commonly used in reliability modeling. The single parameter is the mean of the distribution, and it is also equal to its standard deviation. The parameters for these distributions can be established by analyzing failure and repair data stored in the database of a fab's computer-integrated manufacturing system.

In the model, these distributions are sampled by using a random-number generator. Whenever a failure or repair is scheduled, the model generates a random number, which is then transformed into a sample from the appropriate distribution.

With random equipment failure, the factory is modeled as a stochastic dynamic system. Stochastic dynamic systems are difficult to analyze (see below), and consequently this mode of analysis is usually broached with care.

Case 4: RANDOM **AND** *FIXED EQUIPMENT FAILURES*

With this availability model, equipment at a workstation will fail in a complex manner governed by all four parameters in the workstation table. In the case of two-parameter distributions, six parameters govern the model. Random failure or repair is thus superimposed upon fixed failure or repair.

Consider the example above, where the fixed-failure model parameters are 90 and 10. The equipment will then fail at a time calculated by adding the 90 hours to a time determined by sampling the appropriate exponential distribution. Similarly, the time for repair will be calculated by adding 10 to a number determined by sampling a time-to-repair distribution.

With this model, the factory is again modeled as a stochastic dynamic system. This is an especially complex failure model, and is used only with extreme care. The complexities of stochastic simulation are discussed below.

STOCHASTIC SIMULATION

With deterministic models, the results of a single simulation are easy to interpret. With stochastic models, the meaning of a single simulation or experiment is unclear [7.14, 7.15, 7.16].

In stochastic simulation, **all** output variables are now represented by distributions. Cycle time and throughput are not single variables, but are given by a distribution of values. We can no longer say that the cycle time for process CMOS-1 is 125 hours. To characterize cycle time of the stochastic factory model requires characterizing the distribution of cycle time. For example, cycle time has a mean value of 125 hours and standard deviation of 50 hours.

To characterize the distribution of an output value like cycle time requires a large number of simulations, perhaps over 100. For each of those 100 simulations, the output will be different, and the results will appear in an apparently random order.

Performing stochastic simulations, and generating 100 or more random factory experiments will, needless to say, result in an extremely large set of output. Thus, determining the distributions may require feeding the stochastic model's output into a statistical analysis package. Such a package may then provide plots of the output distributions, and may even provide a statistical analysis of each distribution.

Regardless of the form of the final output, interpretation of factory performance will be difficult and generally will require someone skilled in statistics. Furthermore, validation of a stochastic simulation against fab operating data is challenging.

7.2.5 The Process Flow(s)

The second fundamental entity in the fab are the process flows. Each flow uses some portion of the factory's capacity to process its lots. A factory model, therefore, must be provided with a clear description of how the various flows use the equipment set. Thus, each recipe step in a process flow is assigned to a workstation, given a specific process and setup time, and assigned an expected yield.

Table 7.6 provides an example of the type of process flow information that is needed by a factory-specific model. As a cautionary note, it must be pointed out that the relationship of process step to workstation must be correct in the model, just as it is in the fab. If a step is assigned to the wrong workstation, an unrealistic factory representation will be modeled. Such a blunder in the fab will result in a Class 2 yield crash, and cost $8-15mm.

Table 7.6

********************** PROCESS FLOW **********************
***** PROCESS # 1: CHAP6_P1 *****

PROCESS STEP #	PROCESS STEP NAME	WORKSTATION #	WORKSTATION NAME	SETUP TIME	PROCESS TIME	YIELD (ratio)
1	LASER MASK	1	MISC	.10	.80	1.000
2	INITIAL OX	2	DIFFUSION 1	.10	2.00	1.000
3	P-TUB MASK	3	MASK/ETCH 1	.10	1.00	1.000
4	P-TUB IM	4	IMPLANT	.10	.50	1.000
5	NITRIDE DEP	5	DIFFUSION 2	.10	2.00	1.000
6	FIELD MASK	6	MASK/ETCH 2	.10	1.00	1.000
7	FIELD DIFF/O	5	DIFFUSION 2	.10	2.00	1.000
8	SRC/DRAIN MS	6	METAL ETCH 2	.10	1.00	1.000
9	S/D IM 1	4	IMPLANT	.10	.50	1.000
10	S/D OX 1	2	DIFFUSION 1	.10	2.00	1.000
11	S/D MASK 2	6	MASK/ETCH 2	.10	1.00	1.000
12	S/D IM 2	4	IMPLANT	.10	.50	1.000
13	S/D OX 2	2	DIFFUSION 1	.10	2.00	1.000
14	NITRIDE STRI	6	MASK/ETCH 2	.10	1.00	1.000
15	GATE OX	2	DIFFUSION 1	.10	2.00	1.000
16	POLY DEP/DOP	7	DIFFUSION 3	.10	2.00	1.000
17	POLY MASK 1	3	MASK/ETCH 1	.10	1.00	1.000
18	POLY DEP/DOP	7	DIFFUSION 3	.10	2.00	1.000
19	POLY MASK 2	3	MASK/ETCH 1	.10	1.00	1.000
20	CONTACT MASK	3	MASK/ETCH 1	.10	1.00	1.000
21	METAL DEP 1	8	METAL DEP 1	.10	1.10	1.000
22	METAL MASK	9	MASK/ETCH 3	.10	1.00	1.000
23	DIELECTRIC	10	DIEL-FAB	.10	2.00	1.000
24	METAL DEP 2	8	METAL DEP 1	.10	1.10	1.000
25	METAL MASK	9	MASK/ETCH 3	.10	1.00	1.000
26	TOPSIDE DEP	7	DIFFUSION 3	.10	2.00	1.000
27	TOPSIDE MASK	3	MASK/ETCH 1	.10	1.00	1.000

Wafer Yield

The yield parameter shown in Table 7.6 refers to the wafer or mechanical yield, not the die yield. During manufacturing wafers from a lot are lost due to mechanical breakage or mis-processing. Thus, at each process step, on average, a fraction of good wafers will result. This is the wafer yield of the process step. The yield after a sequence of steps will be the product of the step yields. This yield is called a partial cumulative yield. The yield after all the steps in a process flow is the process cum yield. Die yield is measured on completed wafers after mechanical losses.

Historically, cumulative yields as low as 60 percent were not unknown. With automated wafer handling, most fabs have improved cumulative yields to over 90 percent. Some world-class fabs experience only occasional losses.

7.2.6 Batching and Cycle Time Analyses

Before a lot is ever started through the factory or through the dynamic model, it is possible to perform static analyses.

Table 7.7

******************** BATCHING AND LOAD ANALYSIS ********************
***** PROCESS # 1: CHAP6_P1 *****

PROCESS STEP #	NAME	CUM YIELD (ratio)	LOT SIZE (wafers)	MACHINE CAPACITY (wafers)	BATCH SIZE (wafers)	BATCH/ LOT	BATCH MODE
1	LASER MASK	1.000	25	25	25	1	1
2	INITIAL OX	1.000	25	100	100	1	3
3	P-TUB MASK	1.000	25	25	25	1	1
4	P-TUB IM	1.000	25	25	25	1	1
5	NITRIDE DEP	1.000	25	100	100	1	3
6	FIELD MASK	1.000	25	25	25	1	1
7	FIELD DIFF/OX	1.000	25	100	100	1	3
8	SRC/DRAIN MSK 1	1.000	25	25	25	1	1
9	S/D IM 1	1.000	25	25	25	1	1
10	S/D OX 1	1.000	25	100	100	1	3
11	S/D MASK 2	1.000	25	25	25	1	1
12	S/D IM 2	1.000	25	25	25	1	1
13	S/D OX 2	1.000	25	100	100	1	3
14	NITRIDE STRIP	1.000	25	25	25	1	1
15	GATE OX	1.000	25	100	100	1	3
16	POLY DEP/DOPE 1	1.000	25	100	100	1	3
17	POLY MASK 1	1.000	25	25	25	1	1
18	POLY DEP/DOPE 2	1.000	25	100	100	1	3
19	POLY MASK 2	1.000	25	25	25	1	1
20	CONTACT MASK	1.000	25	25	25	1	1
21	METAL DEP 1	1.000	25	16	16	2	2
22	METAL MASK	1.000	25	25	25	1	1
23	DIELECTRIC	1.000	25	25	25	1	1
24	METAL DEP 2	1.000	25	16	16	2	2
25	METAL MASK	1.000	25	25	25	1	1
26	TOPSIDE DEP	1.000	25	100	100	1	3
27	TOPSIDE MASK	1.000	25	25	25	1	1

BATCHING MODES: 1--NORMAL 2--BATCHING DOWN 3--BATCHING UP

These static analyses are based solely on the structure of the factory, i.e., how the processes fit the available equipment. Such an analysis provides valuable information about the physical limits of the factory for each process. Statics bound dynamics.

Table 7.7, for example, shows the type of static batching and load information that might be calculated for each process. For each recipe step, this table presents the partial cumulative yield ratio as lots move through the processing sequence, as well as the actual number of wafers that yield represents. The table also presents how the equipment batches for each step, which directly impacts the flow's cycle time.

Another example of a static analysis is shown in Table 7.8. For each process, a static cycle-time analysis determines the residence time and the cumulative theoretical cycle time as we march through the process sequence.

Table 7.8

******************** CYCLE TIME ANALYSIS ********************

***** PROCESS # 1: CHAP6_P1 *****

PROCESS STEP #	NAME	PROCESS TIME	SET-UP TIME	RESIDENCE TIME (w/ setup)	CUMULATIVE CYCLE TIME (w/ setup)
1	LASER MASK	.800	.100	.900	.900
2	INITIAL OX	2.000	.100	2.100	3.000
3	P-TUB MASK	1.000	.100	1.100	4.100
4	P-TUB IM	.500	.100	.600	4.700
5	NITRIDE DEP	2.000	.100	2.100	6.800
6	FIELD MASK	1.000	.100	1.100	7.900
7	FIELD DIFF/OX	2.000	.100	2.100	10.000
8	SRC/DRAIN MSK 1	1.000	.100	1.100	11.100
9	S/D IM 1	.500	.100	.600	11.700
10	S/D OX 1	2.000	.100	2.100	13.800
11	S/D MASK 2	1.000	.100	1.100	14.900
12	S/D IM 2	.500	.100	.600	15.500
13	S/D OX 2	2.000	.100	2.100	17.600
14	NITRIDE STRIP	1.000	.100	1.100	18.700
15	GATE OX	2.000	.100	2.100	20.800
16	POLY DEP/DOPE 1	2.000	.100	2.100	22.900
17	POLY MASK 1	1.000	.100	1.100	24.000
18	POLY DEP/DOPE 2	2.000	.100	2.100	26.100
19	POLY MASK 2	1.000	.100	1.100	27.200
20	CONTACT MASK	1.000	.100	1.100	28.300
21	METAL DEP 1	1.100	.100	2.300	30.600
22	METAL MASK	1.000	.100	1.100	31.700
23	DIELECTRIC	2.000	.100	2.100	33.800
24	METAL DEP 2	1.100	.100	2.300	36.100
25	METAL MASK	1.000	.100	1.100	37.200
26	TOPSIDE DEP	2.000	.100	2.100	39.300
27	TOPSIDE MASK	1.000	.100	1.100	40.400

MIN CYCLE TIME = 37.700 MIN CYCLE TIME W/ SETUP = 40.400

Residence time in Table 7.8 is the total time that a lot spends at a particular process step, including setup time.

RT = (# of Batches *(Process Time)) + Setup

The residence time helps identify the impact a process step or steps may have on a workstation. For example, in the sample data above, Process Steps #21 and #24, METAL DEP 1 and 2, would have a hefty impact on Workstation #8. In fact, under dynamic conditions, where there is competition for resources, these steps could cause a bottleneck in the fab, resulting in long wait times and large queues.

The cumulative cycle time is another valuable parameter resulting from an analysis of process structure. This information from wafer statics provides a perspective on the requirements of each process on the factory. It also provides bounds on the best a factory will be able to do in terms of cycle time by process.

7.3 Discrete Event Simulation

Returning briefly to our earlier discussion of fab graphs, each node in the graph represents a workstation or a set of like equipment. At the workstation, machines perform operations on lots by following a set of rules governing loading, batching, setup, processing, yield, and availability. Following these rules, the workstation becomes a center of rich dynamic activity__activity which is described to the model in terms of events. Each fundamental event is defined in terms of changes to the data structures.

Discrete-event simulation [7.15] is a numerical method for obtaining results from a model of discrete events. A wafer fab model that simulates such discrete events could theoretically be implemented using pencil, paper, and calculator, but a computer simulation will be considerably more efficient, and less prone to errors. In fact, de-bugging such simulations for simple cases involves manual solutions.

Three concepts are involved in a discrete event simulation of a wafer fab: the state of the fab, the generation of future events, and the determination of the next event. The state of the fab at any point in time is described by the data structures for fundamental entities. The occurrence of an event changes the state of the fab and thus the data tables.

Future events are generated by applying the dynamic rules, as the model dynamically moves wafer-lots through the various process flows on the factory floor, taking into account the loadings of lots into machines. For example, an initial event could be a lot start. The lot moves into the queue for process step one at the indicated workstation. If a machine is available, the lot is loaded into that machine. At the loading, an unload is scheduled for a future time when the process recipe is complete. At any point in time, the state of the factory contains information as to the next scheduled event.

Thus the key to a discrete-event simulation of a model is the ordered list of events. At the top of the list is the next event scheduled to occur in the factory. As this event is implemented, additional events are scheduled by being added to the list. For example, if a machine is loaded, then a corresponding unload must be scheduled. When the unload is completed, then the lot is moved to the queue of the next workstation of the process flow.

Hidden in this description is an explanation of the speed of discrete-event simulation. Each time a new event is scheduled, the event list must be searched to locate its position relative to the events already scheduled. We are, therefore, concerned with sorting and searching, a basic concept in computer science [7.17].

The Classical Contrast

The wafer fab is by its very nature a man-made, discrete system. In this regard, it is quite different from the continuum systems often modeled in physics, applied science, and parts of continuous systems is often that of differential equations or partial differential equations (PDEs). For example, in electrical engineering (see Chapter 3), circuits and devices are modeled using these mathematical theories. Finite difference is a numerical method of solution for continuous PDE models.

In direct contrast (Table 7.9), the discrete world of wafer fabs requires theories of discrete mathematics. The model is given in terms of discrete structures, fab graphs, 'If-Then' rules, logic statements, and functional relationships. The method of numerical solution for these discrete models is a technique called discrete-event simulation.

Table 7.9: Discrete vs Continuous Models

	Discrete Model	Continuous Model
System	Wafer Fab	Circuit/Device
Mathematics	If-Then Rules Logic Stmts Algebraic Fns	Differential Eqns & PDEs
Method of Solution	Discrete Event	Finite Difference

7.4 Lot Starts

The third, and final fundamental entity in the wafer fab is the lots that are being processed. For each flow, lots are started into the factory at a specified rate, over a specified time intervals. Lots are discrete and are started at discrete points in time over this interval, despite the tendency to speak of a continuous *lots per hour*. Consequently, the parameters for lot starts are the number of lots in each starting group, one or more, and the interval between groups.

Ideally, the cumulative lot-start rates would be within the fab's dynamic capacity. However, this information is rarely available; hence the need for a dynamic model.

As was seen in Chapter 6, the manner in which lots are started into the fab can have a significant impact on whether performance is stable or chaotic. Thus, a model must provide the capability for independent control of each flow's lot-start interval, the number of lots starts, and the size of the lots started.

As lots are started by the model, they are entered into the data structure for lots. This data structure tracks the lots as they move through the process flows, experience yield losses, and are loaded into and unloaded from individual machines. At any time during a simulation, the lot data structure should be able to give the complete status of all lots.

7.5 Factory System Behavior

Using discrete-event simulation as a numerical technique, the dynamic model may be implemented in computer software as a production simulation. The output of the model (Fig. 7.1) includes numerical measures of fab performance such as dynamic capacity, product cycle times, process throughputs, equipment utilization, and scheduled lot completions.

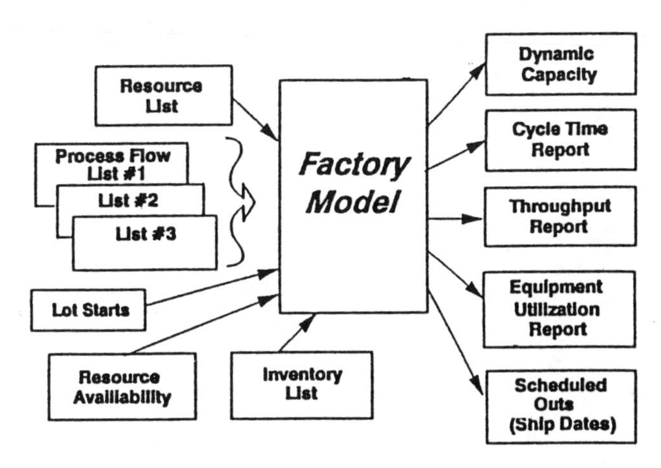

Figure 7.1: Factory-Specific Model

Once a factory-specific model is built, it is then used for a wide variety of simulations to gain fundamental understanding and synthetic experience. The results from these simulations should encompass: a global picture of factory system behavior, all the pertinent fundamental performance analyses, and detailed information on workstation dynamics.

The global-factory system behavior is in effect an executive or high-level overview of the factory's performance during the simulated time horizon. It contains summary results from a process perspective and from an equipment perspective. (See Table 7.10.)

Table 7.10

EXECUTIVE SUMMARY OF FACTORY STATUS

CYCLE TIME AND THROUGHPUT REPORT

PROCESS # NAME	MIN. PROCESS TIME	MAX. SET-UP TIME	TOT.AVG WAIT TIME	CYCLE TIME	LOTS-COMPLETED INVENTORY	STARTS	TOTAL
1 PR1-DMP1	41.5	8.5	183.3	.0	18	0	18
2 PR2-DMP2	47.9	8.9	29.2	83.8	25	22	47
3 PR3-DMP3	83.5	15.4	13.4	111.5	29	31	60

EQUIPMENT PERFORMANCE REPORT

WORKSTATION # NAME	LOTS PROCESSED	THRUPUT (WAF/HR)	AVAIL RATIO	UTIL. RATIO	AVERAGED CYCLE TIME	WIP (LOTS)
1 SUBSTRATE	171	23.7	.8988	.5126	3.23	1
2 WELL FORMATI	172	25.1	.9876	.5167	5.90	2
3 ISOLATION	245	39.8	.9340	.8608	10.96	16
4 ELECTRON EJE	141	23.1	.9397	.1594	5.00	1
5 TUNNEL OXIDE	138	23.5	.9220	.6711	8.63	6
6 GATE OXIDE	247	40.1	.9850	.6105	4.68	2
7 CHANNEL DOPI	247	41.1	.9836	.6076	4.50	5
8 GATE ELECTRO	255	42.7	.9300	.5212	2.96	1
9 INTERPOLY OX	45	7.9	.9400	.3867	8.41	1
10 DUAL LEVEL G	46	8.3	.9600	.2805	4.32	1
11 N+/P+ BURIED	199	36.5	.3575	.3575	22.00	62
12 P+ BURIED LA	14	3.5	.9505	.0395	2.44	0
13 POLY LOAD	15	3.7	.8803	.0970	4.72	0
14 POLY WIRING	15	4.2	.8800	.0354	2.93	2
15 CONTACT	201	36.5	.9087	.4749	2.68	2
16 REFLOW	200	36.6	.9672	.4090	2.15	3
17 1ST METAL	199	36.7	.9500	.4342	2.25	3
18 2ND METAL	259	47.6	.9500	.5412	2.54	1
19 PASSIVATION	124	24.6	.9800	.3320	2.13	3
20 TEST	65	13.3	.9900	.0053	.58	0
21 OPS POOL 1	76	12.9	.8575	.0098	.15	0
22 OPS POOL 2	67	11.9	.8575	.0000	--	0
23 OPS POOL 3	60	11.0	.8575	.0054	.21	0

The process summary results present dynamic cycle time and throughput performance for each flow in the plant. The cycle times that are generated during the dynamic simulation are the result of the model directly counting factory activities. Thus, the numbers accurately reflect all processing, setup, waits, and perhaps transport that befall all starting lots that complete during the simulation. The cycle-time information is presented in terms of the theoretical (minimum processing) and actual cycle times. Statistics are separate for lots completing from starts and for lots processed from inventory or WIP.

The summary factory performance results provided from an equipment perspective include for each workstation: the throughput, the availability ratio, the utilization ratio, an averaged cycle time (averaged across all lots), and the ending WIP. These performance statistics are derived based on all of the processes that are using the equipment.

The distinction between *availability* and *process utilization* is subtle but also critical for understanding the inherent flaw of tying factory performance to equipment usage. The availability ratio is the ratio of time that all equipment at a workstation is available for processing (i.e., not out of service as downed equipment). The process utilization ratio is the ratio of time that all equipment at a workstation is actually either processing, in setup, or down (i.e., not idle). Therefore, unavailable equipment is the fault of the equipment; unused equipment is the fault of the system (i.e., the entire wafer fab).

Witness Workstation #4, ELECTRON EJECTION, in Table 7.10. The equipment has a reasonable availability (0.9397), but it is simply not used very much (0.1594). Improving availability at this workstatiun will in no way improve factory performance.

One final note before moving on: the 'Averaged Cycle Time' is a calculated value, derived from the fundamental statistics generated by the model. The values are the average time that lots from *all* processes spend at each workstation. For a new process, these calculated times, when summed for the new processing sequence, may be used to give a 'ballpark' estimate of cycle time. The accuracy of the estimate, however, may not be particularly good (because of the averaging of old processes).

7.6 Fundamental Performance Analyses

The dynamic behavior of the wafer fab can be very complex. This complexity combined with the sheer volume of model outputs means that obtaining insights from a particular exercise of the model can be difficult and time-consuming. Thus, a useful production model should be prepared to answer the critical questions:

- *Do the flows fit within the fab's dynamic capacity?*
- *If not, which workstation is the bottleneck?*
- *How may the flows be reduced to fit the capacity?*

Or,

- *How much equipment must be added to accommodate?*

7.6.1 Expert System

To answer these questions, the entire dynamic results of the simulation must be knowledgeably analyzed, including factors such as workstation utilizations, queueing dynamics, inventory build rates, etc. Automated analysis and interpretation by the software is possible because of the fundamental understanding of fab structure derived from fab-graph analysis.

Such automated analysis is often called an expert system. The expert system reports most often needed by a wafer fab are for bottleneck identification and fab capacity.

The place to start analyzing the dynamic results is to observe the inventory build or depletion rate during the time period studied. If the fab is doing either to an excessive degree, factory operation is unstable and can not be sustained. A build rate generally is indicative of a bottleneck which should be identified for corrective action. (Long-range corrective action is either buying equipment or backing off on lot starts.)

A WIP depletion rate is generally indicative of excess capacity. The fab may either need to increase lot starts or institute lot-move controls. (See Chapters 5 and 6.)

If the fab has multiple processes, where some are building WIP and others depleting it, then stable, long-term operation may call for a combination of adding equipment, adjusting lot starts, **and** controlling lot moves. There are a few other alternative, though they are usually less amenable to overall fab objectives. Altering process priority may provide a sufficient shift in capacity demands. Also, smoothing the line by adjusting maintenance schedules at key workstations may provide some degree of additional capacity. (Of course, these multi-solution and/or sophisticated approaches will not be answered by one or two simulations.)

The next indicator that something may be unstable in the fab comes when an activity ratio for a workstation is too high. The definition of too high will most likely vary from fab to fab, but a ratio of '1.0', as seen for Workstation #11 in Table 7.11, is usually a dead giveaway that there is a bottleneck. Since *activity ratio* is the fraction of time that the workstation is not idle, a ratio of '1.0' means it is busy all the time. The situation where there is *NO* idle time on equipment is a guarantee that there is a large queue of work sitting there__waiting.

7.6.2 Bottleneck Identification

Indeed, from Table 7.11, we see that our suspected bottleneck has a large queue of work and excessive wait times for Step #7 of Process Flow #1. This step seems to be the culprit as far as the inventory build rate is concerned as well.

Based on the bottleneck identification, remedial actions should be recommended for study in future simulations.

It should be noted that bottleneck identification and removal may need to be an iterative process. In fabs running grossly beyond their capacity, only the first bottleneck in the process sequences may be identified. After its removal, we will then know whether there are other, downstream bottlenecks. A fab capacity analysis, however, should give clues as to whether the fab is indeed running well beyond capacity.

Table 7.11

```
-----------------------------------------------------------
BOTTLENECK IDENTIFICATION:
-----------------------------------------------------------

THE IDLE RATIO OF EQUIPMENT AT THE FOLLOWING WORKSTATION(S) IS
TOO LOW (i.e., the workstation activity level, including down
time, is too high):

        WORKSTATION              IDLE       ACTIVITY
         #    NAME               RATIO        RATIO
        ----------------         -----     -----------
         3  ISOLATION            .0732        .9268
        11  N+/P+ BURIED         .0000       1.0000

AVERAGE QUEUE LENGTHS AND WAIT TIMES AT THESE WORKSTATIONS, ARE:

 WORKSTATION     PROCESS      PROCESS-STEP      AVG.WAIT  AVG.QUEUE
 #   NAME        #  NAME       #    NAME           TIME     LENGTH
 ------------    ---------    --------------    --------  ---------
 3 ISOLATION     1 PR1-DMP1    3 ISOLATION          4.24        .56
                 2 PR2-DMP2    3 ISOLATION          1.42        .10
                 3 PR3-DMP3    3 ISOLATION           .76        .23
                              14 ISOLATION           .36        .10
11 N+/P+ BURIED  1 PR1-DMP1    7 N+/P+ JUNCTI     172.20      43.96
                 2 PR2-DMP2    9 N+/P+ JUNCTI      22.72       4.23
                 3 PR3-DMP3    9 N+/P+ JUNCTI       3.54       1.08
                              20 N+/P+ JUNCTI       1.77        .65

INVENTORY IS BUILDING FOR THESE WORKSTATIONS AT THE FOLLOWING
PROCESS STEPS:
                                                          INCREASED
   WORKSTATION         PROCESS         PROCESS-STEP       INVENTORY
    #   NAME          #  NAME          #    NAME            (LOTS)
   --------------    ----------      ------------       ---------
    3 ISOLATION       3 PR3-DMP3     14 ISOLATION               6
   11 N+/P+ BURIED    1 PR1-DMP1      7 N+/P+ JUNCTI           57
                      3 PR3-DMP3      9 N+/P+ JUNCTI            2
    3 ISOLATION       3 PR3-DMP3     14 ISOLATION               4

THESE WORKSTATIONS MAY BE CANDIDATES FOR FACTORY BOTTLENECKS.
POSSIBLE REMEDIAL ACTIONS ARE:

        (1) ADD MACHINES AT THE BOTTLENECK WORKSTATION(S)
        (2) SLOW LOT STARTS
        (3) ALTER PROCESS PRIORITIES
        (4) MINIMIZE SETUP TIMES
        (5) RESCHEDULE PREVENTATIVE MAINTENANCE
        (6) ADD PERSONNEL
```

7.6.3 Fab Capacity Analysis

The fab-capacity analysis (Table 7.12) indicates whether the fab has been asked to operate beyond its capacity. The report indicates which process flows are building inventory, at their assigned lot-start rates, and then it gives feasible start rates that are within the fab's capacity.

Other remedial actions are also indicated. Specifically, the number of additional machines required for the assigned start rate. Finally, possible additional bottlenecks are indicated.

Table 7.12

FAB CAPACITY:

FACTORY CAPACITY HAS BEEN EXCEEDED. INVENTORY IS BUILDING FOR THE FOLLOWING PROCESS(ES) INDICATING LOT-START RATE IS TOO HIGH:

#	NAME	WIP BUILDUP	RATE	LOT-START RATE
1	PR1-DMP1	46 LOTS/ 200. HRS	.230	2 LOTS/ 6.3 HRS
2	PR2-DMP2	13 LOTS/ 200. HRS	.065	1 LOT/ 5.9 HRS
3	PR3-DMP3	5 LOTS/ 200. HRS	.025	1 LOT/ 3.1 HRS

START RATES THAT MAY BE STABLY MAINTAINED WITHIN THE FAB'S CAPACITY (i.e., CURRENT EQUIPMENT SET) ARE:

#	NAME	LOT-START RATE
1	PR1-DMP1	1 LOT/ 2.9 HRS
2	PR2-DMP2	1 LOT/ 5.2 HRS
3	PR3-DMP3	1 LOT/ 3.0 HRS

IF MORE PRODUCTION IS NEEDED, EQUIPMENT MUST BE ADDED. THE CURRENT BOTTLENECK IS:

BOTTLENECK WKST #	NAME	ACTIVITY RATIO	CURRENT MACHINES
11	N+/P+ BURIED	1.0000	5

THE ADDITION OF 2 MACHINES SHOULD BE SUFFICIENT TO DEBOTTLENECK THIS WORKSTATION.

CAUTION: SECONDARY BOTTLENECK(S) MAY APPEAR AFTER DEBOTTLENECKING! STATIC ANALYSIS INDICATES THE FOLLOWING WORKSTATION MAY ALSO BE UNDERSIZED:

WORKSTATION #	NAME	CURRENT MACHINES	STATIC ESTIMATE
17	1ST METAL	4	3.9
18	2ND METAL	4	4.1

ALSO, INITIAL INVENTORY (WIP) MAY BE TOO HIGH FOR STABLE, LONG-TERM OPERATION. INVENTORY SHEDDING MAY BE WORTH EXPLORING.

7.7 Workstation Dynamics

Although there are masses of fab performance measures to be obtained from a dynamic factory model, some of the more useful are concerned with workstation dynamics. Statistics on workstation utilization explain how the various members of the equipment set actually spend their time. Batching statistics indicate how effectively equipment is used when it is in operation.

7.7.1 Workstation Utilization

The dynamic factory model provides a clear breakdown of how the equipment is actually being used (Table 7.13). The table list lots processed and the fraction of time the equipment spent processing, in setup, unavailable (down), or idle.

Table 7.13

********** RESOURCE UTILIZATION REPORT **********

WORKSTATION #	NAME	LOTS PROCESSED	--------RESOURCE USAGE RATIOS-------- PROCESS	SETUP	DOWN	IDLE	ACTIVITY
1	SUBSTRATE	171	.4281	.0845	.1013	.3861	.6139
2	WELL FORMATI	172	.4749	.0419	.0125	.4708	.5292
3	ISOLATION	245	.7728	.0880	.0660	.0732	.9268
4	ELECTRON EJE	141	.1126	.0468	.0603	.7802	.2198
5	TUNNEL OXIDE	138	.6257	.0454	.0780	.2509	.7491
6	GATE OXIDE	247	.4596	.1509	.0150	.3745	.6255
7	CHANNEL DOPI	247	.5347	.0729	.0164	.3760	.6240
8	GATE ELECTRO	255	.4506	.0705	.0700	.4088	.5912
9	INTERPOLY OX	45	.3455	.0412	.0600	.5533	.4467
10	DUAL LEVEL G	46	.2422	.0382	.0400	.6795	.3205
11	N+/P+ BURIED	199	.2633	.0942	.6425	.0000	1.0000
12	P+ BURIED LA	14	.0257	.0138	.0495	.9111	.0890
13	POLY LOAD	15	.0873	.0097	.1197	.7833	.2167
14	POLY WIRING	15	.0255	.0099	.1200	.8446	.1554
15	CONTACT	201	.4159	.0590	.0913	.4338	.5662
16	REFLOW	200	.3939	.0151	.0328	.5581	.4419
17	1ST METAL	199	.3568	.0774	.0500	.5158	.4842
18	2ND METAL	259	.4721	.0691	.0500	.4088	.5912
19	PASSIVATION	124	.2819	.0501	.0200	.6480	.3520
20	TEST	65	.0028	.0025	.0100	.9847	.0153

As we mentioned earlier, just because equipment is available does not necessarily mean the fab has a need for all that availability. A usage breakdown will provide an operations manager with a clearer picture of where: maintenance dollars should be spent (high process and setup), more operators may be needed (high setup), longer process times may be used for better yield (low process), etc.

The final parameter, total activity, is a measure of machine utilization and excess capacity. Consequently, as discussed above, experience has shown that the sum of the process, setup, and down ratios (i.e., one minus the idle ratio) is a good indicator of possible bottlenecks.

7.7.2 Workstation Batching Statistics

Workstation batching statistics do two things: they present a clear picture of a workstation's throughput performance (over all processes) by batches, lots, and parts; and they determine how completely the equipment is being used while it is processing. (See Table 7.14)

Table 7.14

********** WORKSTATION SUMMARY AND BATCHING STATISTICS **********

WORKSTATION #	NAME	LOTS_IN MACHINE	INVENTORY (lots)	-----PROCESSED------ BATCHES	LOTS	PARTS	LOAD FACTOR ratio
1	SUBSTRATE	1	0	137	171	4743	.2308
2	WELL FORMATI	2	0	145	172	5018	.4614
3	ISOLATION	12	4	188	245	7959	.4234
4	ELECTRON EJE	1	0	124	141	4625	.4973
5	TUNNEL OXIDE	6	0	120	138	4700	.5222
6	GATE OXIDE	2	0	195	247	8019	.5483
7	CHANNEL DOPI	5	0	197	247	8219	.5563
8	GATE ELECTRO	1	0	206	255	8546	.8297
9	INTERPOLY OX	1	0	42	45	1575	.5000
10	DUAL LEVEL G	1	0	43	46	1650	.7674
11	N+/P+ BURIED	7	55	148	199	7290	.6568
12	P+ BURIED LA	0	0	14	14	694	.9914
13	POLY LOAD	0	0	15	15	742	.6596
14	POLY WIRING	2	0	17	15	842	.9906
15	CONTACT	2	0	166	201	7296	.8790
16	REFLOW	3	0	165	200	7321	.8874
17	1ST METAL	3	0	165	199	7346	.8904
18	2ND METAL	1	0	211	259	9521	.9025
19	PASSIVATION	3	0	105	124	4921	.9373
20	TEST	0	0	57	65	2662	.9340
21	OPS POOL 1	0	0	59	76	2575	.2910
22	OPS POOL 2	0	0	67	67	2375	.2363
23	OPS POOL 3	0	0	43	60	2200	.3411

This latter statistic, the *load-factor ratio*, is the averaged comparison of lots actually processed by the machines at a work center, as opposed to the total lots which could have been handled by the same number of batches if the full machine capacities had been used. Thus, load factor is a measure of effective loading. A load factor of 50%, for example, indicates that, on average, the machine is half full during processing.

For reasons of depreciation and wafer-cost allocation, it is generally desireable to have high load factors, i.e., more effective loading of expensive equipment. Large-batch equipment tends, however, to be an exception. Having lots wait in order to fill large-batch equipment can lead to 'chugging' behavior in the fab, which in turn leads to less effective use of the bottleneck or constraining workstation. Fab performance will generally suffer, sometimes dramatically, with ineffective utilization of the bottleneck. (See Section 6.3.)

7.8 Queueing Dynamics

Since it is typical that wafers spend two-thirds to three-quarters of their cycle time sitting in queues, waiting, some of the more valuable model results are answers to the questions of 'Where?', 'When?', and 'Why?' Information, such as that shown in Table 7.15, provides insight on: how the workstations accommodate the process sequence for each flow; which steps in which flows are causing problems; where more racks may be needed; where more equipment may be needed; whether inventory is a fab-wide or workstation-specific problem, etc.

Table 7.15

********** QUEUE LENGTHS AND WAITING TIMES **********
***** PROCESS # 1: PR1-DMP1 *****

PROCESS STEP #	NAME	WORKSTATION #	NAME	AVG.QUEUE LENGTH (LOTS)	AVG.WAIT TIME (IN QUEUE)
1	SUBSTRATE PREP	1	SUBSTRATE	.02	.34
2	WELL FORMATION	2	WELL FORMATI	.00	.20
3	ISOLATION	3	ISOLATION	.56	4.24
4	GATE OXIDE	6	GATE OXIDE	.04	.37
5	CHANNEL DOPING	7	CHANNEL DOPI	.04	.23
6	GATE ELECTRODE	8	GATE ELECTRO	.37	1.17
7	N+/P+ JUNCTION	11	N+/P+ JUNCTI	43.96	172.20
8	P+ BURIED LAY	12	P+ BURIED LA	.08	.65
9	POLY LOAD	13	POLY LOAD	.10	.78
10	POLY WIRING	14	POLY WIRING	.23	1.66
11	CONTACT	15	CONTACT	.03	.24
12	REFLOW	16	REFLOW	.01	.09
13	1ST METAL	17	1ST METAL	.01	.12
14	2ND METAL	18	2ND METAL	.01	.04
15	PASSIVATION	19	PASSIVATION	.07	.75
16	TEST	20	TEST	.04	.21

*** SUM OF AVERAGE WAIT TIMES = 183.29 HRS ***

Also, this information could be invaluable to a fab that has decided to reduce or shed WIP in an effort to improve cycle times.

Obviously, this information provides a quick perusal for identifying potential bottlenecks. But, however strong, this evidence alone is not sufficient to establish a bottleneck's existence. As done by the expert system, one must also review the equipment-utilization reports, as well as the complete in-line inventory picture.

7.9 Lot Dynamics

The final pieces of the puzzle that are needed for analysis of fab performance are the lot dynamics. Lot dynamics may be viewed in two different way: first from a process sequence perspective, and secondly from a workstation perspective. The former provides useful information on how lots are staged throughout each process sequence at the end of the simulation; the latter indicates how lots are staged at equipment at the end of the simulation.

7.9.1 By Process

The lots left in the queues or remaining 'in' equipment at the end of the simulation are a natural output of a dynamic model. After all, the model has been marching these lots through the process sequences, and into and out of specific equipment throughout the simulation. When the simulation concludes, the music stops, and lots are where they are. Thus, for each process flow, it is possible to count the work-in-progress (WIP) waiting in queue at each workstation or actually loaded in machines.

As may be seen from Table 7.16, such a final-inventory status report may be useful for spotting potential bottlenecks. But the table also illustrates one of the inherent flaws with bottleneck-removal efforts. If bottlenecks are identified and removed based solely on fab operating data, only the first one in the process sequence is likely to be found.

In the example in Table 7.16, lots are progressing smoothly through the flow (as evidenced by the throughput) until Step #7 at Workstation #11. At this step, throughput drops from roughly 70 to 14 lots. The remaining lots have joined WIP (at the work center) and are simply waiting. Downstream steps can not process more lots than are getting through this choking step. It is impossible, therefore, to ascertain whether there are more bottlenecks until something is done about the one that has been identified.

Table 7.16

***** FINAL LOT STATUS REPORT BY PROCESS *****
***** PROCESS # 1: PR1-DMP1 *****

PROCESS STEP #	NAME	WORKSTATION #	NAME	LOTS THROUGHPUT	LOTS IN MACHINES	IN-LINE INVENTORY
1	SUBSTRATE PREP	1	SUBSTRATE	67	0	0
2	WELL FORMATION	2	WELL FORMATI	67	2	0
3	ISOLATION	3	ISOLATION	67	2	2
4	GATE OXIDE	6	GATE OXIDE	68	0	0
5	CHANNEL DOPING	7	CHANNEL DOPI	68	2	0
6	GATE ELECTRODE	8	GATE ELECTRO	71	0	0
7	N+/P+ JUNCTION	11	N+/P+ JUNCTI	14	4	55
8	P+ BURIED LAY	12	P+ BURIED LA	14	0	0
9	POLY LOAD	13	POLY LOAD	15	0	0
10	POLY WIRING	14	POLY WIRING	15	2	0
11	CONTACT	15	CONTACT	15	0	0
12	REFLOW	16	REFLOW	15	0	0
13	1ST METAL	17	1ST METAL	15	0	0
14	2ND METAL	18	2ND METAL	16	0	0
15	PASSIVATION	19	PASSIVATION	17	0	0
16	TEST	20	TEST	18	0	0

*** SUM OF LOTS IN MACHINES = 12 ***

7.9.2 By Workstation

The lot dynamics presented in Table 7.17 give a slightly different perspective than those presented in Table 7.16. Above, the final inventory picture was broken down by process sequence; here, all steps at a given workstation (for a given flow) are lumped together to give the total final WIP at each workstation. Again, this information is useful in indicating equipment sets that may be causing first-order bottlenecks.

Table 7.17

********** LOTS PROCESSED BY WORKSTATION **********
***** PROCESS # 1: PR1-DMP1 *****

WORKSTATION #	NAME	TOTAL LOTS PROCESSED	IN-LINE INVENTORY (not workstation)
1	SUBSTRATE	67	0
2	WELL FORMATION	67	0
3	ISOLATION	67	2
6	GATE OXIDE	68	0
7	CHANNEL DOPING	68	0
8	GATE ELECTRODE	71	0
11	N+/P+ BURIED	14	55
12	P+ BURIED LAY	14	0
13	POLY LOAD	15	0
14	POLY WIRING	15	0
15	CONTACT	15	0
16	REFLOW	15	0
17	1ST METAL	15	0
18	2ND METAL	16	0
19	PASSIVATION	17	0
20	TEST	18	0

Using either or both of these two final WIP reports, in conjunction with information on the quantity and staging of initial inventory, provides intricate and abundant detail on factory behavior.

7.10 Lot Completions

The lot dynamics above provide the pertinent data on lots that are still in the factory awaiting additional processing, but what of those that have completed?

A factory model should, by definition, be able to provide a complete and detailed list of: the specific lots that have finished, the flow to which each lot belongs, the number of parts (wafers) in the lot, and the time the lot completes. (Table 7.18.)

Such a lot-completion schedule may be useful for verifying the feasibility of meeting ship dates, based on actual factory capacity and performance. If the projected ship dates are not being met, then some 'what-if' simulations could be performed (changing process priorities, rescheduling maintenance, etc.) to explore some other operating options to meet those ship dates.

Table 7.18

```
------------ LOT COMPLETION SCHEDULE ------------

THE FOLLOWING TABLE GIVES THE LOT COMPLETIONS, OCCURRING BY
SHIFT, FROM THE CURRENT WIP AND FROM STARTING LOTS

  LOT I.D.                 PROCESS        # OF     COMPLETION
  #     NAME             #     NAME      PARTS       TIME
----------------        ---------        -----    ----------
               ********  SHIFT    1 ********

 77 lot   77             1 PR1-DMP1        50            .04
 74 lot   74             2 PR2-DMP2        50           3.49
 75 lot   75             2 PR2-DMP2        50           3.49
 12 lot   12             2 PR2-DMP2        50           3.51
 73 lot   73             1 PR1-DMP1        48           6.94
  7 lot    7             1 PR1-DMP1        48           6.94
 71 lot   71             2 PR2-DMP2        50           7.77

               ********  SHIFT    2 ********

 69 lot   69             2 PR2-DMP2        50          10.39
 66 lot   66             1 PR1-DMP1        48          13.41
 65 lot   65             1 PR1-DMP1        48          14.47
 62 lot   62             2 PR2-DMP2        50          15.48
```

Before leaving the topic of factory modeling, we would like to make one final observation. From thousands of years of simulated experience, as well as observing many years of real-world data, we have often been struck by the natural, wide variability in the number of lots completing in a shift or in a day. This variability is perhaps another indication of the chaotic behavior of this complex dynamic system otherwise known as a wafer fab.

7.11 Synthetic Experience

Following Hamming [7.18], the purpose of performance analysis is insight. Insight, however, does not come from unorganized computational experiments. Sensitivity analysis [7.19] and signature analysis [7.20] are useful in planning performance analysis projects.

At its most effective, performance analysis provides synthetic experience and knowledge. Detailed specific rules result that are independent of the numerical details resulting from the simulations.

The calibration-validation exercises discussed earlier provide excellent opportunities for developing structured knowledge of wafer fab operations. Since these projects deal with data from real factories, the studies establish a strong baseline of reality.

The highly-focused performance studies of Chapters 5 and 6 are powerful examples of the development of synthetic experience. The end results from the analyses are more than just numbers; they are rules of thumb, heuristics, and calculation procedures that can be implemented without a detailed simulation model. However, these results were developed with the insight derived from a large knowledge base acquired by extensive simulations, and validated against real-world experience. Furthermore, the heuristics, once proposed, could be checked by detailed simulation studies.

Carefully designed plots called signatures are extremely useful in developing synthetic experience. A *signature* is a plot specifically organized to illuminate an aspect of fab performance. Figs 5.6(c), 5.7(c), 5.8(c), and 5.9(c) of Chapter 5 are signatures. They show in detail the rate of inventory relaxation for a fixed throughput. A quick glance at the signature provides an indication of whether the fab has any burst capacity at that throughput rate. A well-designed signature provides a specific piece of information at a glance.

7.12 Sensitivity Analysis

Sensitivity analysis is a powerful, general method for performance analysis [7.19] that has been applied across a wide range of applications. Essentially, *sensitivity analysis* is a generalized partial derivative in that, the rate of change of a performance measure, like throughput, is determined relative to the rate of change of a control variable, like availability of a workstation.

Thus, if we let $S(o_J,p_K)$ be the sensitivity of output variable o_J relative to parameter p_K, then,

$$S(o_J,p_K) = (o_J' - o_J)/(p_K' - p_K) = \triangle o_J/p_K$$

The 'delta' notation as usual indicates we are looking at a change. In practice, the sensitivity is evaluated by running the model for a base case giving the output variables o_J. A perturbation $\triangle p_K$ is chosen and a second simulation is run for the perturbed set of parameters, where the new p_K is given by

$$p_K' = p_K + \triangle p_K$$

In simple, direct sensitivity analysis, only one parameter is changed for perturbed simulation runs. More general methods allow multiple perturbations.

A sensitivity analysis looks at changes in factory behavior as factory model and operations parameters change. Thus, the sensitivity measures **how much** the factory changes. The greater the sensitivity, the more important is the parameter to factory operation.

7.12.1 Wafer Fab Sensitivity

Table 7.19 lists a number of possible sensitivity studies.

TABLE 7.19: EXAMPLES OF SENESITIVUTY ANALYSES
• Throughput vs Availability
• Cycle Time vs Inventory
• Dynamic Capacity vs Process Times
• Dynamic Capacity vs Setup Times
• Bottlenecks vs Product Mix
• Bottlenecks vs Process Priority

One of the more interesting aspects of wafer fab performance is that it is characterized by regions of flat sensitivity (low sensitivity) followed by regions of high sensitivity. This behavior is illustrated by the data in Table 7.20, where the sensitivity of throughput and cycle time to an availability parameter at a constraining workstation is studied. Throughput shows no sensitivity for the first four percent decline in availability. However, it has a high sensitivity to the last two percent decline. Cycle time, on the other hand, is more sensitive to every change.

TABLE 7.20: SENSITIVITY TO EQUIPMENT AVAILABILITY

Availability Case:	Throughput (wafers)	Cycle Time (hr)
Base Case	295	59.2
- 4% on WKST #3	295	61.6
- 5% on WKST #3	294	65.3
- 7% on WKST #3	249	81.1

Widget factory with constraining workstation at WKST #3.

When studying availability, increases in sensitivity are indicative of an impending shift in bottleneck status. Similarly, flat sensitivity to availability indicates that the workstation is not a bottleneck or constraining workstation. Clearly, the sensitivity of availability at a bottleneck workstation will be quite large.

7.12.2 Sensitivity and Fab Sizing

We consider now the capital costs of a full set of equipment to implement a given process flow for a required throughput. We will use N different types of equipment, and each type j costs C_J. There are K_J of each type required. Specifying the number of each type of equipment is the fab design problem. Thus, the total capital cost C is given by:

$$C = \sum K_J C_J$$

This simple function involving multiplications and sums becomes more complicated when we recognize that the K_J's are complex functions of a large set of parameters (see Table 7.21), and the parameters are uncertain.

TABLE 7.21: EQUIPMENT SIZING PROBLEM

The number of required pieces of equipment, K_J, is a function of:

P_I : process times performed @ equipment

S_I : setup times performed @ equipment

L_J : load size of equipment

A_k : availability parameters of equipment

T : throughput of equipment defined by K

G_k : interaction parameters (due to discreteness)

Thus,

$K_J = K_J(P_I, S_I, L_J, A_k, T, G_k)$

If the design is to be robust over a range of uncertainty [7.21], then a large set of sensitivities must be considered (see Table 7.22).

TABLE 7.22: SENSITIVITY OF EQUIPMENT SIZING

The sensitivity of equipment sizing, K_J, to various process or operations measures:

T: $\frac{\Delta K_J}{\Delta T}$

L_J: $\frac{\Delta K_J}{\Delta L_J}$

P_I: $\frac{\Delta K_J}{\Delta P_I}$

A_1: $\frac{\Delta K_J}{\Delta A_1}$

S_I: $\frac{\Delta K_J}{\Delta S_I}$

A_2: $\frac{\Delta K_J}{\Delta A_2}$

G_1: $\frac{\Delta K_J}{\Delta G_1}$

A comprehensive sensitivity analysis will involve centuries worth of simulations to gain synthetic experience. But once the simulations are made, and the experience understood, the engineer or analyst can literally write the book on the fab.

7.13 Modeling Methods in General

The purpose of fab performance analysis is to learn something about wafer fabrication dynamics and hopefully to operate the fabs better. We have presented a variety of tools for performance analysis based on the fab graph. The fab graph was chosen because it represents the essential structure of these factories and their dynamics.

There are a variety of other methods available for analyzing manufacturing operations, including general purpose methods such as spreadsheets and simulation languages, mathematical theories [7.23]-[7.25], and empirical methods.

7.13.1 General Purpose Methods

General purpose methods for fab modeling include computer spreadsheets and simulation languages. While mathematical theories may take years to learn, general purpose methods are usually quick and easy.

Spreadsheets are useful for getting numerical results fast for relatively simple models. Spreadsheets have been used extensively for fab capacity planning based on static models. Some of the shortcomings of static models, however, have been discussed elsewhere (see Chapter 2).

Spreadsheet models suffer two major drawbacks. First, with these models, it is difficult and impractical to include the full detail of a wafer fab in a spreadsheet. Secondly, spreadsheets simply do not support production simulations.

Simulation languages [7.14, 7.15] also allow rapid implementation of models. These are more powerful tools than spreadsheets in that they naturally permit more complete modeling structures. A *simulation language* is a special purpose modeling language built upon a programming language like Fortran or C. Examples of simulation languages include Siman, GPSS, and Autosim. In principle, the fab-graph dynamics model described above could be implemented using a simulation language.

There are several drawbacks, however, associated with using simulation languages. First, the overhead associated with using a simulation language slows down production simulations. Secondly, model maintenance, model stability, and code maintenance are all difficult issues when a simulation language is used as the basis for a production simulation.

In addition, simulation languages are not content free. Each simulation language contains built-in, preferred modeling concepts, which encourage a tendency to make the problem fit the concepts of the simulation language, whether such concepts apply to wafer fabrication or not. When the concepts of the simulation are not a good fit to the problem, it is very difficult to achieve successful performance analysis projects.

7.13.2 Mathematical Theories

Mathematical theories that can be applied to performance analysis of wafer fabrication include queueing theory [7.22], petri nets [7.23], and differential equations [7.25]. Gershwin [7.24] and Desrochers [7.25] present several other methods. Such methods are useful in general and very useful in specific applications.

However, they have some serious drawbacks as tools for intensive application in fab performance analysis. Mathematical theories can be difficult to learn. They often require an expert to apply them appropriately, and they can be difficult to use even as part of a production simulation.

Learning a mathematical theory to use as a modeling tool can require considerable effort. In the case of queueing theory, it could take perhaps six months or more, even with appropriate background. Fab operations personnel, making this investment and not getting useful results, can feel a great deal of frustration.

On the other hand, an expert can do surprising things with a powerful mathematical theory; even represent a wafer fab. After spending several years getting facility with a theory like petri nets or queueing networks, expert practitioners can transmogrify a wafer fab so that to them it looks like a queueing network. They then can calculate the properties of this fab-like queueing network, and, most miraculously of all, they can transform the results back to a real world situation.

However, if the mathematical theory is incorporated into a production simulation, things do not go so well. The users are not experts in the theory, and they have difficulty making the fab look like a petri net or queueing network. They have even more difficulty relating simulation results to fab production data.

7.13.3 Empirical Methods

Empirical methods are extremely powerful when coupled with an appropriate model. For example, we have already discussed the use of fab operations data in calibration and validation with fab graph dynamics.

7.14 Cluster Tool Modeling

So far, this chapter has discussed factory modeling as it applies to the entire wafer fab. But the same principles apply when we model some portion of the system.

Integrated processing equipment, also known as modular equipment, cluster tools, or production integrated processing equipment (PIPEs) are machines capable of performing multiple, independent processing steps without leaving a controlled environment. By performing sequences of processing steps in a single machine, the equipment is in effect a mini factory. It exhibits the same complexities and dynamics of a full-blown factory. Modeling these marvels of modern engineering, however, is more difficult than doing the entire factory [7.26]-[7.29].

The key issues in the detailed modeling of PIPEs are how to deal with buffer dynamics and lockup. The few detailed-simulation models that have been used successfully for PIPE performance analysis have followed the same methodologies used in full factory models: representing, in detail, the complex interactions of the fundamental entities. In the case of PIPEs, the fundamental entities are wafers and the interconnected modules for processing, transport, and storage. Otherwise, the analogies with the full factory follow almost one-on-one.

Wafers enter the PIPE and then attempt to follow a process recipe. The process steps in the recipe are, however, only a fraction of the total operations necessary to complete the wafer. Like the factory, there are also transports, waits, and simultaneous processing. Hence, one requirement of a PIPE model is that it account for all operations steps.

Performance analysis of a PIPE predicts the throughput and cycle time of the machine as a function of equipment design, process recipe, and operating conditions. The process recipe that is being performed can strongly affect the cycle time of a cassette of wafers.

Integrated processing equipment has complex, and often poorly understood, internal material movement. For example, a four-step recipe may require nearly 30 operational steps, each having a time element associated with it. Thus, the cycle time for a cassette may be over twice the cycle time expected from processing times alone.

In extending factory models to integrated equipment, it is necessary to extend the theoretical understanding of the buffer dynamics associated with finite storage in production systems. When finite storage capacity at a resource is considered, qualitatively different phenomena arise. Buffer dynamics often becomes dominant, and lockup may occur. *Lockup* is the phenomena where the system freezes (or gets lost) because all acceptable 'next events' are physically blocked or impossible.

This phenomena is not only observed in software, but in the machines as well.

In order to design effective PIPEs, engineers are having to develop an understanding of buffer dynamics and of the competition for processing resources. In one PIPE design, for example, wafers may spend time in the process chamber waiting for the next processing step. In a second design, an internal buffer may be provided, and wafers may move to this internal buffer if no processing resource is available. In a third design, multiple process chambers may be capable of performing the same process step. This last design may be used in conjunction with either waiting in the chamber or in the internal buffer.

Further complexity in analysis of PIPEs results if the path of the wafer through the machine is dependent on the dynamic condition of the machine. For example, in the design with an internal buffer, a wafer may spend additional time in the buffer if the next processing chamber is unavailable. However, use of the buffer will also acquire additional transport time in addition to the wait time. Thus, the additional transport times in the PIPE can be critical in establishing cycle-time performance.

The impetus for introducing PIPEs into wafer fabrication has been the substantial yield improvements that are expected. Because the wafers never leave a controlled environment, contamination and unnecessary human handling are reduced. The complexity of this integrated equipment, however, introduces new problems in equipment performance analysis. Chapter 8 covers the topic of integrated equipment in some detail.

Further Reading and Notes

1. Graph theory is quite useful in factory modeling and performance analysis. References [7.3] and [7.12] provide an introduction to a large literature.

2. Practical aspects of stochastic simulation are covered in [7.15] and [7.16].

3. The ACHILLES software package [7.8] provides a production simulation of wafer fabs in terms of fab graph dynamics.

4. The THOR software package [7.27] provides a production simulation of several cluster tool designs. Cluster tool dynamics are discussed in detail in Chapter 8.

5. Systems thinking and systems analysis provide a sound foundation for performance analysis. Weinberg [7.32] is a highly recommended introduction. The Boulding quotation of Chapter 2 is from Weinberg. See also [7.33].

6. Simon's views on the sciences of the artificial [7.2] provides an interesting perspective for viewing manufacturing. More on computation and synthetic experience is given in Pagels [7.31].

7. Expert systems are covered in [7.34, 7.35].

References

[7.1] Karlin, S., **Mathematical Methods and Theory in Games, Programming, and Economics**, Dover, New York, 1959, 1992.

[7.2] Simon, Herbert A., **The Sciences of the Artificial**, (2ed.), MIT Press, Cambridge, MA, 1981.

[7.3] Beineke, L. W., and R.J. Wilson, (eds.), **Selected Topics in Graph Theory**, Academic Press, 1978.

[7.4] Atherton, R. W., A. H. Miller, and J. E. Dayhoff, "Operations Management and Analysis in the Management of Electronic Testing," **1983 International Test Conference Proceedings**, IEEE Computer Society Press, Los Angeles, pp. 418-426, 1983.

[7.5] Dayhoff, J. E., and R. W. Atherton, "Simulation of VLSI Manufacturing Areas," **VLSI Design**, 84, Dec. 1984.

[7.6] R. W. Atherton and J. E. Dayhoff, "An Introduction to Fab Graph Structures," ECS Abstracts, vol. 85-1, Princeton, N.J., pp. 146-147, Princeton, N.J., 1985.

[7.7] Dayhoff, J. E. and R. W. Atherton, "A Model for Wafer Fabrication Dynamics in Integrated Circuit Manufacturing," **IEEE Trans. Systems, Man, Cybernetics**, vol. SMC-17, 91-100, 1987.

[7.8] Atherton, Linda F., and R. W. Atherton, **ACHILLES User's Manual**, In-Motion Technology, Los Altos, CA, 1987-1995.

[7.9] Atherton, R. W., **Real World Modeling and Control Process**, U.S. Patent, No. 4,796,194, 1989.

[7.10] Atherton, R. W., L.F. Atherton, M.A. Pool, "Detailed Simulation for Semiconductor Manufacturing," **1990 Winter Simulation Conference Proceedings**, Society for Computer Simulation, December, 1990.

[7.11] Atherton, Linda, and R. W. Atherton, **Factory Dynamics of Wafer Fabs and Cluster Tools**, course notes, In-Motion Technology, Los Altos, California, 1990-1994.

[7.12] Deo, N., **Graph Theory with Applications to Engineering and Applied Science**, Prentice-Hall, Englewood Cliffs, NJ, 1974.

[7.13] Horowitz, Ellis, and S. Sahni, **Fundamentals of Data Structures**, Computer Science Press, Rockville, MD, 1981.

[7.14] Fishman, George S., **Concepts and Methods in Discrete Event Simulation**, Wiley, New York, 1973.

[7.15] Bratley, Paul, B.L. Fox, L. E. Schrage, **A Guide to Simulation**, Springer-Verlag, New York, 1983.

[7.16] Bury, Karl V., **Statistical Models in Applied Science**, Wiley, New York, 1975.

[7.17] Aho, A.V., J.E. Hopcroft, and J.D. Ullman, **The Design and Analysis of Computer Algorithms**, Addison-Wesley, Menlo Park, CA, 1974.

[7.18] Hamming, R.W., **Numerical Methods for Scientists and Engineers**, McGraw-Hill, New York, 1973.

[7.19] Frank, Paul M., **Introduction to System Sensitivity Theory**, Academic Press, New York, 1978.

[7.20] Atherton, R.W., and J. E. Dayhoff, "Signature Analysis: Simulation of Inventory, Cycle Time, and Throughput Trade-Offs in Wafer Fabrication," **IEEE Trans. Components, Hybrids, Manufacturing Technology**, vol. CHMT-9, 498-507, 1986.

[7.21] Tribus, Myron, **Rational Descriptions Decisions and Designs**, Pergamon, New York, 1969.

[7.22] Buzacott, John A., and J. George Shanthikumar, **Stochastic Models of Manufacturing Systems**, Prentice-Hall, Englewood Cliffs, 1993.

[7.23] Desrochers, A. A. (ed.), **Modeling and Control of Automated Manufacturing Systems**, IEEE Computer Society Press, Washington, D.C., 1990.

[7.24] Gerschwin, S.B., **Manufacturing Systems Engineering**, Prentice-Hall, Englewood Cliffs, NJ, 1994.

[7.25] Jacoby, S.L.S., **Mathematical Modeling with Computers**, Prentice-Hall, N.J., 1980.

[7.26] Atherton, R. W., **Real World Modeling and Control Process for Integrated Manufacturing Equipment**, U.S. Patent, No. 5,305,221, 1994.

[7.27] Atherton, Linda F., **THOR User's Manual**, In-Motion Technology, Los Altos, CA, 1989-1993.

[7.28] Atherton, R. W., F. T. Turner, L. F. Atherton, M.A . Pool, "Performance Analysis of Multi-Process Semiconductor Manufacturing Equipment", **Proceedings of Advanced Semiconductor Manufacturing Conference**, September, 1990.

[7.29] Pool, M.A. and L.F. Atherton, "Cluster Tools Need Performance Analysis", **Semiconductor International**, September, 1990, pp. 98-99.

[7.31] Pagels, H., **The Dreams of Reason**, Simon and Schuster, New York, 1988.

[7.32] Weinberg, G. M., **An Introduction to General Systems Thinking**, Wiley, New York, 1975.

[7.33] Warfield, J.N., **Societal Systems: Planning, Policy, and Complexity**, Wiley, New York, 1976.

[7.34] Martin, James, and S. Oxman, **Building Expert Systems: A Tutorial**, Prentice-Hall, Englewood Cliffs, N.J., 1988.

[7.35] Rich, Elaine, **Artificial Intelligence**, McGraw-Hill, New York, 1983.

Exercises

1. Design a data table (data structure) for wafer fab equipment.

2. Design a data table for lots.

3. Lots with size L are to be processed at a set of workstations. The load size of the equipment sets allows three cases: two batches are required to process the lot, one batch handles exactly a lot, or a batch may hold two lots. Write the *If-Then* rules to determine the appropriate case at an arbitrary workstation.

4. Continuing problem three, for each of the three case, list the changes in the data tables for lots and for machines to specify the loading of lots into machines.

5. Write an expression for the cumulative wafer yield after N steps.

6. What must be the average wafer yield such that the cumulative yield after 400 steps is 0.98?

7. What parameters of queueing theory are difficult, if not impossible, to measure in the typical wafer fab?

8. What parameters of petri net theory are difficult to determine for a wafer fab?

9. After lengthy operations, tracking data for BiCMOS-1 contains sufficient information to make an informed identification of the most likely bottleneck (Wkst 8), as well as the next most heavily-used workstation (Wkst 29). Further, over this extended period of operation, equipment availability for the two workstations varied significantly, all the way from up-ratios of 1.0 to ratios of 0.65. When fab capacity is plotted against the availabilities of these two work centers, the following graph is obtained.

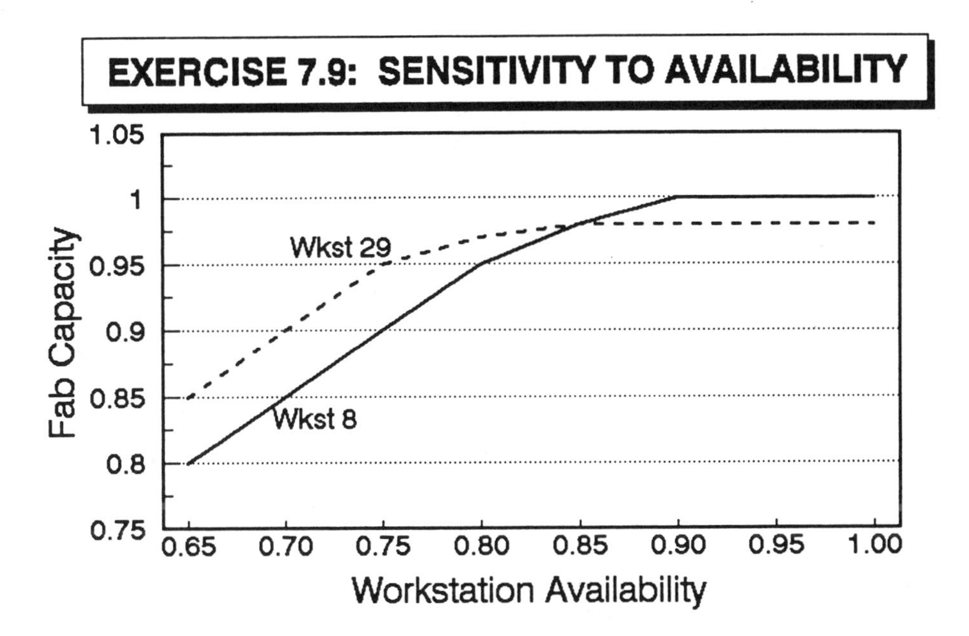

Discuss the meaning of these curves in terms of bottlenecks, bottleneck shifts, fab sensitivity to constraining workstation, and the relationship between bottleneck and fab capacity.

10. In Exercise #9, when should time and effort on improving availability cease to be a high priority for the two workstations?

The whole is greater than the sum of the parts.
-Mystical Saying

Chapter 8
Cluster Tools

8.0 Introduction

A quiet revolution is happening in the semiconductor industry in the design and use of manufacturing equipment, and, hence, in the design and operation of wafer fabs. Individual process steps that were traditionally performed in discrete equipment are being grouped, or clustered, into a single machine. The driving force behind this remarkable innovation has been the pursuit of good or improved process performance in the face of rapidly-advancing and increasingly-complex technologies. With each new generation, chip makers are cramming more transistors onto their chips, making the feature sizes smaller, and looking towards a future of larger wafers.

Mainly for process reasons, clustering of photolithography began taking shape in the early 1980's, and by about 1987 photo track systems were pretty much the industry standard. For much the same reasons (i.e., process achievability and yield), fabs have been slowly introducing integrated equipment for a variety of other processes including plasma etching, plasma deposition, physical-vapor deposition, reactive ion etching, rapid thermal processing, etc.

Often, the clustering of steps occurs because the process engineers make it happen. Whether their objective is the latest-and-greatest design, control of deadly particle deposition, wafer-to-wafer uniformity, die yield across large wafers, inter-step process control, or any of a number of other formidable tasks, the engineers find cluster tools to be versatile platforms for process development.

Despite the processing prowess delivered by cluster tools, their acceptance into wafer fabs has been slow and grudging. As is the case with most new, cutting-edge technology, these machines are exceptionally complex, and consequently struggle with reliability issues. Only rigorous, extended use in a manufacturing environment will lead to tool refinement and better operations. But since wafer fabs are in the business of making chips, their preferred mode of operation is with proven-reliable equipment.

Thus, as a result of the fundamental mismatch in fab operating philosophy and equipment development necessity, the emergence of cluster tools into the fab has been seriously resisted. In fact, each new acceptance has generally come about because, at some degree of chip complexity, the advanced component can not be downgraded to previous-generation tools. So integrated equipment has been making its way into wafer fabs, and in so doing has provided major opportunities for process innovation. But industry penetration has been slow.

This slow penetration is perhaps the primary reason that one of clustering's potentially-significant contributions has yet to be well recognized. Cluster tools offer the means for serious manufacturing simplification. By clustering several process steps into one manufacturing operation (Fig. 8.1), the number of manufacturing operations is reduced, thus offering important economic advantages over fabs using conventional equipment.

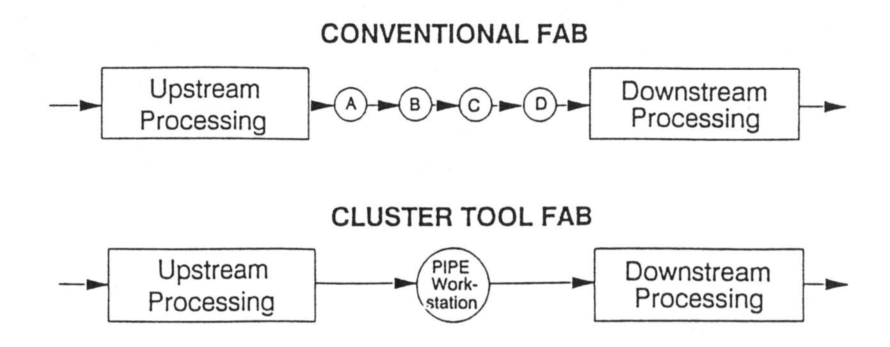

Figure 8.1: Simplification through Reduction of Manufacturing Operations

In addition to savings in cycle time, inventory, and labor costs, cluster tools provide a dramatic opportunity for yield improvement. Not simply for the obvious processing reasons of performing sequences of steps in one machine without exposure to atmosphere, but because fast cycle times mean more turns per year which translates directly into yield.

So, when we talk about cluster tools, we are talking about potentially-tremendous advantages in yield improvement, in manufacturing simplification, in cost reduction, and in process innovation. But, the key word here is *potentially*. Even were reliability no longer an issue, the equipment still has the major drawback of being much more complicated to configure and use than previous-generation, single-process equipment. For any given set of process steps, the clustering options are numerous, and someone must decide on which steps to cluster, how many different tools should do the job, and how to configure those tools for best process capability and cost effectiveness.

In making such clustering decisions, it is necessary to evaluate the various tools, and their configurations, to determine if they are capable of providing both the promised process results as well as 'good' cassette cycle times. For, it is the cycle time per cassette that determines the throughput of the machine, a number which must be known for workstation sizing and capacity planning.

To ascertain how a cluster tool will perform, from both a process and a manufacturing perspective, a performance analysis will need to be made to determine both overall and detailed tool behavior. The overall behavior of a tool will include its cycle time and throughput, as a function of equipment configuration, control rules, and process recipe. The detailed behavior will include the individual wafer routings, wait times in specific resources, and any special sequencing effects, again as a function of configuration, control rules, and process recipe.

However, the very complexity of integrated equipment makes performance analysis difficult, if not impossible, unless one has a fundamental understanding of how these machines work and how they fit into the fab. Gaining this fundamental understanding will be a necessary step for engineers and fab managers if they are to effectively design and operate the fabs of the future.

One final note before proceeding: these complex tools are known by a variety of names, such as cluster tools, multi-chamber or modular equipment, integrated equipment, etc. We have found the acronym PIPE for Production Integrated Processing Equipment to be useful in discussing these machines.

8.1 Mini-factories

The complexities inherent in PIPEs may be addressed directly by comparing them to wafer fabs. In performing multiple process steps without leaving a controlled environment, the PIPE is no longer a simple machine, but is in reality a mini-factory. In essence a segment of the factory has been reduced in size, but not in scope, and moved inside a single piece of gear (Fig. 8.2). Like the factory itself, more than just processing is taking place within its walls. Wafers are transported from place to place, loaded into chambers, processed, unloaded from chambers, and sometimes even left to wait. Thus, the cycle times of individual wafers, and consequently of lots and cassettes, have time components that include transport, wait, and processing.

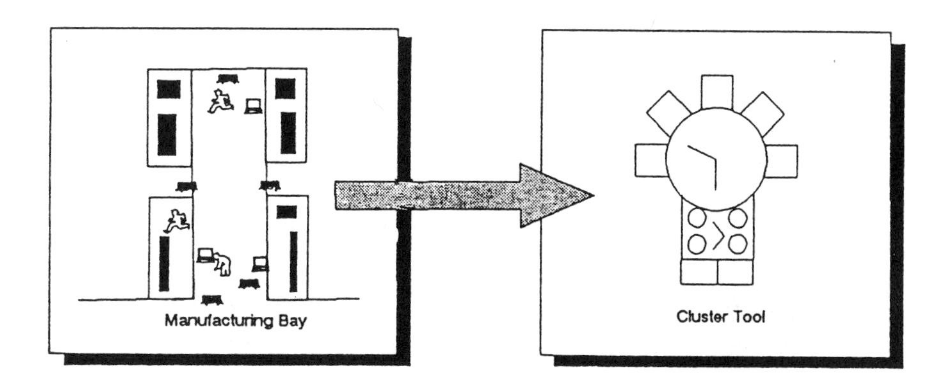

Figure 8.2: Cluster Tool as a Mini-Factory

Because the wafers are in a single piece of integrated equipment, a natural, but erroneous, assumption is to equate cycle time with processing time. Processing time, however, is only one element of the time the wafer spends in the machine, and even it is not straightforward. Since a cluster tool, in general, has multiple process modules, simultaneous processing is usually taking place. Hence, a summation of the process times, multiplied by the number of wafers in the cassette, has little relationship to the true cassette cycle time.

As a case in point, consider the clustering of a tungsten CVD followed by a tungsten etch. For processing reasons, the W-CVD takes considerably longer than the W-etch (150 versus 60 seconds, respectively, for single-wafer processing). Therefore, the PIPE configuration that will be used for the integration of these two steps is shown in Fig. 8.3. Two process modules will be provided for the CVD step, and one for the etch step.

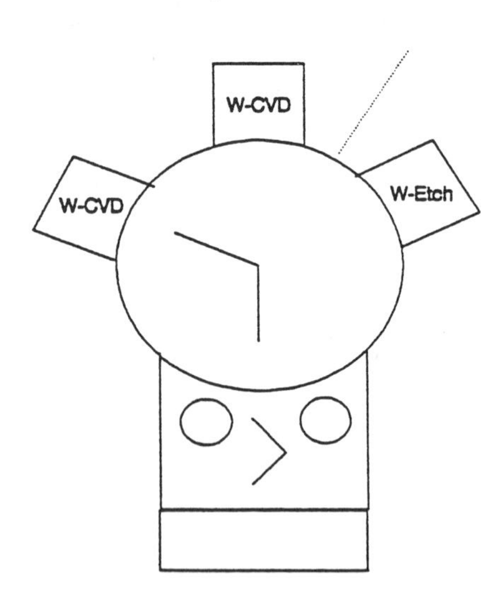

Figure 8.3: Clustering of W-CVD and W-Etch

If cassette cycle time were indeed equivalent to the sum of the process times multiplied by the number of wafers in the cassette (usually 25), then it would take 1.46 hrs to complete the cassette (25 x (150 + 60)/3660). In actuality, because of the simultaneous processing and other simultaneous activities that are taking place, the cassette cycle time in this particular machine is 0.90 hrs.

Because the PIPE is a small factory, there are many other components to cassette cycle time besides just processing (as is the case for the entire fab). Table 8.1 shows some of the other contributing factors and times. Since so many of these things are occurring simultaneously, the performance of the cluster tool is difficult to anticipate without some understanding of the dynamic behavior of the machine.

TABLE 8.1: CONTRIBUTORS TO PIPE CYCLE TIME

- PROCESS MODULE(S)
  - TYPE OF PROCESSING IN THE MODULE
  - QUANTITY OF MODULES (parallel capability)
- PROCESS FLOW
  - PROCESS SEQUENCE
  - PROCESS MODULE ASSIGNMENT(S)
  - PROCESS AND SETUP TIMES
- TRANSPORT MODULE
  - BI-DIRECTIONAL TRANSPORT TIMES
- PROCESS OVERHEAD
  - PIN UP / DOWN TIMES
  - HEAT / ROTATE TIMES
  - CLAMP UP / DOWN TIMES
  - I/O DELAY TIMES
  - TRANSPORT EXTEND/ RETRACT TIMES
- CASSETTE MODULE
  - PUMP / VENT TIMES
  - I/O DELAY TIMES
  - PRE / POST PROCESS TIMES
  - CASSETTE LOAD / UNLOAD TIMES
- BUFFER(S), WAITS, AND CONTROL RULES
  - CAPACITY OF BUFFER
  - BUFFER AND/OR WAIT CONTROL RULES
  - TRANSPORT MODULE PRIORITY

Although machine dynamics will be covered later, we can nonetheless say categorically that cassette cycle time is <u>**not**</u> the sum of the wafer processing times; it is instead some complex function of the process times, the transport times, the wait times, the machine overhead times, and the simultaneous processing that is taking place.

Thus, even without getting into the nitty, gritty intricacies of PIPE dynamics, it is still possible to understand where the bulk of the fab-level performance advantages are occurring. The most obvious improvement, the reduction of cassette cycle time from 1.46 to 0.90 hours, is the direct result of the simultaneous operations taking place inside the cluster tool. But such intra-tool savings are, more often than not, swamped by the fab-level savings that result from manufacturing simplification. In this case, the simplification resulting from replacing two manufacturing operations with one.

Fig. 8.4(a) shows fab graphs for the W-CVD and W-etch steps in both a conventional and a clustered fab. Fig. 8.4(b) shows the same pertinent tools sets, except this time with the fab's buffers (or queues) taken into consideration. *Buffers* are spaces/areas/racks in the fab, near each equipment set, where the work-in-process (WIP) is stored until it can be loaded into machines. Generally, buffer capacity at each workstation is large, and sometimes even considered to be unbounded.

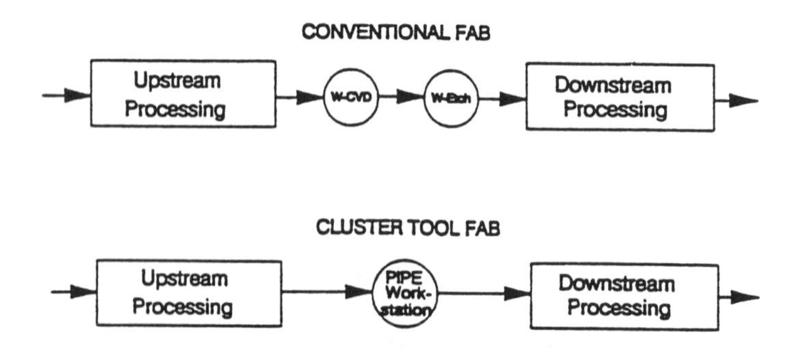

Figure 8.4(a): Conventional vs Clustered Fab

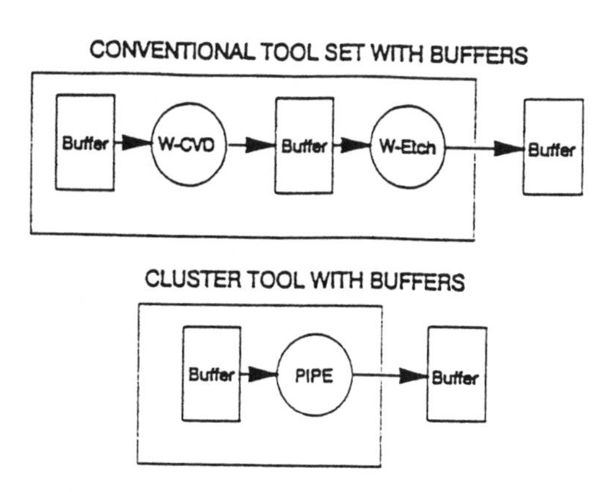

Figure 8.4(b): Tool Sets with Buffers

With Fig. 8.4(b) in mind, consider all the components that contribute to product cycle time in the fab: transport time from work center to work center, wait time in buffers or queues at every workstation, time needed to unload the cassettes into machines, processing time, time to reload wafers back into the cassettes, and, in the case of the cluster tool, simultaneous processing. Of all these components, waiting is the largest contributor to the fab's cycle-time performance. Having a virtually infinite-capacity buffer before every process step, coupled with the tendency to run with comfortable inventory levels, insures that products spend three to four times the processing time simply waiting__even in world-class fabs.

To demonstrate the impact of manufacturing simplification through the reduction of operations, Table 8.2 assigns some representative values to the obvious cycle-time components. A simple one-step reduction cuts the cycle time for this portion of the process flow by **more** than half. By eliminating an operation, we eliminates a work center, any waiting in the buffer at that work center, the transport between the two former work centers, and a round of cassette unloading and reloading. For this simple clustering of two process steps, the flow's cycle time has been reduced by 2.7 hrs. If this example is representative, **any** reduction of manufacturing operations will have a significant impact on fab performance.

TABLE 8.2: REPLACING CONVENTIONAL TOOLS			
	W-CVD	W-Etch	Cluster Tool *
Transport Time	----	10 min	----
Wait Time	60 min	60 min	60 min
Unload Time	5 min	5 min	5 min
Process Time	86 min	50 min	54 min
Reload Time	5 min	5 min	5 min
Total per Tool	156 min	130 min	124 min
Total for Tool Set		286 min	124 min

8.2 Manufacturing Simplification

Since the early 1980's, wafer complexity has been feeding on itself, growing increasingly-more complicated with each generation, and each generation coming faster than the one before it. Increased chip complexity not only necessitates more process steps in the manufacture, but the steps themselves are more difficult. As a result of the rapid escalation in complexity, the cost of semiconductor equipment has increased at a much faster rate than the cost of the building and other fab-related costs (Fig. 8.5). By the late 1980's, enough historical data existed to see where the industry was headed: billion-dollar-plus fabs, and counting. Yet, there were no ready answers on how to curtail the soaring costs.

Then, in 1990, along comes IBM with the concept of using clustering as a method to cap complexity [8.1].

It is not clear whether IBM knew going in that clustering could lead to manufacturing simplification, or whether their push for cluster tool development was based strictly on process needs. But, at some point during the development of their 16Mb DRAM line, manufacturing advantages became at least as much of a driving force as process capability.

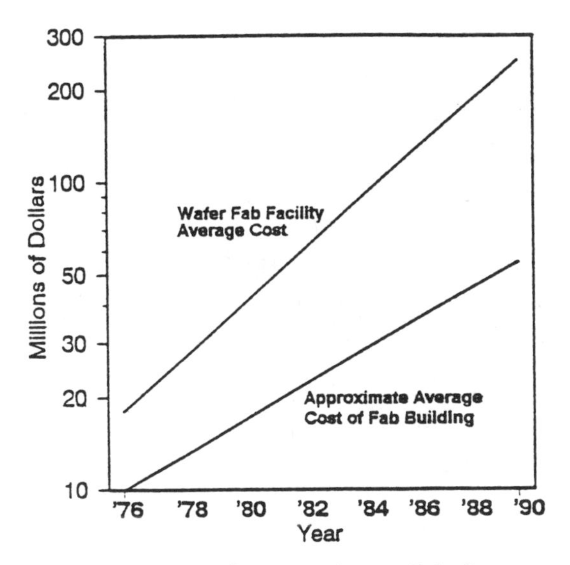

Figure 8.5: Semiconductor Fab Costs

For perspective, Fig. 8.6 shows the increase in the number of manufacturing steps with each new generation of DRAM. IBM postulated that by clustering, putting multiple process steps together into one manufacturing step throughout the fab, it would be possible to essentially break the back of the complexity curve, and hence control costs.

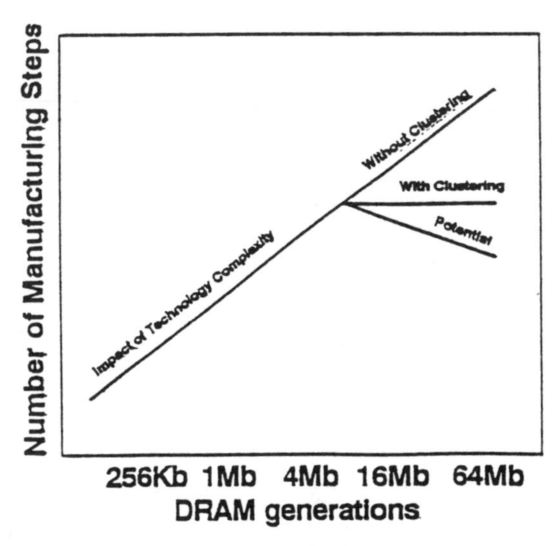

Figure 8.6: Impact of Clustering

Of course, critics would say that eliminating an operation here and there could not possibly have such a dramatic impact. But one has only to look at the most obvious opportunities for clustering to understand how truly large the potential is.

To illustrate, consider IBM's estimates of where cluster tools could be used in a 16Mb DRAM fab (Table 8.3). In addition to showing the number of steps that could be clustered together, we have also given the typical hub number, i.e., the number of process steps performed at the cluster candidates. The product of the two numbers, '# of steps clustered' minus one for the replacement PIPE and 'typical hub #', determines the number of manufacturing operations that clustering would eliminated. The total for the 16Mb DRAM fab is: 160 to 207.

TABLE 8.3: OPPORTUNITIES FOR CLUSTERING

Cluster class	Description	# of steps clustered	Typical hub # *	Eliminated Mfg opns
Photo	Mid uv	6-8	12	60-84
	Deep uv	7-10	6	36-54
Hot process	Tube furnaces	4	9	27
	Other	4	4	12
Dep/etch	Deep trench	3	1	2
	Recess etch/collar dep/etch	3	1	2
	Isolation trench etch	2-3	1	1-2
	Planarization etch	2	1	1
	Gate conductor etch	5	1	4
	Sidewall spacer dep/etch	4	1	3
	Surface strap etch	3	1	2
	Contact etch	2-3	2	2-4
	Metal 1 and 2 etch	2-3	2	2-4
	Interlevel dielectric dep	5	1	4
	Interlevel dielectric etch	2	1	1
	Terminal via etch	2	1	1
			TOTAL	160-207

* September 1990

A conservative estimate of the cycle times savings for each eliminated step would be:

60 minutes of wait
10 minutes of transport
5 minutes of cassette unload
5 minutes of cassette reload
80 minutes of cycle time savings

This savings estimate is conservative because it totally ignores the simultaneous processing taking place within the PIPEs.

Multiplying the cycle time reduction per step by the number of eliminated steps translates into an overall reduction in fab cycle time of 213 to 276 hours (about 1.33 to 1.75 weeks). Such a substantial reduction in product cycle times should translate directly into improved yields, both because of more learning cycles per year as well as the fact that wafers will spend considerably less time simply sitting around waiting.

In addition to reduction in cycle time and improvement in yield, a clustered fab should require less capital than a non-clustered one. Although cluster tools are pricey compared to single-step machines, they generally are no where near as costly as the sum of the equipment they are replacing. For instance, a radial cluster tool performing three steps might cost $4.5mm as compared to three separate machines at $2mm each, for a total of $6mm.

The bottom line on reducing manufacturing operations appears to be that it is an all-around win-win situation, leading to improved cycle times, higher yields, and reduced capital costs.

But, if such reasoning is valid, why hasn't the industry reaped the promised benefits? Because, as we know, much of the clustering shown in Table 8.3 has already occurred in wafer fabrication; photolithography tracks are the industry standard. Why, then, have cycle times not dropped dramatically, and why are plant costs moving into the multi-billion-dollar stratosphere?

The answer is that, despite the photolithography clustering, the number of process steps is growing, not shrinking. To understand why, we must look at what has been happening in the back end of wafer fabrication, more specifically in the interconnect area.

Regardless of product type, the fabrication of wafers is broken down into two parts: formation of transistors (*device fabrication*) and connecting the transistors together via metal 'wiring' (*interconnect*). (See Fig. 8.7.) The processing involved in interconnect, included in Table 8.3 under 'Dep/etch', has yet to be widely clustered, primarily because of industry conservatism, but also because of equipment availability and reliability issues.

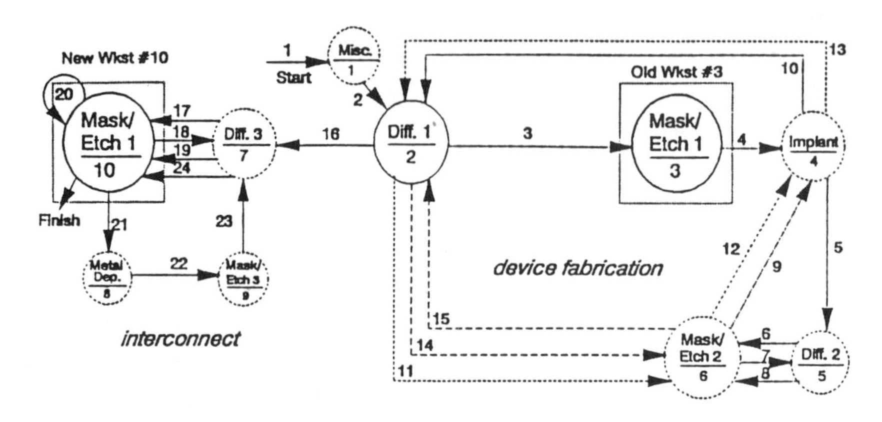

Figure 8.7: Two Areas of Wafer Fabrication

But there is another reason the number of process steps is growing and not shrinking. In addition to little clustering taking place in the 'Dep/etch' area, it is also fair to say that Table 8.3, which was first compiled in September 1990, no longer presents an accurate picture of wafer fabrication. If anything, the information in this table is a distractor because it indicates that clustering 'Dep/etch' would only eliminate about 30 operations. In 1990, that might have been so, but it certainly is no longer the case.

Two things occurred at about the same time in the late 1980's, and one of them quite effectively obscured the impact of the other. At about the same time that photolithography was being widely clustered, semiconductor technology took a giant leap on the complexity curve: the density of transistors on chips had finally necessitated the stacking of one level of metal interconnections over another in order to save silicon *real estate*. The industry had quietly moved from one to two metal layers, and a whole new technology was born: multilevel interconnect.

So, even though photolithography had been clustered, vastly reducing the number of manufacturing operations, the rest of the processing (i.e., metallization and multilevel interconnect) had taken its first strides toward process-step proliferation. Not only did multilevel interconnect involve more steps than single level, but the steps were more complex as well. But, because photo clustering caused such a huge reduction in manufacturing operations, the increase passed almost unnoticed.

As technology has continued to advance, the density and functionality of microprocessors has become space-limited every few years, requiring even more, denser levels of interconnections. Each additional level, ramps up the number of interconnect processing step for **every** layer, not just the latest one.

Currently, for each layer of metal interconnect on a chip, the following process steps are needed:

- dielectric deposit (5 process steps)
- pattern (full photo/litho, 6-8 steps unclustered)
- dielectric etch (2 step)
- resist strip (2 steps, dry/wet)
- planarization (2 step)
- metal deposit (6-8 steps)
- pattern metal (full photo/litho, 6-8 steps)
- metal etch (2-3 step)
- resist strip (2 steps, dry/wet)
- plus possibly others

Thus, the number of process steps to make each metal layer ranges from 33 to 40 (or more). With 1995 state-of-the-art chips having four metal layers, it can take 132 to 160 process steps for metallization and interconnect.

If we consider the next generation of chips, already in development, the number of process steps to make the five or six metal layers could well be nearing 250. It's easy to see how the escalation in component complexity has more than swamped the manufacturing simplification that was gained by the clustering of photolithography. Process flows are now in the range of 500 process steps, over half of which are involved in interconnect. Fig. 8.8 gives a more realistic picture of the opportunities for clustering than does Fig. 8.6.

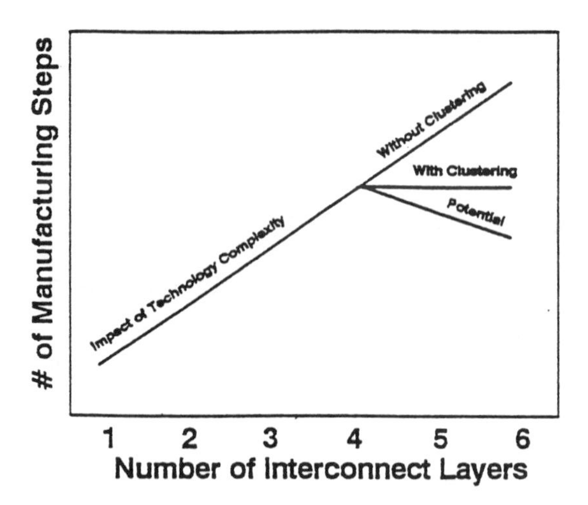

Figure 8.8: Impact of Clustering Metals

To see how new fab cost have soared into the multi-billion dollar strata, assigns some fairly conservative estimates to the equipment used in each metal layer (Table 8.4). The capital price tag for one round of machines is in the vicinity of $40 million.

TABLE 8.4: EQUIPMENT COSTS PER LAYER OF METALS

Description	# of Steps	Cost/Step ($ MM)	Total Cost ($ MM)
Dielectric dep	5	2.0	10.0
Pattern	1	1.0	1.0
Dielectic etch	2	2.0	4.0
Resist strip	2	0.2	0.4
Planarization	2	2.0	4.0
Metal dep	6	2.5	15.0
Metal pattern *	1	1.0	1.0
Metal etch	2	2.0	4.0
Resist strip	2	0.2	0.4
TOTALS	23		$39.4

* Photolothography performed in track system.

But one machine at each workstation is unrealistic; first, because each work center is at least a fourth-order hub (four metal layers), and second, because these new, costly fabs will be starting on the order of 20,000-24,000 wafers a month. The sizing for this number of wafer starts will call for roughly 15 to 20 machines at each workstation. So, a more credible cost estimate for interconnect is obtained by multiplying the $40mm by the 15 to 20 machines at each workstation, for a whopping capital cost of $600 to 800 million dollars. Suddenly, it is not difficult to understand why there is a billion-dollar-plus price tag for a new, four-metals wafer fab.

It is ironic then that, despite the fact that the number of steps in the metals/interconnect area is growing by leaps and bounds, and despite the fact that the equipment to perform those steps is expensive and becoming more so, wafer fabs are still not pursuing clustering as a means to control costs, much less manufacturing simplification. Eventually, though, they will most likely be driven to it.

Multilevel-interconnect technology is, and will be for the foreseeable future, the gating technical hurdle for high-volume submicron IC fabrication. As chips get denser and geometries get smaller, even Class-1 clean room particles take on the dimensions of boulders. Someday soon, technical advancement in interconnect may 'hit the wall' because of process limitations; and the only way around it, or through it, may be to cluster. It is only a matter of time, and the clock is ticking.

Like photolithography before it, clustering will be done for reasons of component capability and/or yield. It will be almost a serendipitous side benefit that there will be a considerable simplification in manufacturing that goes along with it, as well as possibly a substantial reduction in cost.

But, then again, perhaps not. Such benefits will probably permit (or perhaps encourage) process engineers to find the next, higher level of complexity to scale.

Finally, one last point about clustering and manufacturing simplification before moving on: along with reducing the number of manufacturing operations, there will also be a concurrent reduction in the physical space requirements of a clustered fab. And not just any space. All the eliminated equipment, as well as the queuing areas near each workstation, would normally require a Class-1 clean room. Truly expensive real estate, and another opportunity to whittle away at the high cost of wafer fabrication.

Throughout the remainder of this chapter, keep in mind that, although the clustering revolution began, and has been carried forth as a means of breaking technical barriers, the coincidental simplification of the manufacturing itself may far overshadow the advantages gained in processing.

8.3 Types of Cluster Tools

Cluster tools are automated systems for wafer processing, transport, and process control. While a variety of equipment designs are possible, such designs are fundamentally limited by the physics and chemistry of the processing steps (see Chapter 3). Consequently, proposed integrated-equipment designs have also been strongly influenced by traditional equipment concepts for each type of processing.

Below we will discuss design concepts for cluster tools for photoresist processing, chemical vapor deposition, furnace processing, the generic radial cluster, and wet processing. With the exception of the radial cluster, it will be clear that the proposed designs are extrapolations of existing equipment.

For a host of reasons, each design should be subjected to a performance analysis (see Chapter 7 and Section 8.5) before it is ever becomes a piece of hardware. The throughput of the cluster tool as a system may not be as expected. The system bottleneck may not be obvious. Like many automated systems, the tool may be transport limited, and control rules may have subtle, but large impacts. Conversely, a simple correction in a control rule could have a significant impact on throughput.

Of course, the ultimate reason for performance analysis of cluster tool designs is that any tool not meeting throughput expectations can result in bottlenecks for the entire wafer fab. Furthermore, poorly designed cluster tools are difficult to operate and maintain.

In the discussion below, we will indicate designs that present obvious problems.

8.3.1 Photolithography

Fig. 8.9 shows four designs for automated photoresist processing systems. The major variation in the four designs is the type of transport system and the number of transport modules.

Fig. 8.9(a) shows the popular wafer track transport system. The track moves a single wafer from one module to any other module. THe speed of the track limits the number of modules that can be effectively supported.

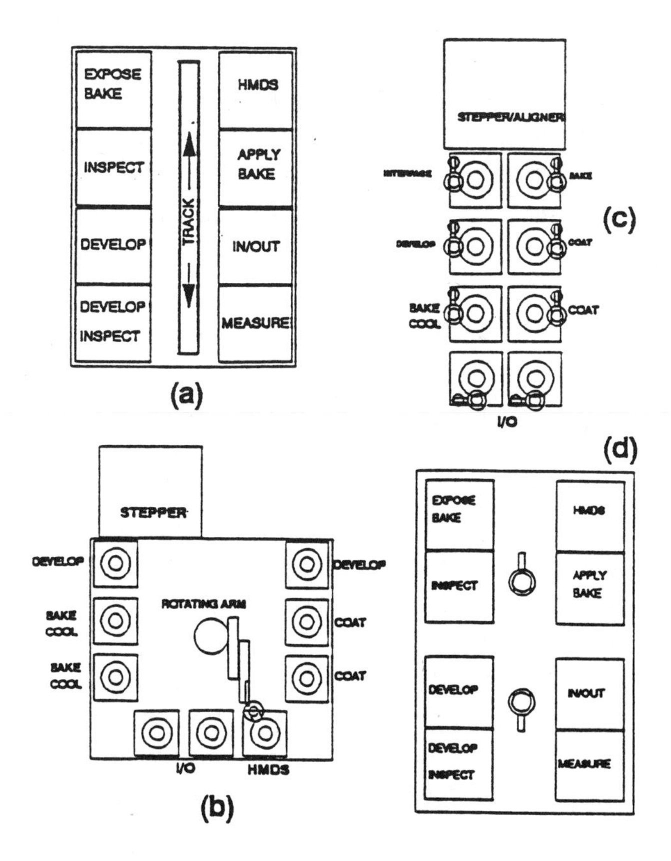

Figure 8.9: Integrated Photolithography Tools

The integrated system in Fig. 8.9(b) has the stepper, a photolithography tool, as part of the system. The other difference is the use of a single rotating or 'robot' arm as the transport module. The robot arm moves a single wafer from one module to another, and, like the track, its speed of wafer transfer limits system configurations and throughput.

Fig. 8.9(c) presents an attempt to get beyond the transport limitation of the above two designs. Each processing module of this tool has a transport resource that can reach adjacent stations. While more transport resources have been added, the complexity of possible equipment configurations has greatly increased. Designs based on this concept require considerable performance analysis.

Fig. 8.9(d) represents a compromise between the last two designs, (b) and (c). There are two transport arms; each servicing a subset of the processing resources. An additional overhead is involved with the transfer from one transport arm to the other. Performance analyses indicate this design can add throughput in many cases. An interesting question is the value of a third transport arm.

8.3.2 Carousel Clusters

The carousel-type tools shown in Fig. 8.10 have served as the basis for equipment for chemical vapor deposition (CVD). Each wafer position is coincident with a reaction station. The wafer rotates from position to position, and the total deposition is the sum over all stations.

In Fig. 8.10(a), two carousel reactors are integrated with a transport arm. If the transport arm can keep both reactors supplied, this is a reasonable attempt at automation. However, it is not one that is easily extensible to longer process sequences.

Fig. 8.10(b) illustrates some of the difficulties involved. In this design, two additional processing modules have been added for pre-processing and post-processing. These modules are single-wafer chambers. Thus, two large, in-parallel processing systems are coupled to single-wafer processing through a single transportation resource. This design will be quite difficult to operate effectively for good fab performance.

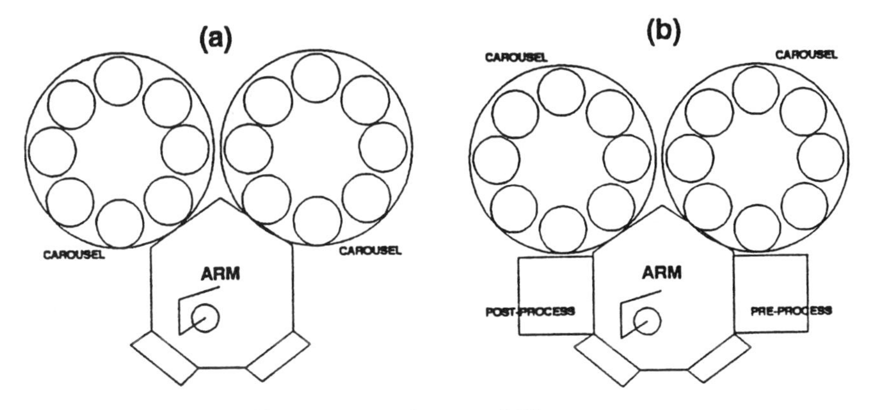

Figure 8.10: Carousel Clusters

8.3.3 Furnaces

The furnaces in Fig. 8.11 illustrate the difficulties in automating old equipment designs. As discussed previously, furnaces can hold several cassettes of wafers (generally 4 to 8).

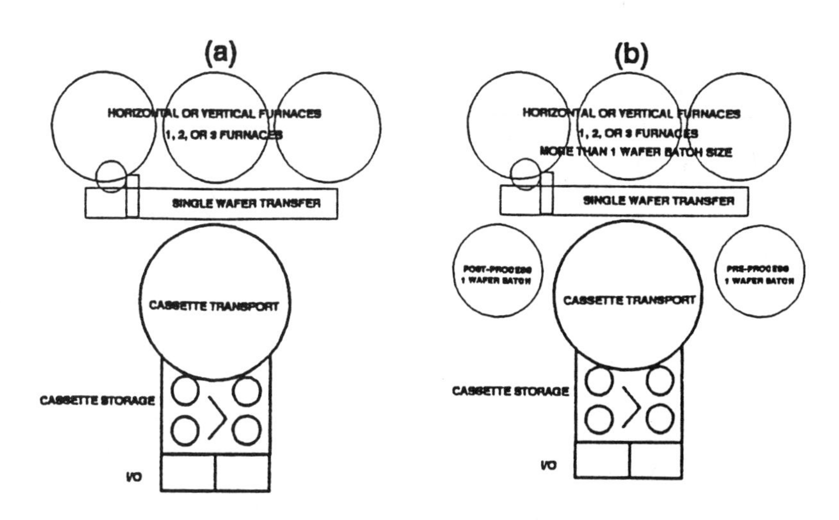

Figure 8.11: Mixed Single-Wafer and Batch Furnaces

Automating the loading and unloading of furnaces involves handling multiple cassettes, removing wafers from them, and later replacing them. Thus, the tool designs show equipment modules for cassette storage, cassette transport, and single-wafer transfer.

Fig. 8.11(a) is a reasonable first attempt at automation, if sufficient transport resources are provided and the throughput of the system is balanced. Balancing the throughput of the system again becomes problematic as the sequence of process steps is lengthened.

Fig. 8.11(b) shows a second configuration with single-wafer processing modules for both pre-processing and post-processing. The resulting complex configuration with this design will be difficult to operate.

8.3.4 Radial Clusters

The generic radial cluster tool design provides a more flexible solution for longer process sequences (i.e., greater than one) than either carousels or large batch systems (furnaces). In the radial cluster tool, all processing chambers provide single wafer processing. This simplification allows developing designs and operating rules for multiple-step process sequences. The radial cluster tool is the subject of much of the performance analysis in the remainder of this chapter.

Fig. 8.12 shows successively more complex radial tools. The tool in Fig. 8.12(a) is a conventional radial with two transport resources and five processing chambers. The tool in Fig. 8.12(b) provides a second group of chambers for additional processing flexibility. In Fig. 8.12(c), a third transport resource has been added.

With the third transport arm and docking modules in Fig. 8.12(c), there now exists the possibility of direct interfaces to other cluster tools performing the sequences before as well as after those of the current tool. Connecting cluster tools in this manner provides additional opportunities to reduce contamination and cycle time.

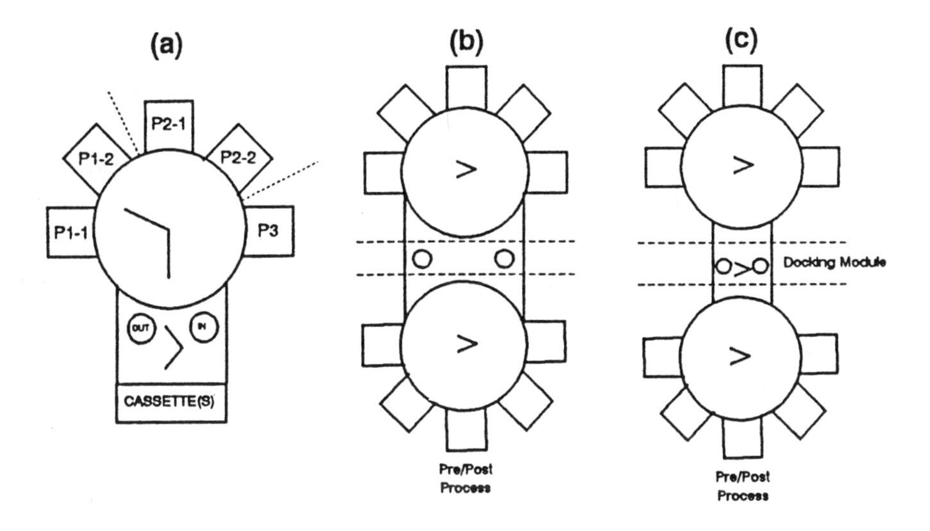

Figure 8.12: Radial Cluster Tools

8.3.5 Wet Etch Station

Fig. 8.13 shows an automated sequence of wet etching and cleaning steps. Transport is provided by a robot arm. This design is similar to 8.9(b) for photolithography.

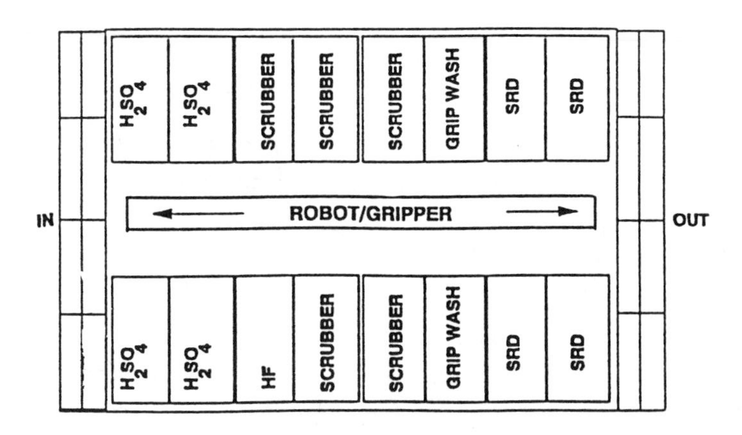

Figure 8.13: Wet Etch Station

8.3.6 Automated Bays and Automated Fabs

In an automated bay a sequence of process steps is tied together by automation, but not placed inside the same machine. Performance analysis of an automated bay is similar in complexity to analyzing a cluster tool like the one shown in Fig. 8.13.

Fabs have also now been built with automated material handling between each workstation. The dynamics of coupling transport with processing, for a thousand or more steps, can be awesome. Not surprisingly, the operation and design issues of such a factory are quite challenging.

8.4 Configuring Cluster Tools

By now, it should be obvious that cluster tools are not off-the-shelf items. Not only are there many types of tools, from a growing number of vendors, but every platform that has multiple-chamber or processing options must be **configured** to best meet the processing and manufacturing needs of a particular wafer fab. For instance, for a 3-step process sequence, is it better to buy three 3-chamber radial cluster tools, where:

- every chamber does exactly the same processing

or

- each chamber on the tool does a different step?

As may be seen in Figs. 8.14(a) and (b), each process step has exactly the same number of chambers or modules available to perform the processing, though they are dispersed quite differently among the three machines. So which equipment set is better?

Life just got **real** complicated!

First, define better. Do we want better processing results, better manufacturing results, or the best mix of the two? What is the measuring stick for 'better'? Higher yield, shorter cassette cycle times, higher machine throughput, lowest capital expenditure, most manufacturing simplification, highest equipment utilization, or what?

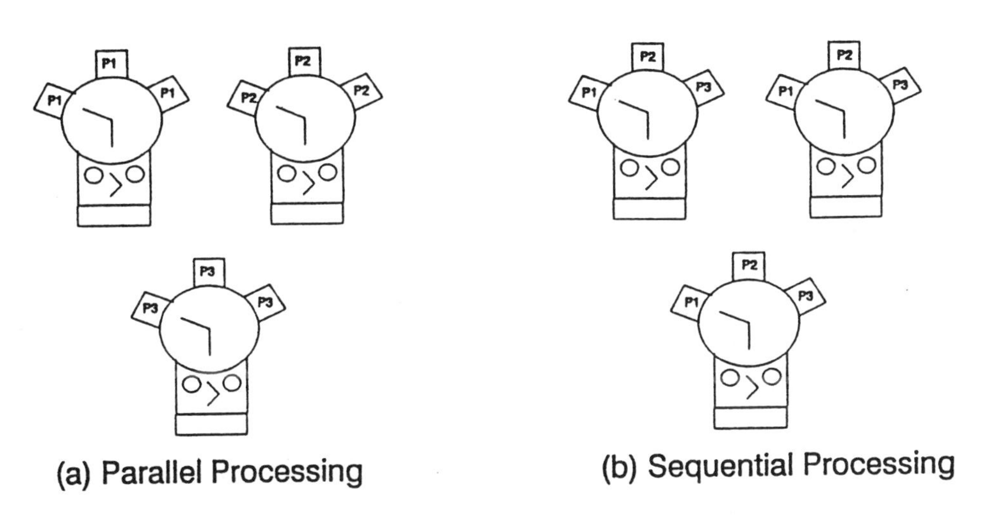

Figure 8.14: Clustering Options for 3 Process Steps

To complicate matters further, once the pertinent performance measures are selected, we must ask ourselves if we have considered all the possible clustering options.

Answering the last question first: NO. Not even close.

Clustering one or more process steps turns out to be a problem of combinatorial enumeration, i.e., ascertaining and/or listing all possible combinations of steps and tool resources. A PIPE has a platform which can support multiple process modules (four, six, ten, etc.), and each of the process modules can perform the same step, a different step, or need not be used at all. Thus, there are many different ways that a sequence of steps may be assigned to one or more tools.

For the hypothetical 3-chamber platform in this example, one, two, or all three of the process-module docking sites may be used for any one, or combination, of steps. Therefore, on the way to selecting the 'best' PIPE configuration for a given sequence of process steps, we must consider all of the possible options.

8.4.1 Combinatorial Enumeration of Configuration Options

Fig. 8.15 contains two examples of what are known as cluster diagrams, one for a heavily-clustered area of a wafer fab, and the other for an area where only photolithography is clustered. As may be seen, these diagrams basically list the individual steps in a process flow and block out those steps which are candidates for clustering. Such schematics are valuable not only because they help to visualize the reduction in manufacturing operations that result from clustering, but also because they act as a reality check on which steps may viably be clustered. (For instance, it would take a stretch of the imagination to want to cluster steps #10 and 37.)

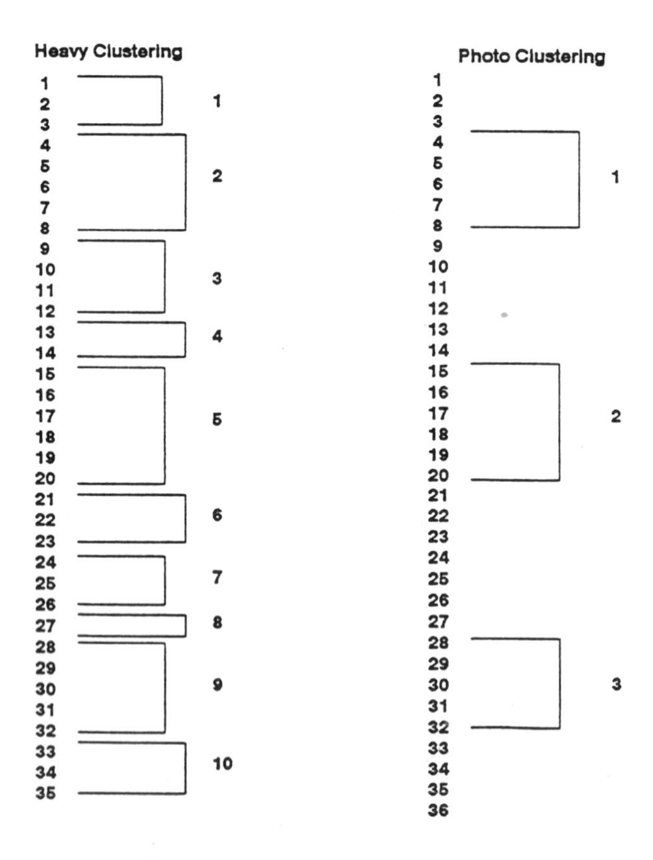

Figure 8.15: Clustering Diagram

Perhaps, most importantly, the cluster diagram is useful in defining: the subset of the flow (i.e., the process sequence) that may be clustered from a process perspective; the order in which the sequence will be performed inside the PIPE (i.e., the recipe); and the relationship of the PIPE to other equipment (or other PIPEs).

First enumeration: Determining recipe sequence for clustering.

For the little 3-step example above, assume that the first three process steps, P1, P2, and P3, are candidates for clustering. From the cluster diagram, we see that P3 must always follow P2 which must always follow P1. The recipe is set.

Second enumeration: Listing the feasible grouping of steps.

The best way to determine potential groupings is to draw the matrix of possibilities, without restriction to anything other than recipe order. Fig. 8.16 shows the potential grouping for Process Steps #1-3.

NON-CLUSTERED STEPS	STEP-CLUSTERING OPTIONS			
1	1	1	1	1
2	–	2	–	2
3	2	–	2	3
	–	3	3	
	3			

Figure 8.16: Possible Groupings of Steps

Third enumeration: Deciding upon the tool/hardware options.

The next step is to consider the hardware and its possibilities. For this 3-step example, the cluster tools being considered have ports for up to 3 process modules on their platforms. Thus, Fig. 8.17 shows the various hardware choices. These tools may be equipped with one, two, or three process modules.

Figure 8.17: Hardware Options for 3-Chamber Platform

In light of the chamber options available, the feasible grouping of steps shown in Fig. 8.16, may now be expanded to take advantage of multiple chambers for one or more steps. The expanded options are portrayed in Fig. 8.18. This matrix is actually a condensed representation of all possible clustering alternatives; multiple chambers are indicated where feasible, even though they may not necessarily be used.

NON-CLUSTERED STEPS	STEP-CLUSTERING OPTIONS					
1	1,1,1	1,1	1	1,1,1	1,1,1	1
2	2,2,2	2	2,2	2,2	2	2
3	3,3,3	3,3,3	3,3,3	3	3,3	3

Figure 8.18: Possible Groupings with Multiple Chambers

For instance, an entry such as '2,2,2' means that there are three possible machine configurations that can perform process step #2: a machine with one chamber, a machine with two chambers, or a machine with three chambers, all dedicated to step #2.

Similarly, the entry '2,2 and 3' means that there are two possible machine configurations that can perform process steps #2 and 3: a two chamber machine with one chamber dedicated to each step, or a three chamber machine with two chambers for process step #2 and one chamber for process step #3.

Fig. 8.19 shows **all** possible machine configurations that can perform the three process steps on a three-module platform.

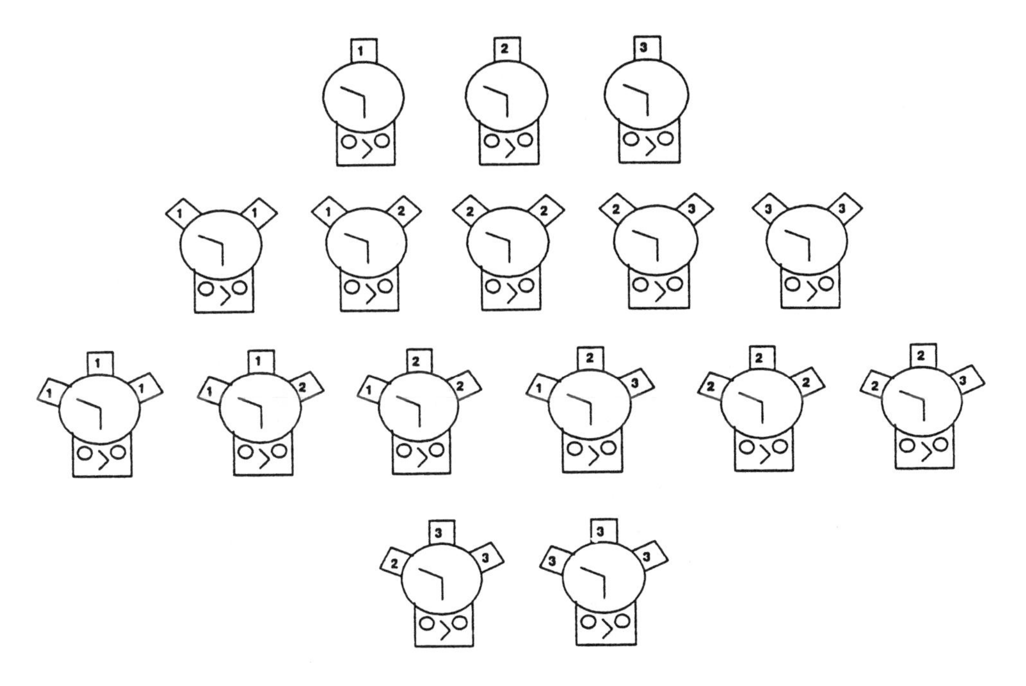

Figure 8.19: Combinatorial Enumeration of Configurations

Fourth enumeration: Matching feasible-step groupings with multiple-chamber options.

At this point, the enumeration of the possible combinations is complete. But, from the numerous hardware configurations found, how does one narrow the choices down to a manageable few, and what selection criteria should be used?

As always, tool selection is based first on good process performance, followed as closely as possible by good (or at least adequate) manufacturing performance.

8.4.2 Process Performance in Cluster Tools

Achieving good process performance in a cluster tool involves at least three things:

- selecting the appropriate steps to cluster
- correctly configuring the tool for those steps, **and**
- knowing the configured tool's intrinsic dynamics and how it handles waits.

For virtually every clustering effort, especially in the yield-sensitive metals/interconnect area, serious effort must go into all three aspects in order to attain good process results, and hence good yield.

Table 8.5 gives some general rules of thumb on which process steps are amenable to clustering, and which are not.

TABLE 8.5: RULES OF THUMB ON SELECTING STEPS PROCESS CONSIDERATIONS
■ Choose consecutive process steps.
■ Except for photolithography, clustering wet steps with gaseous is difficult.
– Some wet steps, however, like clean and etch, can be replaced with vapor steps, and then be clustered.
■ Large batch-size processing is not conducive to clustering due to the intrinsic mismatch between single-wafer and batch processing.
– Some large-batch processing may be replaced with single-wafer processing, such as thermal oxidation with rapid-thermal processing (RTP).
■ In metals/interconnect, clustering should be considered whenever:
– Delay is a problem because of surface aging.
– Contamination will result because of exposure to atmosphere (O_2 /H_2O vapor).
– It is desireable to avoid cool down/heat up between steps.
– Steps are involved in a particular feature (dielectric, W-plug, metal wire, etc).

Of course, the metals area is still virgin territory when it comes to pushing the bounds of clustering, and, until it is extensively explored by the process engineers, progress is likely to be slow. There are some process sequences, however, that are naturals for clustering. For example, W-CVD followed by W-etch (used in making a tungsten plug) are two steps that would clearly benefit by not being separated and subjected to an inter-step delay.

From a manufacturing and cycle-time perspective, it is advantageous to cluster as many steps as are possible while still achieving good yield. However, the cautious nature of the industry says that, at least in the near future, steps will only be clustered on the manufacturing floor when:

- formidable process advantages will be gained, or
- there is no other way to perform the processing.

In either scenario, process engineers will be extensively involved, and only appropriate steps will be clustered.

Which brings up the second strategic element for good process performance: correctly configuring the tool for the chosen steps. As was seen in Fig. 8.19, even a simple recipe will have many hardware options. Selecting the best option will be a matter of identifying the configuration or configurations which come closest to matching cluster tool resources to process recipe, and, in doing so, achieving a balanced PIPE. Table 8.6 provides some general guidelines.

TABLE 8.6: RULES OF THUMB ON CONFIGURING PIPES PROCESS CONSIDERATIONS
▪ Evaluate recipe in various PIPE configurations: - parallel - sequential - mixed ▪ Determine if waits (and locations of waits) will impact yield. If yes, how does PIPE handle waits? - buffer(s) - waiting in chamber(s) - multiple process chambers ▪ Determine if the PIPE is balanced (i.e., resources match recipe). If not, then: - can it be balanced by adding multiple chambers, or - are control rules flexible enough to compensate?

8.4.3 Balanced PIPEs

Exactly what is a balanced PIPE?

Perhaps the best way to describe a balanced PIPE is to trigger a mental picture of something with which we are all familiar. Remember movies of the old-time fire brigades, where a string of men are passing buckets of water down the line, from man to man. When everyone is in sync, each man swings his bucket forward into the hands of the man in front of him, and on the backward swing takes the bucket from the man behind him. Arms never stop swinging, and buckets never have to wait—as would happen if a man stopped to scratch his nose.

When a PIPE is balanced, it has this type of perfect synchronization. Wafers move from resource to resource, through the prescribed recipe, never incurring waits. In other words, PIPE resources are available when needed; they are not busy doing something else. But note that a PIPE's *resources* are not just its process modules, but those chambers in addition to transport arm(s), pre-/post-processing modules, and every other machine component.

Simply stated, the key factor in balancing a PIPE is the elimination of waits. In the cluster tool, as in the fab itself, it is generally not a good result to have wafers sitting around waiting, especially in the highly-reactive metals/interconnect area. In an unbalanced PIPE, not only will wafers experience waits, but the length and location of the waits will vary from wafer to wafer. Needless to say, continued, and usually erratic, exposure to temperature and/or gases may have processing ramifications.

One could argue that the primary purpose for clustering in the first place is to eliminate random waiting and its deleterious side effects. So if a PIPE does not, or can not, eliminate waits, then one must question the advisability of the clustering decision or at least the competence with which the decision is carried out. (Manufacturing ramifications will be discussed later.)

In most tools with two or more chambers, the best way to achieve a balanced PIPE is to have all processing times be of approximately the same length. When the tool is used for *parallel processing*, i.e., all chambers performing exactly the same step, this criteria is met automatically.

But the big 'win' from both a process and a manufacturing perspective comes from performing *sequential processing* in a cluster tool, where every chamber does a completely different process step. However, in this case, process times can be quite disparate.

When several process steps are to be clustered, it may be necessary to do what is known as *mixed processing* in order to balance the PIPE. In mixed processing, the tool performs a sequence of steps, but, for at least one of the steps, multiple (or parallel) chambers are provided. Fig. 8.20 illustrates the three different types of cluster-tool processing.

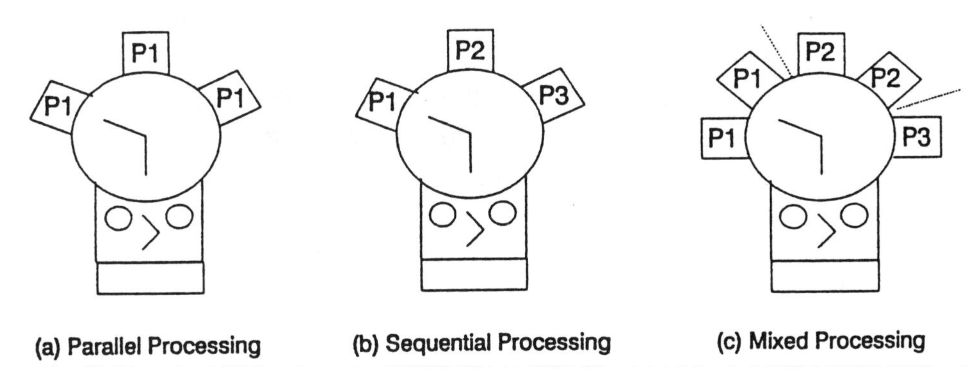

Figure 8.20: Types of Cluster Tool Processing

Using Buffers for Balance

Of course, there are other ways to configure and/or design for wait states in a cluster tool. Some equipment vendors have opted to provide an internal buffer where wafers may move to wait if the next processing resource is unavailable. The reasoning that went into this design innovation, however, was flawed. Just because a wafer is ready to be moved, whether it is to the next process chamber or to the buffer, does not necessarily mean that the transport mechanism is available to move it. In fact, with machines having more than three process modules, and performing standard-type single-wafer processing (i.e., 60 seconds or less), the transport mechanism is generally the system bottleneck; waits are almost inevitable. The presence of an internal buffer in such a case exacerbates the wait problem.

To elaborate, once a wafer has finished a process step, the control system on the machine will look ahead to the next process chamber to see if it is available. Since only the first wafer in a cassette will have a clear, unobstructed path through the machine, the next chamber may very well be tied up with another wafer. Accordingly, the control system will route the wafer to the buffer to wait.

However, with the transport arm being the system bottleneck, it is probably busy moving other wafers around the machine. Thus, the wafer in question will end up waiting for transport in the process module, despite the presence of the buffer. When the transport arm is finally free to move the wafer, its journey to the buffer will further stress the bottlenecked resource (the arm itself) for a non-productive move of an 'already-cooked' wafer. While the bottlenecked arm is making this non-productive move to the buffer, other wafers will end up waiting randomly around the system. The arm will be further over-taxed when it has to move the wafer out of the buffer and back into production.

The usual consequences of having a buffer in a cluster tool is that wafers will end up waiting anyway, and the bottlenecked arm will be even more bottlenecked, making the number, length, and frequency of waits increase. A corollary to this observation is that the wait problem worsens with each additional buffer slot that is made available.

To illustrate, Table 8.7 presents snapshots of the wafer waits in a failed radial-cluster-tool design. With a buffer capacity of 8 wafers and a fast transport arm, it was expected that the tool would be able to eliminate wafers waiting in process chambers after they completed processing. It was a noble but misguided effort. For just about every recipe that did not have sequentially-decreasing process times, waits occurred in process chambers, even for simple two-step recipes. For strict sequentially-decreasing process times, the buffer was not needed.

TABLE 8.7: WAITING IN A BUFFERED PIPE

----Wafer Locations----

time	549.3	598.0	625.5	719.5	741.5	816.2	911.0	979.8
P1	-	-	5	7	8	-	12	14
P2	2	-	3	4	-	8	10	10
P3	1	1	2	-	3	-	-	6
P4	-	-	1	1	1	2	3	3
Buffer	3,4	2,3,4	4	2,3,5,6	2,4,5,6,7	3,4,5,6,7,9	4,5,6,7,8,9,11	4,5,7,8,9,11,12,13

Process Recipe: 30/60/90/120 sec
8 Buffer Slots

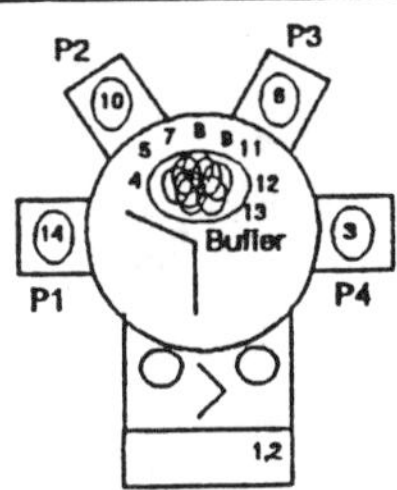

This particular cluster-tool design never made it out of development. Of the buffered tools that are on the market, wafer fabs generally do not like to use them with recipes that incur trips to the buffer.

Balance through Understanding and Control

Rather than providing buffers, a much more desireable way to balance a PIPE is to control the start of wafers into the machine. If wafer starts are gated by the slowest recipe step, plus a small increment of time for machine overhead, then wafers should never have to wait. (There are recipe-specific exceptions to this rule, but they are rare and usually involve extremely rapid processing.) Thus, controlled wafer starts looks to be one of those equipment-design innovations that will ratchet up the process performance of cluster tools.

Unfortunately, most cluster-tool control software is not yet 'smart' enough to have this capability. Processing hardware has far out-distanced the software in regard to meeting performance requirements. Equipment vendors do not realize that an entire portion of the fab has moved inside their gear. Until they do, the software will not be able to support advanced process sequences where timed moves and waits are critical.

Understanding the performance of a cluster tool requires knowing exactly what occurs inside the entire machine, not just what happens inside the various process chambers. Because wafers in general do not follow the same operations sequence through the PIPE, wafer-to-wafer uniformity is a critical yield concern. Unfortunately, detailed performance information (such as in Table 8.1) is neither understood, nor readily available, on a recipe-by-recipe basis. Little wonder then that fab management is leery of voluntarily clustering more than is absolutely necessary.

Perhaps, even more remarkable, few vendors and still fewer fab managers truly understand what a good cluster-tool performance analysis entails. Contrary to popular opinion, it is not merely providing the cassette cycle time for a given recipe.

PIPE performance analysis determines the cassette cycle time and throughput for a recipe, as well as the exact operations sequence for every wafer in the cassette. The *operations sequence* consists of everything that happens to a wafer as it follows its own particular path through the machine.

Not only will wafers in general not follow the same operations sequence through the PIPE, but when and where the next move occurs depends on the dynamic state of the machine. The dynamic state of the machine can only be determined **at** the time that a control decision is made. Of course, in order to do an accurate performance analysis, that deals with the complex dynamics of a cluster tool, a machine-specific computer model is a necessity. A first-level analysis, however, is possible with an operations or OPS graph.

8.4.4 Operations Graphs

An operations or OPS graph is a mapping of a process recipe, and all other tool operations required to perform that recipe, onto the resources of the cluster tool. Thus, the OPS graph shows the tool's resources, such as transport module(s), process modules(s), pre- and post-processing modules, cassette loader/unloader, buffer, etc. It also shows the moves the wafers make amongst the tool's resources. Finally, it shows any possible waiting.

Like the fab graph does for the entire factory, the OPS graph provides a basic understanding of process and resource interactions. Also like the fab graph, it identifies hub structure and provides a traffic analysis of what is happening inside the tool, showing wafer paths and possible wait states.

To represent the tool's resources in an OPS graph, any geometric symbol may be used. The wafer routings, including the process recipe, are represented by numbered arrows going from resource to resource (node to node). Possible wait states are indicated by numbers at the appropriate resource.

Examples of several OPS graphs follow.

Example 1: Sequentially-Decreasing Recipe in a Buffered PIPE

The first example of an OPS graph will be for the simple two-step metallization recipe given in Table 8.8. The cluster tool to be used will have two process chambers, as well as a buffer for any possible waiting. The other tool resources include a transport arm and a cassette loader.

TABLE 8.8: METALLIZATION IN BUFFERED PIPE

Process Step		Process Time	PIPE Resource	
1	CVD	30 sec	1	Chamber A
2	PVD	24 sec	2	Chamber B

For some machines, the drawing of the cassette loader may not be required if it does not impact the tool's dynamics. For this particular tool, however, the transport arm services the cassette loader, as well as the process chambers and the buffer. Its presence, therefore, does directly affect machine dynamics.

The cluster tool's resources are shown in Fig. 8.21 as geometric symbols (circles, ovals, nodes). Wafer movement, from resource to resource, is reflected by the arrows. The numbers at the resources and near the arrows indicate the operations sequence of a particular wafer, i.e., everything that happens to it as it follows its own path through the machine. These events include processing, transports, and waits. For this simple two-step recipe, the operations graph has 5 nodes (resources), 8 arrows (transports), and 14 times (process, transport, and wait).

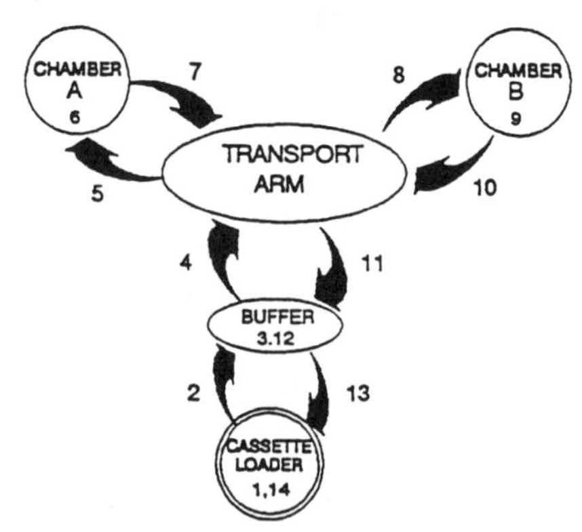

Figure 8.21: OPS Graph for Metallization in Buffered PIPE

As might be expected, the recipe is the primary determinant of whether there will be waits, and whether the buffer will be used. Since this two-step recipe has sequentially-decreasing process times (30 and 24 seconds, respectively), the unavailability of the downstream process chamber will never be a cause for an upstream wafer to wait. It is possible to envision a scenario where a wafer might wait because the transport arm is busy with another wafer. But under no circumstances would the buffer be used with this recipe.

Only detailed knowledge of a machine's timing parameters permits a comprehensive and accurate OPS graph for every wafer in the cassette. Without such data, even for this simple case, it is impossible to know whether an occasional wafer ends up with a small wait while the transport arm is engaged with another wafer.

While an occasional one- or two-second wait may not sound like much, it still means that the particular wafer has followed a different operations sequence through the machine, and is therefore different from the other wafers. How much different is relative to what is being done. But at least one large wafer fab is known to run dummy wafers for every fourth wafer through a particular type of gear because every fourth wafer follows a different operations sequence.

Finally, before moving on to another example, it should be noted that, even though there are only two process steps involved in this simple recipe, the number of operations occurring inside the mini-factory is fourteen:

(1) waiting in cassette loader
(2) move to buffer/in-bound site
(3) waiting at in-bound site
(4) pickup by transport module
(5) transport to Chamber A
(6) CVD process step (30 sec)
(7) pickup by transport module
(8) transport to Chamber B
(9) PVD process step (24 sec)
(10) pickup by transport module
(11) transport to buffer/out-bound site
(12) waiting in out-bound site
(13) move to cassette loader
(14) waiting in cassette loader

There are, of course, other representations of the operations sequence which are also correct. Potential waiting in process chambers for the transport module could be indicated by an additional wait number as seen in Fig. 8.22(a). Another option is to show the pickup and move by the transport module as one step (one arrow) as indicated in Fig. 8.22(b). Similarly, the movement of the wafer from the cassette loader to the process chamber could be represented as one operation, Fig. 8.22(c), since the wafer only moves **through** the buffer for this recipe (and visa versa for the return). In fact, since the buffer is never used for this recipe, the operations graph in Fig. 8.22(d) with no buffer is also technically correct.

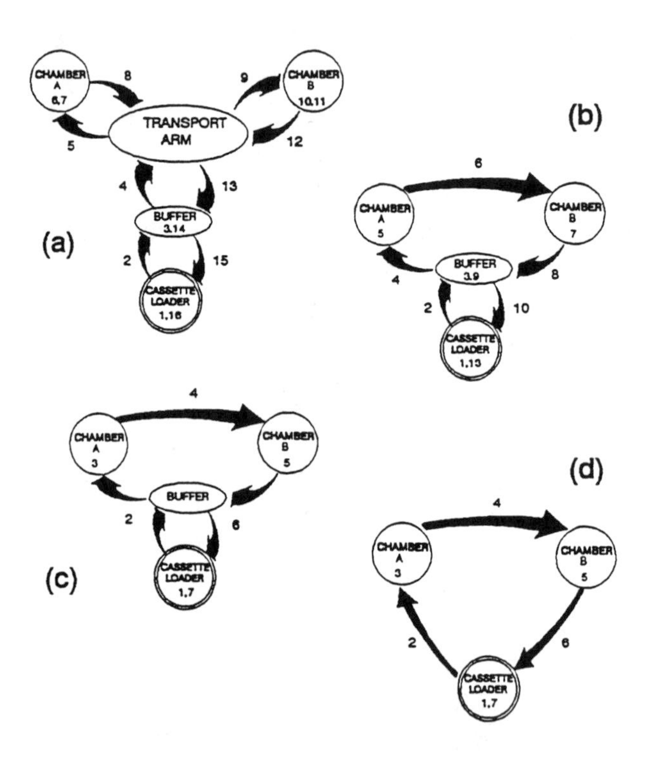

Figure 8.22: Alternate OPS Graphs for Example 1

The bottom line is that, like the fab graph for the entire factory, the operations graph may take many forms and still be correct. The OPS graph is simply a first-level visual model to aid in cluster-tool analysis.

Example 2: Sequentially-Increasing Recipe in a Buffered PIPE

Since, from a processing point of view, unexpected wafer waits may be the most deleterious aspect of clustering, consider the recipe in Table 8.9 where the steps are not only substantially different, but sequentially increasing as well.

TABLE 8.9: SEQUENTIALLY-INCREASING RECIPE
Buffered PIPE

Process Step		Process Time	PIPE Resource	
1	Etch Clean	60 sec	1	Chamber A
2	CVD	320 sec	2	Chamber B

For such a disparate recipe, the natural inclination for clustering is either to provide multiple process chambers for the longest step or to use a tool with a buffer. The OPS graph in Fig. 8.23 explores this latter option.

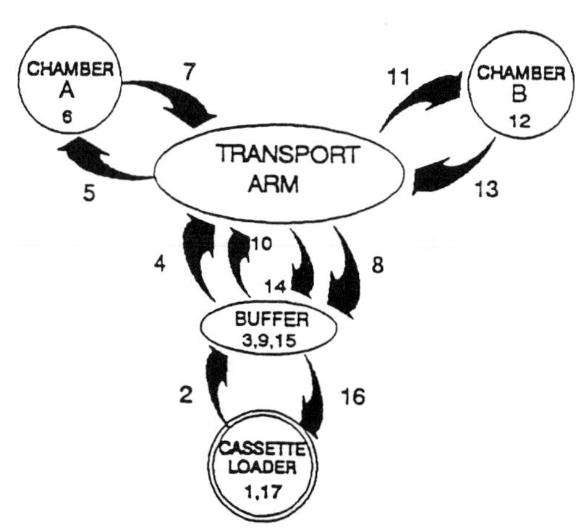

Figure 8.23: OPS Graph for Sequentially-Increasing Recipe

The operations sequence presented has the wafers returning to the buffer after the first process step to wait for the second chamber to clear. But not only will the second wafer end up in the buffer waiting for wafer #1 to complete its CVD processing, but also wafers #3 through #5 (Fig. 8.24(a)). Furthermore, when wafer #2 finally does make it to Chamber B for 320 seconds worth of chemical-vapor deposition, the waiting wafers #3-5 will continue to wait, and will be joined by wafers #6-10, if that many buffer slots are available (Fig. 8.24(b)). That's eight wafers waiting; only two in production or completing.

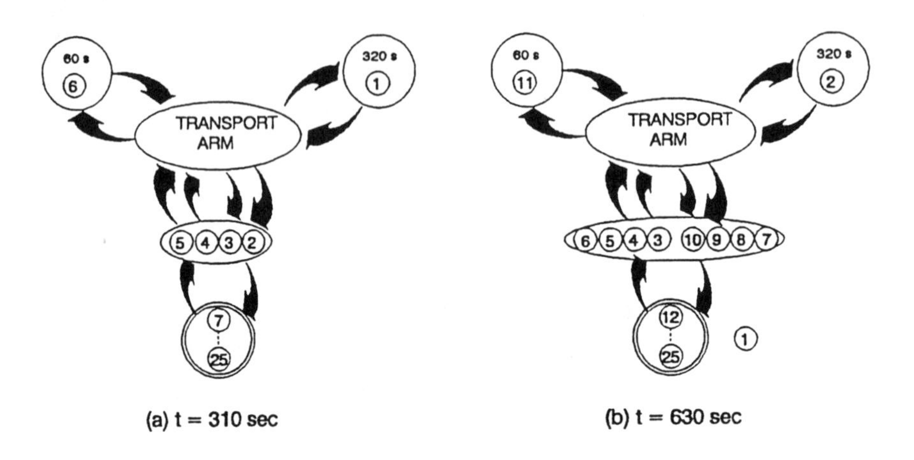

Figure 8.24: Buffer Waits with Long Downstream Step

The progression associated with a very long downstream process time is clear: at some point, the buffer will fill, and all subsequent wafers will wait in Chamber A until Chamber B is free. When Chamber B does clear, the wafer that went into the buffer first will be put into B, the wafer waiting in Chamber A will move to the buffer, and Chamber A will be filled with a new wafer to process, then wait.

Once the buffer has filled, all wafers will spend 260 seconds (320 minus 60) waiting in Chamber A (plus transport and overhead times). The buffer's sole function will be to add to the burden of the transport arm, moving wafers in and out of it **after** they have already spent the long wait time in the process chamber. Furthermore, until the buffer has filled, the wafers will spend varying wait times in the process chamber, which may be more of a problem than long, but consistent wait times.

Example 3: 'Seemingly-Random' Results in a Buffered PIPE

Although the widely disparate process times in Example 2 clearly demonstrate the problems that can occur with buffered PIPEs, perhaps a more-dangerous situation arises when the recipe appears to be fairly balanced, as is the case in the example in Table 8.10.

TABLE 8.10: "RELATIVELY-BALANCED" RECIPE
Buffered PIPE

Process Step		Process Time	PIPE Resource	
1	Soft Clean	24 sec	1	Chamber A
2	PVD TiN	30 sec	2	Chamber B
3	CVD	30 sec	3	Chamber C
4	Etchback	35 sec	4	Chamber D

As is always the case, Wafer #1 will have a clear path through the PIPE because it will not have to wait for any other wafers; it is basically starting with an empty machine. The only waits incurred by Wafer #1 will be a result of the transport arm being otherwise occupied when a process step completes.

When Wafer #1 completes processing in Chamber A, it will be collected by the transport arm and deposited into Chamber B. As soon as this happens, Wafer #2 will be collected and moved into Chamber A. Because the process time in Chamber A is shorter than the time in Chamber B, Wafer #2 will be returned to the buffer to wait. Meanwhile, Wafer #3 will be moved into Chamber A (Fig. 8.25(a)).

However, if the transport arm and machine overhead times are extremely fast, Wafer #1 may complete processing while Wafer #2 is being moved to the buffer. In such a case, the PIPE's control rules will mandate that Wafer #1 has priority over Wafer #3, and it will be moved to Chamber C. Wafer #2 will be collected from the buffer and moved to Chamber B; and only then will Wafer #3 be moved to Chamber A (Fig. 8.25(b)).

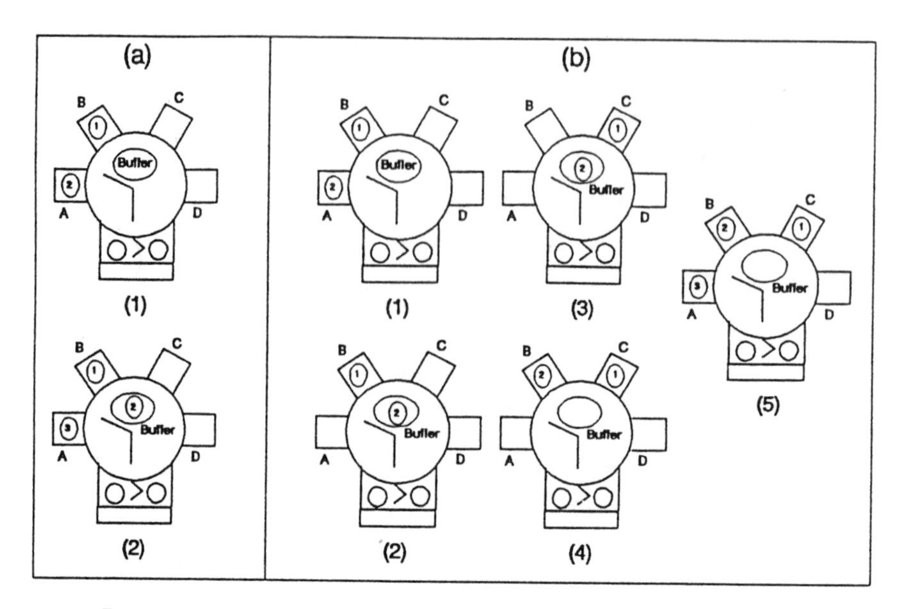

Figure 8.25: Wafer Movement in "Relatively-Balanced" Recipe

A likely scenario with such a closely-balanced recipe is that both 'routes' will be used; some wafers in the cassette following the first, others the second. In fact, many of the eight possible routes could conceivably be used by one or more of the wafers in the cassette. (Fig. 8.26 shows the OPS graphs for the two extremes. See Exercise #1 for the other six possible routes.)

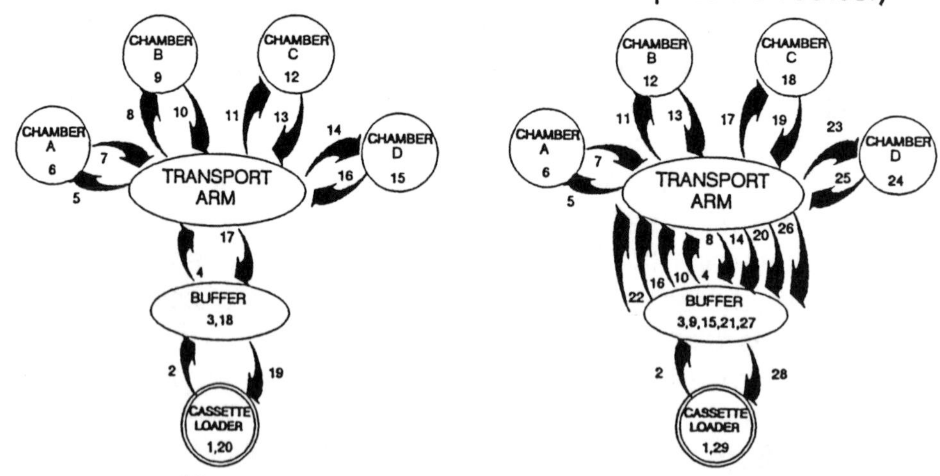

Figure 8.26: Two of Eight Combinatorial Routings

Needless to say, depending on the types of processing being performed, such a variety of operations sequences could have a significant impact on wafer-to-wafer uniformity. The problem may be even more deleterious if it is unclear which wafers follow which paths, or, worse yet, the multiplicity of paths is totally unexpected.

So how does one know for sure exactly what is happening to wafers inside the machine? One method is to draw the operations graphs for each possible operations sequence, know the exact timing parameter that will contribute to wafer processing and wafer movement (Table 8.1), and then, with a pencil and paper, march pennies through a machine schematic (such as those in Fig. 8.20). Such a tedious, but simplistic exercise will eliminate unforeseen surprises and explain seemingly-random processing results.

By far, the most reliable and accurate method to ascertain what is happening inside a PIPE is to use a detailed, machine-specific computer model that can predict exact wafer movement as a function of recipe, machine configuration, timing parameters, and machine control rules. Unfortunately, not only are most timing parameters and control rules unavailable, but most computer models are built with simulation languages and take a 'quick-and-dirty' approach, relying heavily on short-cuts and gross assumptions about machine behavior. Furthermore, even after the models are built, they tend to have limited flexibility when it comes to configuration options and control rules.

Perhaps the best way to comprehend cluster tools, and what happens to wafers inside them, is to develop a fundamental understanding of their *system* dynamics. By knowing the system dynamics of PIPEs, as well as some rudimentary, but specific information on timing parameters and control rules, it is possible to anticipate when a cluster tool will give good or poor process results. This same fundamental understanding makes it possible to knowledgeably configure a tool for a given set of process steps.

8.5 System Dynamics

In production, fab managers have often described cluster tool performance as being random. But nothing could be further from the truth. Because the cluster tool is a mini-factory with a small, finite buffer capacity, it is a very tightly-constrained system, and it must be treated as such. The 'randomness' that has been observed in tool performance is the result of allowing it too many degrees of freedom, i.e., not *capping* the number of possible operations sequences through the system, and limiting when each sequence will be used.

Although cluster tool control software does not permit the programming of operations sequences, or assignment of wafers to sequences, such constraints can *de facto* be imposed by:

- selecting appropriate steps to cluster,
- adjusting process times to eliminate waits,
- configuring the cluster tool to balance times, and
- eliminating the need and/or use of buffers.

PIPE dynamics will vary dramatically depending on the type of processing being performed. Of course, there will be some variation from vendor to vendor, depending on whether the machine's design methodology was plain vanilla, exotica, or *atrocitica*. But, for each type of processing, certain fundamental behavior may be expected.

8.5.1 Parallel Processing

Parallel processing, where all chambers of the cluster tool perform exactly the same process step, should, in principal, not incur waits. The exception arises, however, if the transport arm is bottlenecked. A bottlenecked transport arm may be the result of a short process time, too many process chambers on the tool's platform, and/or an arm that must service pre-/post-processing in addition to the process chambers. Of course, if the arm is bottlenecked or close to being so, the addition of process chambers exacerbates the wait problem.

To illustrate the dynamics of parallel processing, consider the plain-vanilla radial cluster tool in Fig. 8.27.

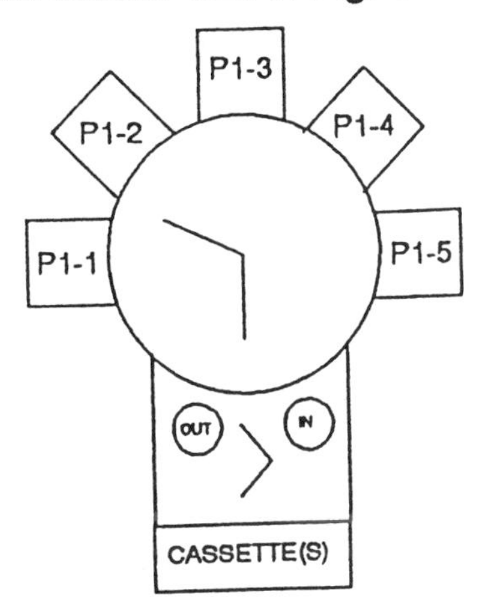

Figure 8.27: Adding Chambers for Parallel Processing

For a representative (though conservative) set of transport and machine overhead times, the tool's performance results are shown in Fig. 8.28. In particular, the graphs show the impact on performance of adding successively more process chambers to the cluster tool's platform. For this particular example, four different processing times are portrayed (all chambers at 30 seconds; all at 60 seconds; at 90 seconds; and at 120 seconds). The four different processing times are within the norm for single-wafer processing.

As may be seen in Figs. 8.28(a) and (b), going from one to two chambers of parallel processing always benefits throughput and cycle time. Addition of the third parallel chamber also has advantages, especially for the longer processing times (60+ seconds). The addition of the fourth chamber may improve throughput and cycle time for long parallel processing. But, as seen in Fig. 8.28(c), the cost of the improved manufacturing performance may be at the expense of long wait times, and hence poorer process performance. For this particular example, addition of the fifth chamber provides little or no benefit. Further, in the 30- and 60-second cases, the fifth chamber is never used.

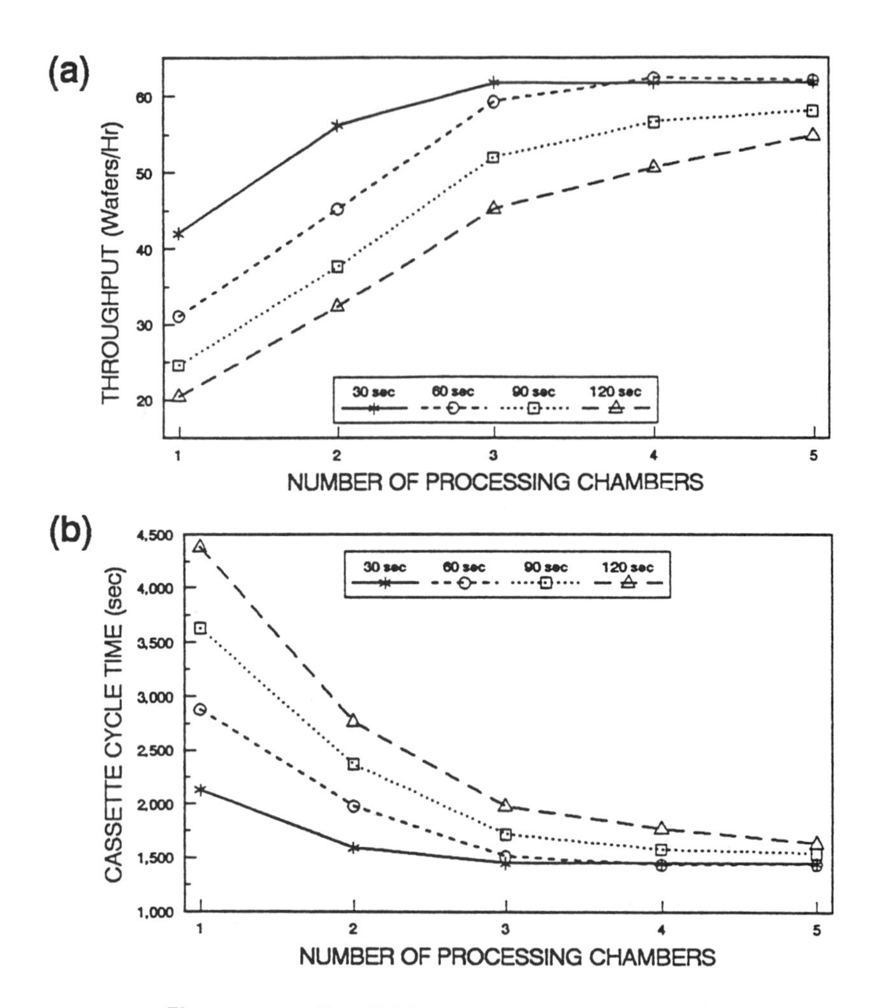

Figure 8.28: Parallel-Processing PIPE Performance

Fig. 8.28(d) explains why manufacturing performance starts leveling off after the addition of the third chamber. Except for the long process times (90 and 120 seconds), the transport arm has become the bottleneck resource. For the shortest process time (30 seconds), the fourth chamber may be added, but it will never be used. The transport arm simply cannot move the wafers in and out of more than three parallel chambers when the process time is this short.

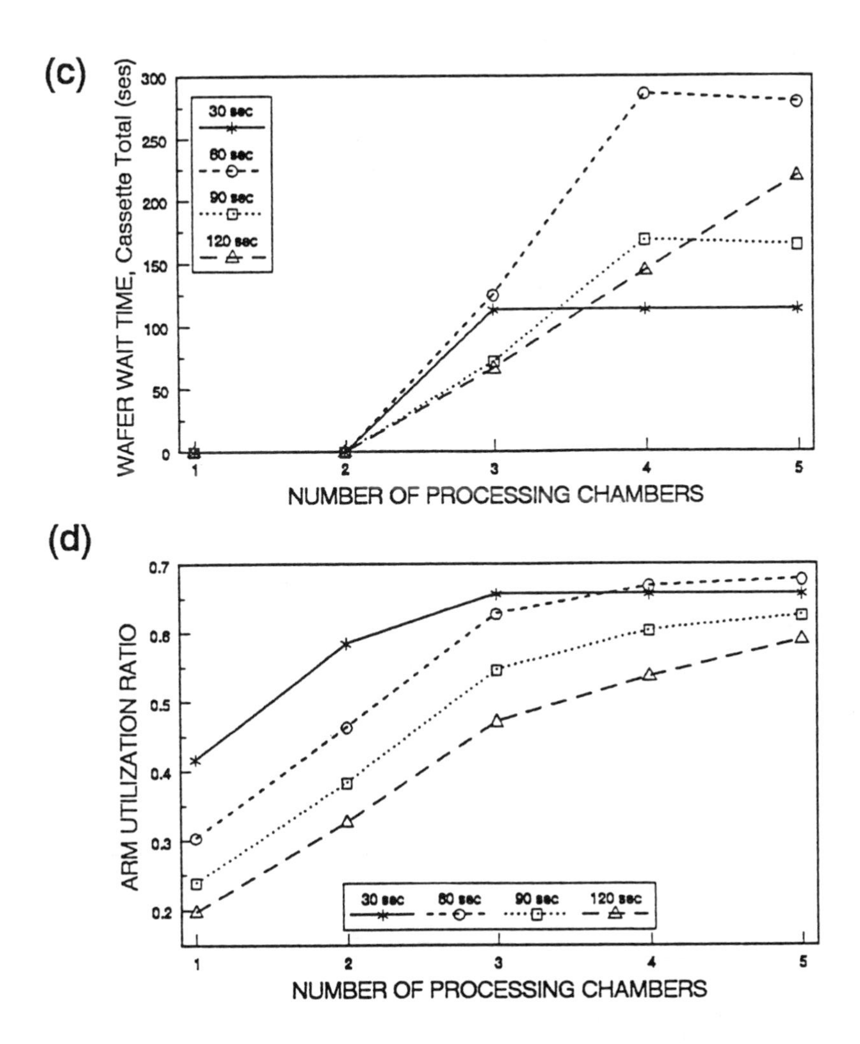

Figure 8.28 (cont): Parallel Processing PIPE Performance

It is interesting to note that the wait-time curves for this type of processing have a characteristic shape. The transitions in the curves are a function of inherent machine characteristics for the configuration and process-type dynamics. The inherent machine characteristics specifically relate to a machine's transport and overhead times.

A different machine, with faster transport and overhead times and/or multiple transport mechanisms, may be able to support more process chambers before transport becomes the tool's bottleneck. However, even though the numbers may shift up or down for the various cluster tools on the market, the trends shown in these figures will usually be representative of the system dynamics of parallel processing.

Of course, for every generalized statement such as the one just made, there is always a good counterexample. As is often the case in cluster-tool analysis, it is easy to overlook or neglect a system component when it is not directly involved in wafer processing. If the overlooked component is a long post-processing step, then the unanticipated wafer waits can be excessive.

Table 8.11 shows the wafer-to-wafer uniformity report for two different cluster tools, each with two chambers performing 60-second parallel processing. The first tool, Machine #1, has a 30-second post-processing step; the other tool, Machine #2, has a 40-second post processing step.

TABLE 8.11: WAFER-TO-WAFER UNIFORMITY REPORT

MACHINE #1 = 30 sec Post-Processing			MACHINE #2 = 40 sec Post-Processing		
WAFER	--ACTUAL\EXPECTED\WAIT--		WAFER	--ACTUAL\EXPECTED\WAIT--	
#	P1-1	P1-2	#	P1-1	P1-2
1	60 /60 /00	---	1	60 /60 /00	---
2	---	60 /60 /00	2	---	63 /60 /03
3	60 /60 /00	---	3	66 /60 /06	---
4	---	60 /60 /00	4	---	70 /60 /10
5	60 /60 /00	---	5	73 /60 /13	---
6	---	60 /60 /00	6	---	76 /60 /16
7	60 /60 /00	---	7	79 /60 /19	---
8	---	60 /60 /00	8	---	80 /60 /20
9	60 /60 /00	---	9	79 /60 /19	---
10	---	60 /60 /00	10	---	80 /60 /20
11	60 /60 /00	---	11	79 /60 /19	---
12	---	60 /60 /00	12	---	80 /60 /20
13	60 /60 /00	---	13	79 /60 /19	---
14	---	60 /60 /00	14	---	80 /60 /20
15	60 /60 /00	---	15	79 /60 /19	---
16	---	60 /60 /00	16	---	80 /60 /20
17	60 /60 /00	---	17	79 /60 /19	---
18	---	60 /60 /00	18	---	80 /60 /20
19	60 /60 /00	---	19	79 /60 /19	---
20	---	60 /60 /00	20	---	80 /60 /20
21	60 /60 /00	---	21	79 /60 /19	---
22	---	60 /60 /00	22	---	80 /60 /20
23	60 /60 /00	---	23	79 /60 /19	---
24	---	60 /60 /00	24	---	80 /60 /20
25	60 /60 /00	---	25	79 /60 /19	---

In either machine, the wafers from both process modules (P1-1 and P1-2) must undergo a post-processing step before exiting the machine. But, whereas there are two physical entities capable of performing the 60-second process operation, there is only one entity for the 30- or 40-second post-processing operation. Thus, for the 60-seconds of parallel load, Machine #1 has 60-seconds of post-processing load, and Machine #2 has 80-seconds. For the latter case, waits are inevitable.

The definition of 'long' when it is applied to post-processing will get shorter as more parallel chambers are added. But be strongly advised: **post processing can cause waits, impacting both process and manufacturing performance.**

Parallel Processing in a Buffered PIPE

Before moving on, we need to briefly address the use of a buffered PIPE in parallel processing. Since only one process step is taking place inside the machine, the use of a buffer is **never** appropriate for this type of processing. When the process step is complete, the wafer should exit the machine. If it cannot do so conveniently, then the machine has not been properly configured.

The only exception to this generalized statement may be the case where a long post-processing step causes wafers to wait. If a buffer is to be used for these waits, then the machine should have a separate transport arm to provide movement between the buffer and the post-processing step.

8.5.2 Inherent Machine Characteristic

Before discussing the system dynamics of sequential and more complex processing, it is important to understand the concept of the inherent machine characteristic. The *inherent machine characteristic* (IMC) is the absolute minimum time between any two process chambers being able to start processing. In most machines, processing can not begin simultaneously because, after a wafer is deposited in one chamber, it takes a certain length of time for a second wafer to be loaded into another chamber and prepared for processing. (The exception occurs with a perfectly-synchronized dual-loading arm.)

Figured in the IMC time is movement of the transport mechanism to collect a second wafer, transport of the wafer to the process chamber, and any wafer/chamber preparation that may be necessary (raise pins, rotate wafer, clamp down, etc). Generally, the IMC is not simply a summation of these times. Depending on the tool's control rules, many of the activities may be performed in parallel.

In addition to the IMC being a function of the control rules of a given cluster tool, it will also vary depending on machine configuration and process recipe. The same 3-chamber machine may have a different IMC depending on whether it is performing strictly sequential or mixed processing. In general, though, the IMC will be equal to or greater than the minimum transport time.

The concept of IMC should become clearer as we examine the system dynamics of sequential processing.

8.5.3 Sequential Processing, Balanced Recipe

Fig. 8.29 shows a typical radial cluster tool that may have anywhere from one to five process modules on its central platform. In strictly-sequential processing, a wafer will move to every module on the platform for a different process step. Some highly advanced wafer fabs, however, have explored returning to process modules for additional steps. Tool operation and system dynamics become extremely complex in such cases, and will not be discussed here. (Advanced topics on cluster tools are covered in [8.2].)

To investigate the impact of adding successively more steps inside a single tool, we are going to be adding both steps and process chambers.

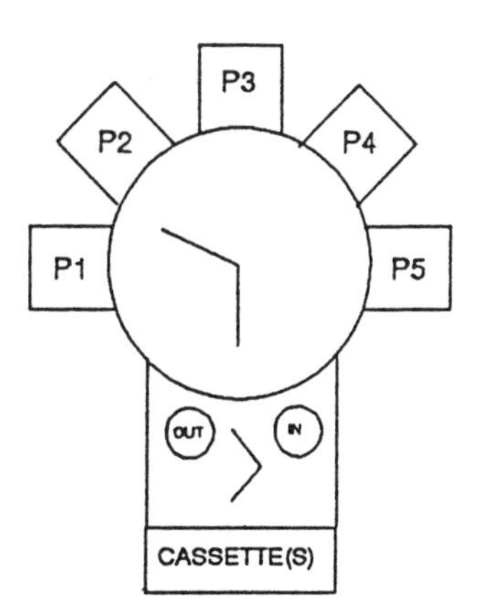

Figure 8.29: Adding Steps and Chambers in Sequential Processing

To simplify this first look at sequential processing, all steps put into the tool will have the same length of process time (i.e., two, three, four, or five 30-second steps, etc.). The process times studied are: 30, 60, 90, and 120 seconds.

Figs. 8.30(a) and (b) show machine throughput and cycle time, respectively. As may be seen, except at very short times, the simultaneous addition of process steps and chambers has almost negligible impact on manufacturing performance; machine throughput goes down very little and machine cycle time sees very little increase. In other words, the additional process steps are virtually 'free' from a manufacturing perspective. This observation makes one of the major industry complaints about cluster tools seem rather frivolous.

An excuse often tendered for not clustering is that no piece of equipment will exceed 30 wafers per hour of throughput. And indeed, at typical process times (60+ seconds), 30 wafers/hr is about the best the equipment can do. **But**, for sequential processing, the throughput of the cluster tool is roughly 30 wafers/hr, regardless of whether it is performing one step or five steps. To compare the throughput over the number of process steps performed, it would be approximately 150 wafers/hr for the five-step case.

Being able to perform a second, third, fourth, etc. process step in a cluster tool without incurring additional

machine cycle time is a tremendous economic benefit. Effective throughput is doubled, tripled, quadrupled, etc.

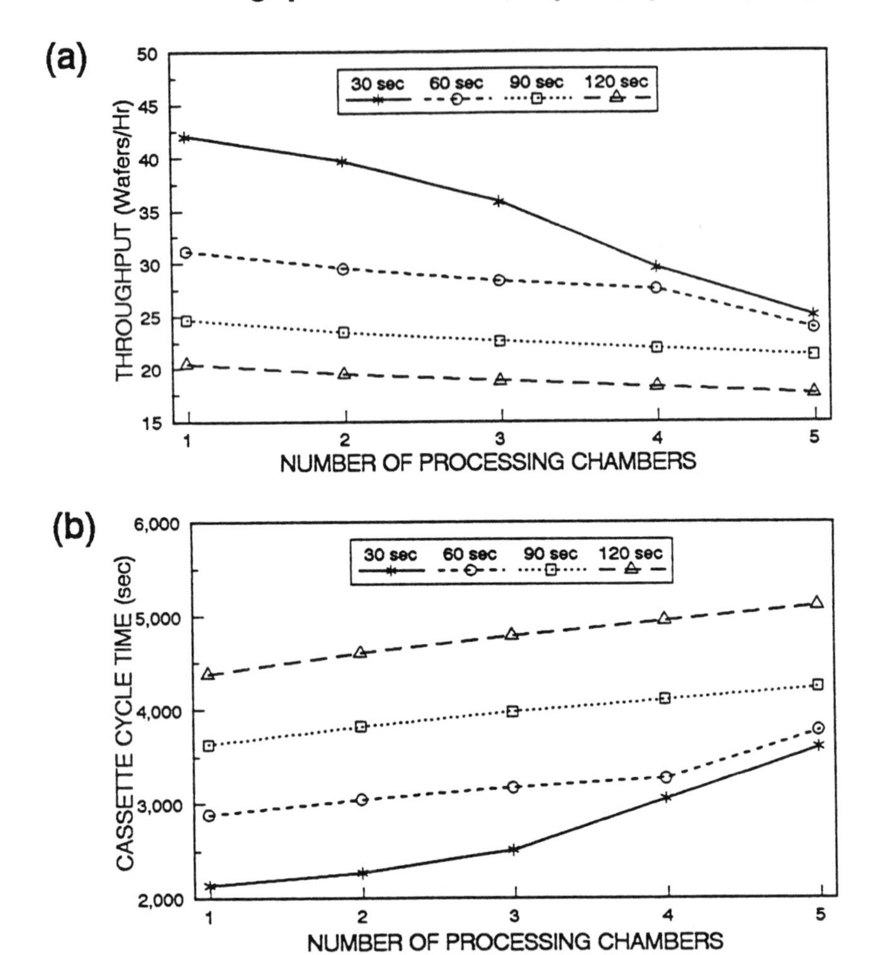

Figure 8.30: Sequential Processing PIPE Performance

But, why are these additional steps essentially 'free'?

The answer is: simultaneous processing. As long as the transport arm does not bottleneck and, thus, is able to keep the chambers stocked with wafers, steps two, three, four, etc. are processing at exactly the same time as step one. Other than a little non-overlapping at the end of the cassette, manufacturing performance is basically what would be expected for just one step. It is the length of that one step that determines the throughput and cycle time performance of the machine.

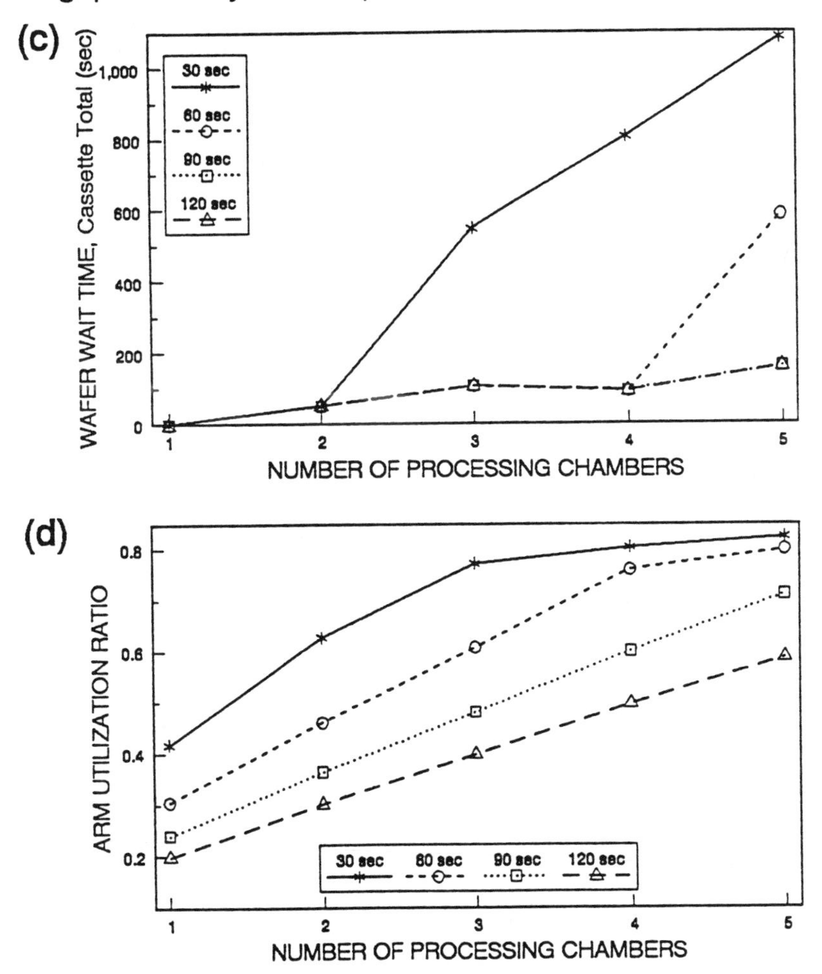

Figure 8.30 (cont): Sequential Processing PIPE Performance

If the transport arm does not bottleneck, the length of the longest sequential process step determines the cassette cycle-time and throughput of the machine. Once the transport arm bottlenecks, the story changes.

As seen in Fig. 8.30(c) for the 30-second case, with the addition of the second chamber, the transport arm becomes quite busy; the addition of the third chamber causes it to bottleneck.

For such a short process time, the arm is constantly in motion trying to keep wafers moving through the process sequence. But, as seen by the excessive waits in Fig. 8.30(d), the arm simply can not keep up with the demands placed upon it. Arm utilization maxes out at about 0.80. Addition of more chambers will not squeeze any more usage out of the arm; it is already busy **all** the time. More chambers will simply incur more waits.

For the longer 60-second recipe, the transport arm bottlenecks with the addition of the fifth chamber. There is a modest drop-off in throughput and cycle time, and substantial waits are starting to accrue. Again, with five chambers, the transport arm is being used virtually all the time (0.80 utilization ratio).

At the longer process times (90 and 120 seconds), the arm does not bottleneck, even with five process chambers. However, if the central platform were capable of supporting more process chamber, there would come a point were the addition of one more chamber would cause arm utilization to reach 0.8 and wafers would begin to wait.

It is a fundamental rule of sequential processing that all steps must be performed, and all chambers must be visited. It is a corollary to this rule then that, if too many steps are added or if the multiple steps performed are too short, the transport arm **will** bottleneck, and waits will occur. The exact definitions of *too many* and *too short* will vary from machine to machine, and will be a function of the speed of the transport mechanism and, to some extent, the machine's IMC.

8.5.4 Sequential Processing, Unbalanced Recipe

So far, the sequential recipes considered have all had steps with exactly the same length of process times. But in the real world, clustered steps will vary in length; some considerably. Some examples of unbalanced recipes are: a tungsten CVD (150 sec) followed by a tungsten etch (60 sec); a titanium-metal sputter (180 sec) followed by W-CVD (90 sec) followed by a short

W-etch (30 sec); and a five-step phospho-silicate glass (PSG) oxide planarization consisting of deposition (30 sec), etch (60 sec), dep (30 sec), etch (90 sec), and dep (120 sec).

As discussed above, a cluster tool's manufacturing performance (i.e., throughput and cycle time) is determined by the length of the longest process step. But it is the sequence or order of the process steps in the recipe that governs the amount of wait times experienced by the wafers. So, if wafer waits degrade process performance by causing lower yields, then the location of the longer process step(s) in the recipe should be considered when making clustering and/or configuring decisions.

Long process steps followed by shorter ones (sequentially decreasing) will generally not incur waits. Conversely, short process steps followed by longer ones (sequentially increasing) will incur waits, usually even if a buffer is present. When the process times go up/down/up/down or up/down/up/up, as in the case of the planarization above, then there will be waits after short steps that are followed by longer ones. Of course, if the transport arm bottlenecks, there could even be waits for a strictly-decreasing recipe.

To study the impact of process-step sequence, consider the recipes given in Table 8.12. The first four entries are 2-step recipes, two balanced and two unbalanced.

Two-Step Recipe

Of these 2-step cases, the best performance (machine throughput and cycle time) is seen with the 60-second/60-second recipe. Since manufacturing performance is rate limited by the slowest (or longest step), and since the other three recipes have at least one 90-second step, little performance difference is seen between them. Once the 90-second process time has been set as the longest step, the length of the other step really does not matter.

The impact of sequence (i.e., the order of the long and short steps) is seen in the wait times, and, therefore, most likely in process performance. The long step followed by the short step has no waits; the short step followed by the long step has considerable waits; the balanced recipes have very small waits that are equivalent to the inherent machine characteristic for a two-module configuration.

TABLE 8.12: IMPACT OF PROCESSING SEQUENCE
Sequential Processing / No Buffer

PROCESS TIME OF EACH CHAMBER	ARM UTILIZATION	THROUGHPUT (wafers/hr)	CASS. CYCLE TIME (sec)	TOTAL WAIT TIME (sec)	WAIT TIME/ WAFER (sec)
60/60	0.4620	29.5	3,049.6	52.8	2.1
60/90	0.3688	23.7	3,896.0	772.8	30.9
90/60	0.3736	24.0	3,746.8	0.0	0.0
90/90	0.3658	23.5	3,829.6	52.8	2.1
60/60/60	0.6075	28.3	3,176.0	107.4	4.3
30/60/90	0.4922	23.1	3,896.0	2,237.4	89.5
90/60/30	0.3736	22.1	4,077.5	224.4	9.0
90/90/90	0.4808	22.6	3,986.0	107.4	4.3
60/60/60/60	0.7606	27.5	3,274.3	94.1	3.8
30/60/90/120	0.5083	18.9	4,774.1	4,362.7	174.5
120/90/60/30	0.5105	19.1	4,716.4	150.0	6.0
90/90/90/90	0.6012	21.9	4,113.2	94.1	3.8

WAIT TIME/WAFER = Total Wait Time/25 Wafers
ARM UTILIZATION includes Pump and Vent Times

Three-Step Recipe

The results for the 3-step recipes are similar. The best manufacturing performance is obtained for the sole recipe that does not have a 90-second step (60/60/60). Throughputs and cycle times for the other recipes, each having at least one 90-second step, are roughly equivalent. Again, the impact of sequence is seen in the wait times.

The sequentially-decreasing recipe has only small waits. These small waits occur only when the non-bottlenecked arm is needed by two wafers at once. The sequentially-increasing recipe has substantial waits associated with it. The two balanced recipes have very small waits, again equivalent to the IMC for the three-module configuration.

Four-Step Recipe

There are no big surprises for the 4-step recipes other than the fact that the transport arm has become incredibly busy for the 60/60/60/60 case (0.7606 utilization ratio).

Balancing Recipes

Perhaps of special note for all the cases in Table 8.12 is that the balanced recipes give the overall best manufacturing **and** process performance, i.e., bétter production and better yield. Furthermore, balancing a clustered recipe will generally give more predictable results, make less demands on the tool's control system, and provide the greatest wafer-to-wafer uniformity. Therefore, it stands to reason that process engineers should give serious consideration to balancing a recipe around the longest step being performed in the tool.

Since the longest step has already predetermined the machine's throughput and cycle time, why not slow the other steps down to improve overall performance? The only roadblock to doing so is the predilection and mindset that 'faster' is better. In reality, slower recipe steps may, in fact, lead to better devices and possibly even higher yield.

For deposition, slower rates may give:

- more uniform thickness,
- better structural integrity, and
- superior film properties.

For etching, slower rates may give:

- better definition,
- more complete clearing, and
- less device damage.

In addition to showing the advantages that go along with balancing a recipe, Table 8.12 also provides a clear illustration of the concept of inherent machine characteristic, and how this number may be used to achieve optimal process implementation.

In particular, note the average wait time per wafer in each of the balanced-recipe cases. For both the 60/60 and 90/90 cases, the average wait time per wafer is 2.1 seconds. For a 2-step recipe in this particular tool, 2.1 seconds corresponds to an inherent transport delay, associated with the design and transport characteristics of the specific machine, regardless of the length of processing time. For a 3-step recipe in this machine, the IMC is 4.3 seconds; and for a 4-step recipe the IMC is 3.8 seconds. (The improved characteristic in going from three to four steps is due to the greater utilization efficiency of the transport arm.)

Process Development for Balance

By identifying the IMC, and knowing beforehand that the recipe should be balanced around the longest step, the logical process development is to make each successive step in the recipe sequence equal to the length of the step before it minus the IMC. Thus, for the 2-step recipe, if the longest step is 60 seconds, the next step is 57.9 seconds. If the longest step is 90 seconds, the next step is 87.9 seconds, etc.

Table 8.13 shows the results of using the IMC to balance the process recipe around the longest process step. Clearly, in every case, optimal sequencing is achieved and process and manufacturing performance are at their best.

Of course, an astute observer will immediately observe that one recipe is conspicuous by its absence: 60/60/60/60 which becomes 60/56.2/52.4/48.6.

If this IMC-adjusted recipe were entered on the table, there would indeed be seen a slight improvement in performance (28.2 wafers/hr throughput and 3195.2 seconds cassette cycle time). But the total wafer wait time goes up (391.7 seconds), not down as expected. The reason for the increase in wait time is to be found in the arm utilization ratio. The arm, which was already very busy for the non-adjusted recipe (0.7606 ratio), becomes bottlenecked (0.8277 ratio) with the faster process times.

TABLE 8.13: OPTIMAL SEQUENCING
Sequential Processing / No Buffer

PROCESS TIME OF EACH CHAMBER	ARM UTILIZATION	THROUGHPUT (wafers/hr)	CASS. CYCLE TIME (sec)	TOTAL WAIT TIME (sec)	WAIT TIME/ WAFER (sec)
60/60	0.4620	29.5	3,049.6	52.8	2.1
60/57.8	0.4708	30.1	2,994.6	0.0	0.0
90/60	0.3736	24.0	3,746.8	0.0	0.0
90/90	0.3658	23.5	3,829.6	52.8	2.1
90/87.9	0.3705	23.8	3,777.1	2.4	0.1
90/87.8	0.3708	23.8	3,774.6	0.0	0.0
60/60/60	0.6075	28.3	3,176.0	107.4	4.3
60/55.7/51.4	0.6286	28.8	3,125.0	0.0	0.0
90/90/90	0.4808	22.6	3,986.0	107.4	4.3
90/85.7/81.4	0.4957	22.9	3,935.0	0.0	0.0
60/60/60/60	0.7600	27.5	3,274.1	93.9	3.8
90/90/90/90	0.6008	21.9	4,114.1	93.9	3.8
90/86.3/82.5/78.8	0.6125	22.3	4,035.6	0.0	0.0

Wafer Traces

To understand how the transport mechanism bottlenecks, consider the partial wafer traces for the two cases. Fig. 8.31(a) shows wafer movement for the 60/60/60/60 process sequence, and Fig. 8.31(b) shows wafer movement for the adjusted case.

A wafer trace shows how each wafer, W_N, utilizes the machine's resources, Arm, P1, P2, Return CM, etc. 'Arm' is the transport mechanism, the 'Pi' are the processing modules, and 'Return CM' is the return to the cassette module. Waits are indicated by the dotted portions of the lines.

Not surprisingly, in the 60/60/60/60 case, the wafers move smoothly through the machine with delays resulting only from the transport characteristic. For the adjusted case, however, after the fourth wafer enters the PIPE, the transport arm bottlenecks; it stays busy <u>all</u> the time. Thus, when a wafer completes a step, it must wait until the arm is free again before it can be moved to the next step.

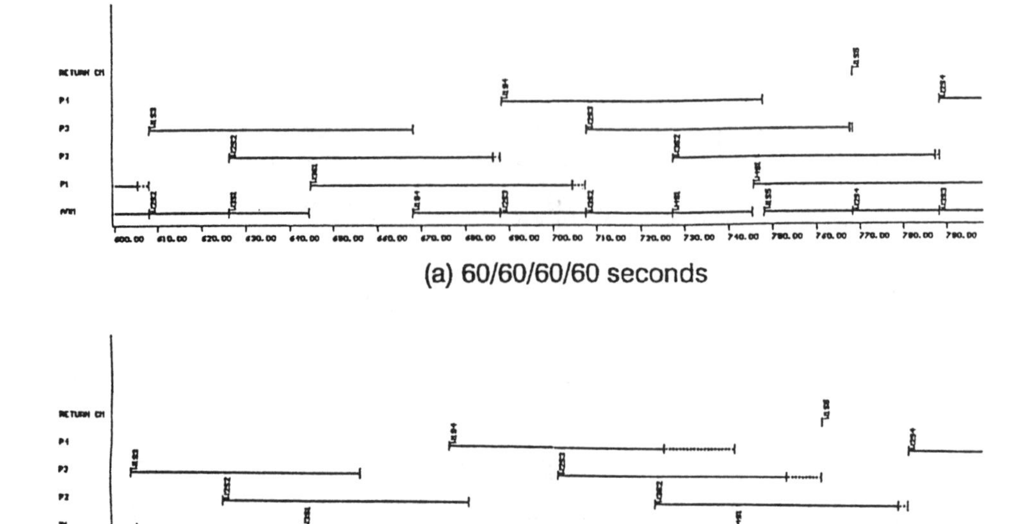

(a) 60/60/60/60 seconds

(b) 60/56.2/52.4/48.6 seconds

Figure 8.31: Wafer Traces for 4-Step Sequential Processing

So, for a 4-step recipe in this particular machine, completely eliminating waits for the 'balanced' 60-second recipe, might require successively increasing upstream times, as opposed to decreasing downstream times. But incrementally increasing process times by the IMC would drop machine throughput and increase cassette cycle time. In the worst case, where the 60-second step is the final one (71.4/67.6/63.8/60), the longest process time would be increased by almost 20%.

Manufacturing vs Process Tradeoffs

In this type of scenario, the process engineer would have some serious tradeoff decisions to make. For short processing times, clustering an additional step may be highly economical in terms of adding an extra chamber to an already-paid-for platform, as well as getting the process step basically 'free' in terms of manufacturing performance. But the real cost of clustering the extra step may show up as substantial and varied waits, possibly at reactor conditions.

Since the major driving force behind clustering is precision processing, such wafer-to-wafer variation is not acceptable in most situations. It is, therefore, often quite critical that the process engineers know exactly what is happening inside the PIPE to each and every wafer in the cassette.

8.5.5 Mixed Processing

In mixed processing, more than one process step will be performed in the PIPE, and at least one of the processing steps will have multiple chambers available that are capable of performing it. Mixed processing, therefore, involves both sequential and parallel processing.

Consider the hypothetical case of a three-step mixed process recipe, where the processing times for the first two steps are 90 seconds each. In order to study the impact of following these two long steps by a short third step, the process time for P3 will be varied from 20 to 60 seconds. But, since the first two steps are long compared to the third step, P1 and P2 will each have two chambers available. P3 will have only one chamber. The schematic of the PIPE is shown in Fig. 8.32.

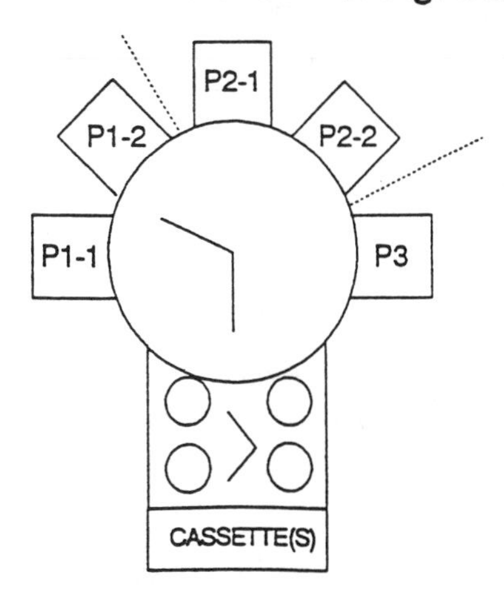

Figure 8.32: 3-Step Mixed Processing

As seen from the schematic, there are four possible chamber sequences for wafers moving through the PIPE:

P1-1, P2-1, P3
P1-1, P2-2, P3
P1-2, P2-1, P3
P1-2, P2-2, P3.

A detailed wafer trace will determine the route followed by each wafer in the cassette. But, as will be seen shortly, the routes the wafers follow through the PIPE, and the wait states associated with each route, will change depending on process recipe.

For the moment, examine the results in Table 8.14 where the process time for the third step is varied from 20 to 60 seconds. Something unusual is obviously happening in the cluster tool when the time in the third chamber goes from 20 to 30 seconds. There is a significant improvement in the tool's manufacturing performance; throughput goes up, cassette cycle time drops, and arm utilization becomes more efficient.

TABLE 8.14: MIXED PROCESSING
Short 3rd Process Step

PROCESS TIME OF CHAMBER 3	ARM UTILIZATION	THROUGHPUT (wafers/hr)	CASS. CYCLE TIME (sec)	TOTAL WAIT TIME (sec)	WAIT TIME/ WAFER (sec)
20	0.6433	29.4	3,062.6	985.6	39.4
30	0.7287	33.2	2,714.5	1,324.4	53.0
40	0.7151	32.4	2,774.2	1,564.9	62.2
50	0.6585	29.8	3,021.7	2,428.0	97.1
60	0.6101	27.5	3,271.7	3,304.9	132.2

Normally, one would expect just the opposite trend in a machine's performance with an increase in process time. Indeed, as the time for the third process step is increased further, from 30 to 60 seconds, the PIPE's performance does again fall off. But why did going from 20 to 30 seconds in the third chamber cause such a dramatic improvement?

After looking at intermediate times between 20 and 30 seconds (Table 8.15), perhaps the more pertinent question is, Why did going from 20 to 21 seconds in the third chamber cause the improvement?

TABLE 8.15: LOCAL OPTIMUM IN PIPE PERFORMANCE
Mixed Processing

PROCESS TIME OF CHAMBER 3	ARM UTILIZATION	THROUGHPUT (wafers/hr)	CASS. CYCLE TIME (sec)	TOTAL WAIT TIME (sec)	WAIT TIME/ WAFER (sec)
20	0.6433	29.4	3,062.6	985.6	39.4
21	0.7390	33.6	2,677.7	1,439.7	57.3
25	0.7354	33.5	2,689.5	1,379.9	55.2
30	0.7287	33.2	2,714.5	1,324.4	53.0

The case where P3 is 21 seconds appears to be a local optimum in every performance category except wait times. In order to understand why the PIPE performs better in terms of cycle time and throughput, even though wafers are spending more time waiting, it is necessary to look at what that one second is doing to the flow of wafers through the PIPE. Table 8.16 gives the detailed chamber waits for each wafer in the cassette for the two recipes, 90/90/20 and 90/90/21, respectively.

For both cases, there is an initial period during which wafer routing and machine operations are coming to an equilibrium. This 'line-out' period takes basically 6 wafers in the 20-second case and 7 wafers in the 21-second case. (If the machine can load more than one cassette, and if cassettes are continuously loaded, this 'line-out' period disappears.)

For the 20-second case, after equilibrium is reached, wafer flow through the PIPE follows one of two paths, each path having an exact waiting pattern. Odd-numbered wafers use P1-1, P2-1, and P3. Each wafer waits in P1-1 for 22.6 seconds, but has no waits after the second or third steps. Even-numbered wafers use P1-2, P2-2, and P3. Each wafer waits 44.1 seconds in P1-2, 20.6 seconds in P2-2, and 0.7 seconds in P3.

TABLE 8.16: WAIT TIMES FOR 3-STEP RECIPE

WAF	20-second case					21-second case				
	P1-1	P1-2	P2-1	P2-2	P3	P1-1	P1-2	P2-1	P2-2	P3
1					18.8					17.8
2		1.6		58.5	0.2		1.6		58.5	18.3
3			6.8					9.2		
4		41.6		8.8	9.6		41.6		12.2	16.7
5						12.8				16.7
6		2.6		21.0	0.7		13.3		19.2	17.7
7	23.0					0.7		29.3		16.7
8		44.5		20.6	0.7		21.9		26.8	17.7
9	22.6					28.4		28.7		16.7
10		44.1		20.6	0.7		26.8		26.8	17.7
11	22.6					27.8		28.7		16.7
12		44.1		20.6	0.7		26.8		26.8	17.7
13	22.6					27.8		28.7		16.7
14		44.1		20.6	0.7		26.8		26.8	17.7
15	22.6					27.8		28.7		16.7
16		44.1		20.6	0.7		26.8		26.8	17.7
17	22.6					27.8		28.7		16.7
18		44.1		20.6	0.7		26.8		26.8	17.7
19	22.6					27.8		28.7		16.7
20		44.1		20.6	0.7		26.8		26.8	17.7
21	22.6					27.8		28.7		16.7
22		44.1		20.6	0.7		26.8		26.8	

For the 21-second case, after line-out, wafers again settle into two distinct patterns. Even-numbered wafers use modules P1-2, P2-2, and P3. The associated waits are 26.8 seconds, 26.8 seconds, and 17.7 seconds, respectively. Odd-numbered wafers use P1-1, P2-1, and P3, with associated waits of 27.8 seconds, 28.7 seconds, and 16.7 seconds, respectively. For this particular example, the uniformity of the waits indicates that the one-second difference in the P3 process time enables more effective utilization of PIPE resources.

In the 20-second case, the processing in P3 is complete when the PIPE looks for the next operation to perform. Thus, the transport arm retrieves the wafer out of P3 and returns it to the cassette. This operation ties up the transport arm for roughly 17 seconds. During that 17 seconds, wafers are not being loaded into the chambers that require the longer processing times.

In the 21-second case, however, processing in P3 is not complete and the transport arm instead keeps the longer-recipe chambers full. Thus, the arm loads and/or moves wafers through the PIPE more uniformly. There will be longer waits associated with each wafer at each step, but the rate of material movement through the PIPE will be considerably higher. The process implications of more, but also more uniform, waits may need serious and careful consideration.

For the moment, assume that cassettes arrive at the cluster tool frequently enough so that the beginning and end wafer-wait effects only occur in unusual circumstances. Normally all wafers will settle down into one of the two wait patterns associated with the respective recipes.

For the 20-second case, wafers will see two markedly different wait patterns. In a case such as this, two distinct PIPE recipes may be more appropriate to insure wafer uniformity. That is, the two distinct operations sequences might be used for different processes, or two different regimes of the same recipe.

For the 21-second case, the waits in any one chamber, for any one step, are within a second of each other. These two distinct operations sequences may be close enough to give acceptable wafer uniformity.

8.5.6 Machine Resonance

The mixed-case example discussed above provides a graphic illustration of some of the complexities of PIPE behavior. Of particular note are the qualitative and quantitative changes in machine behavior that can be caused by even a slight change in a process time (or a machine timing parameter). In mixed processing especially, the last step in a recipe sequence is inordinately influential on PIPE behavior because it gates all upstream operations. Witness the huge impact just one-second had on PIPE performance.

The extreme sensitivities and the rapid transitions to and away from local optima demonstrate the 'resonance' phenomena of PIPE behavior. *Machine resonance* may be defined as the tightly-constrained rhythmic sequences of motion or activity in the PIPE. (Similar tightly-constrained 'resonance' was seen in the old-time bucket brigade when they were in perfect sequence.) In such a tightly-constrained sequence, even a slight perturbation in a timing parameter may shift the resonant behavior to a different operations plateau.

The results seen in Tables 8.15 and 8.16, as well as some of the earlier examples, brings up an interesting and disturbing quandary. If the impact of something so slight as a one-second shift in process time can be like falling off a cliff, then do these cluster tools really belong in a manufacturing environment? Can they be controlled and operated with the needed degree of accuracy? What are the ramifications, both process and manufacturing, if they are not? Finally, why go to so much effort to use cluster tools in the face of such sensitivity and uncertainty?

The answer to the first two questions, about suitability of PIPEs in the fab, is *yes*, but with certain caveats. Although these machines can be extremely sensitive to timing parameters, due to their highly-integrated nature, most of the sensitivity goes away if the tool is configured to operate in a balanced manner. If the tool/recipe cannot be balanced, then clustering may not be wise; a tool operating on the razor's edge will, sooner or later, fall off.

Putting unreliable or unpredictable equipment into the plant is a fab manager's worst nightmare. The degree of trepidation goes up exponentially though if the unreliable equipment turns out to be cluster tools. When an integrated piece of equipment goes down, the fab loses a bigger portion of the line.

An even more serious situation arises if the PIPE continues to operate, but performs in an unanticipated manner. With stand-alone gear, process deviations are usually caught quickly at one of the numerous intermediate inspection stations. Most state-of-the-art cluster tools, however, are not yet sophisticated enough to have in-situ instrumentation for process control. This critical shortcoming will have to be corrected before widespread, heavy clustering occurs in the precision metals/interconnect area. By the time wafers reach this portion of the process flow, they have too much value added to take chances with reliability risks.

But are cluster tools indeed reliability risks? Do they have more than just a novelty future in wafer fabrication? What are the real incentives that keep driving the industry (admittedly at a snail's pace) towards clustering, despite the reliability, complexity, and cost issues?

8.6 Clustering for Manufacturing Performance

Section 8.2 discusses how clustering of process steps means the elimination of manufacturing operations, and hence results in manufacturing simplification. Table 8.2 presents the type of performance gains expected with clustering; and Table 8.4 shows the magnitude of the capital savings that may be achieved with fab simplification. But, after all the positive reasons that were presented for clustering, Section 8.5 delved into the *dark side of the force*, discussing how these complex, tightly-constrained, highly-integrated systems present special design, operational, and reliability challenges. Resistance to clustering is the norm, rather than the exception, bringing the common criticisms that:

- the tools are too complex, hence less reliable;
- the loss of a chamber will bring down the entire tool;
- the loss of a tool will bring down a bigger portion of line;
- in-situ instrumentation is inadequate for process control.

Yet most of these criticisms are valid only when the tools are used for new and/or unusual applications, or when the tools have been improperly configured for manufacturing performance.

Since the introduction of the first radial cluster tool in 1987, these versatile machines have been the tool of choice for the process development engineers. The radials provide the greatest process flexibility for the cheapest cost; they provide the means to accomplish process results that can be obtained **no** other way;

their single-wafer processing capability is the perfect development vehicle; and they are openly flexible to in-situ instrumentation.

But wait! Criticism #4, only two paragraphs back, said 'in-situ instrumentation is not adequate....'

Bingo!

The tools are amenable to in-situ instrumentation, but equipment vendors generally do not receive the necessary feedback on what might improve a tool or its instrumentation. With inadequate feedback, equipment designers must rely on their own best guesses.

By not interacting **with** the equipment vendors during tool design and refinement, the real losers in this guessing game are the chip makers. Rather than getting the exact solutions they need, they instead end up buying pricey gear that is sensitive, of questionable reliability, and may perform abysmally unless configured precisely right.

But, chip sophistication will eventually reach a point that it will require the integration of certain process steps. When that technical wall is reached, stand-alone equipment will no longer be feasible and cluster tools will be used. Hopefully, when that day arrives, the tools that go into production will be co-developed, co-specified, and co-configured by the process-development engineers, the manufacturing engineers, and the equipment designers.

In the mean time, there are some basic insights that will enable wafer fabs to use existing cluster tools and get the best possible manufacturing performance from them.

8.6.1 Clustering Metals Steps

A variation of the simplified CMOS fab used in earlier chapters will provide a not-too-unrealistic entity for illustrating the basics of clustering for good manufacturing performance. The process flow for the unclustered CMOS-1 is given in Table 8.17.

TABLE 8.17: METALS CLUSTERING, CMOS-1

STEP #	PROCESS STEP	WORKSTATION	WKST #
1	LASER MARK	MISC.	1
2	INITIAL OX	DIFF 1	2
3	P-TUB MASK	MASK/ETCH 1	3
4	P-TUB 12	IMPLANT	4
5	NITRIDE DEP	DIFF 2	5
6	FIELD MASK	MASK/ETCH 2	6
7	FIELD DIFF/OX	DIFF 2	5
8	SOURCE/DRAIN MSK 1	MASK/ETCH 2	6
9	S/D 12 1	IMPLANT	4
10	S/D OX 1	DIFF 1	2
11	S/D MASK 2	MASK/ETCH 2	6
12	S/D 12 2	IMPLANT	4
13	S/D OX 2	DIFF 1	2
14	NITRIDE STRIP	MASK/ETCH 2	6
15	GATE OX	DIFF 1	2
16	POLY DEP/DOPE 1	DIFF 3	7
17	POLY MASK 1	MASK/ETCH 1	3
18	POLY DEP/DOPE 2	DIFF 3	7
19	POLY MASK 2	MASK/ETCH 1	3
20	CONTACT MASK	MASK/ETCH 1	3
21	**Ti METAL DEP**	**METAL SPUTTER**	**10**
22	**W-CVD**	**METAL CVD**	**11**
23	**W-ETCH**	**PLASMA ETCH**	**12**
24	**Al METAL DEP**	**METAL SPUTTER A**	**13**
25	METAL MASK	MASK/ETCH 3	9
26	TOPSIDE DEP	DIFF 3	7
27	TOPSIDE MASK	MASK/ETCH 1	3

For the well-balanced line, which is neither building nor depleting inventory, the equipment sizings for the metals workstations (#10-13) are:

Workstation	# of machines
10 Metal Sputter	3
11 Metal CVD	3
12 Plasma Etch	2
13 Metal Sputter-A	2
Total	10

The non-clustered fab was designed for a throughput of 1.0 lot/hour, and an expected cycle time of 80 hours. If the metals steps in this fab are to be clustered, it should be done in such a way that there is no deleterious impact on manufacturing performance. Moreover, because of the eliminated queue waits, cassette unloads and reloads, etc., integration should have a visibly positive impact. If it does not, then the steps have been inappropriately clustered.

Once the decision has been made to cluster the metals steps, then it is time to consider the possible options. Fig. 8.33 shows the enumeration of the possible groupings for the four steps.

NON-CLUSTERED STEPS	STEP-CLUSTERING OPTIONS						
21	21	21	21	21	21	21	21
22	22	22	22	–	–	–	22
23	23	23	–	22	22	22	–
24	24	–	23	23	23	–	23
		24	24	24	–	23	–
					24	24	24

Figure 8.33: Combinatorial Enumeration of Metal Clusterings

Of these groupings, some may be automatically eliminated from consideration for process reasons. For instance, if sequential clustering is to be done at all, then it would be highly unusual **not** to cluster the two tungsten steps (22 and 23). Thus, three of the seven possible groupings are eliminated.

The next phase in the clustering evaluation is to ascertain how the various step clusters will perform from an equipment perspective; i.e., the expected cassette cycle time as a function of clustered steps.

For the hypothetical radial cluster tool, Fig. 8.34 provides the cassette cycle times for the various clusters and machine configurations. The type of tool used here has representative transport, overhead, pre- and post-processing times, and a platform that can accommodate up to four process modules. For a real fab, equipment vendors can generally supply reliable cassette cycle time data if the fab furnishes recipe times.

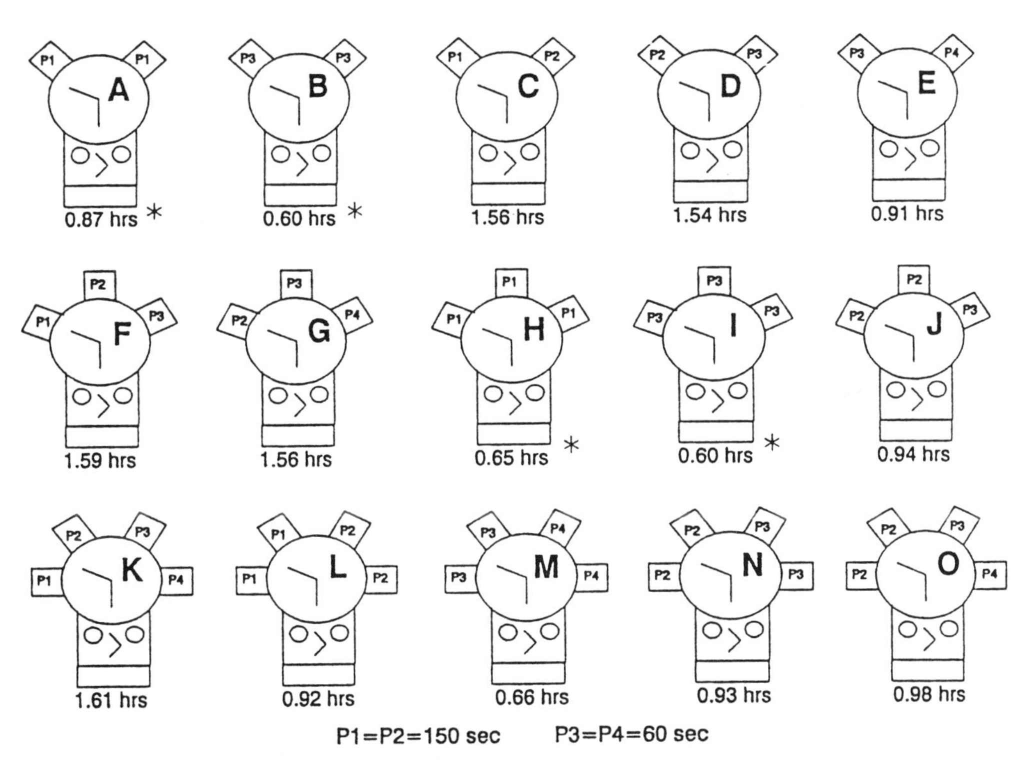

Figure 8.34: Cycle Times of Hypothetical Tool Configurations

It is unfortunate, and purely beyond the equipment vendor's control, that these times are not realized in actual fab operation. But like stand-alone equipment, cluster tools, and thus cluster-tool workstations, are dependent on everything else that happens in the fab, especially upstream. Thus, it is important to remember that these tools are simply one part of the overall system; they only function up to the lowest common denominator of the system (i.e., the fab bottleneck).

As discussed in earlier chapters on factory performance, if a fab has been properly sized, and if it is running reasonably balanced (not waiting for large furnaces to fill, etc.), then workstation utilizations will usually be less than 95%, sometimes much less, for all but the bottleneck or constraining workstation. (So that we will have a basis of comparison, Table 8.18 gives the process times for the stand-alone metallization gear.)

TABLE 8.18: STAND-ALONE METALLIZATION DATA

Step #	Process Step	Workstation #	Workstation Name	Process Time (hrs)
21	Ti Metal Dep	10	Metal Sputter	1.44
22	W-CVD	11	Metal CVD	1.44
23	W-Etch	12	Plasma Etch	0.84
24	Al Metal Dep	13	Metal Sputter-A	0.84

For this clustering exercise, CMOS-1 has been sized to have no true bottleneck; it will be run as a well-balanced line. By having no real bottleneck, the clustering results will not be distorted by other fab behavior.

Table 8.19 shows fab performance as a function of types of clustering, as well as the number of machines needed to maintain the same level of performance as the unclustered fab. In the unclustered fab, throughput is 0.9762 lots/hour, cycle time is 80.72 hours, and the cycle-time-to-process ratio is 2.149.

TABLE 8.19: FAB PERFORMANCE AS A FUNCTION OF METALS CLUSTERING

Metals Workstations	# of Tools per Wkst	$\triangle$Inventory (lots)	Thrupt (lots/hr)	Cycle Time (hrs)	CT/P_{min}
Non-Clust- 21/22/23/24	3 / 3 / 2 / 2	13	0.9762	80.72	2.149
21 / J / 24	3 / 2 / 2	5	0.9921	77.29	2.140
H / J / 24	1 / 2 / 2	3	0.9960	76.90	2.177
21 / J / B	3 / 2 / 1	13	0.9762	76.92	2.144
H / J / B	1 / 2 / 1	5	0.9921	76.81	2.189
F / 24	3 / 2	(6)	1.0139	73.88	2.097
F / B	3 / 1	1	1.0000	73.51	2.101
21 / G	3 / 3	(1)	1.0040	75.81	2.118
H / G	1 / 3	1	1.0000	75.33	2.152
21 / O	3 / 2	(3)	1.0079	74.65	2.119
H / O	1 / 2	(4)	1.0099	74.53	2.165
K	3	(8)	1.0179	72.30	2.107
H / H / B / B	1 / 1 / 1 / 1	5	0.9921	77.82	2.192

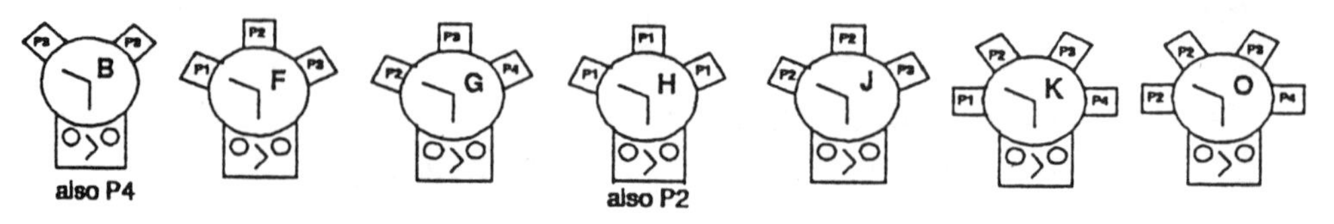

As a reminder, in the non-clustered base case, the four workstations to be considered for clustering, 21/22/23/24, have a total of ten stand-alone machines (3/3/2/2).

Of the four metals steps, the most obvious candidates for clustering are the W-CVD (step 22) and W-Etch (step 23), so these are our first clustering candidates. But since the W-CVD is considerably longer than the W-Etch, tool **J** from Fig. 8.34 will be used. This cluster tool has two chambers available for step 22 and one chamber for step 23. When the fab is re-sized using **J**-type PIPEs in place of the five stand-alone machines, only two tools are needed, taking the total number of machines from ten down to seven.

In addition to reducing the total pieces of gear by clustering, fab performance also improves: throughput rises to 0.9921 lots/hr and cycle time drops to 77.29 hours. These improvements may not sound significant, but a throughput increase of 0.0159 more lots an hour translates into 2.67 more lots a week, or 11.57 more lots a month, or 34.7 more lots a quarter. That's 867.47 more wafers a quarter. If wafers cost roughly $1500 each to make, and sell for $7000, then this slight improvement in fab performance will generate $4.8mm more in profits over a quarter. Well, that was certainly worthwhile, so let's cluster some more.

If, in addition to integrating steps 22 and 23 into **J**-type machines, step 21 is clustered in an **H**-type machine for parallel processing, another slight improvement in fab performance may be achieved (from 0.9762 to 0.9960 lots/hr). Also, clustering this step replaces the three stand-alone machine with one cluster tool. Two tries, two successes.

Leaving step 21 unclustered, but clustering step 24 instead (**B**-type), in conjunction with the **J**-type tools has no positive impact on throughput, though it does offers a slight improvement in cycle time. Clustering this step, however, does allow two stand-alone machines to be replaced by the one PIPE. (For a machine with faster transport and overhead times, some small advantage may be gained by parallel processing.)

The **H**, **J**, and **B** clustering gives the best throughput and cycle time performance so far, as well as the best overall reduction in equipment. The ten stand-alone tools are replaced by four cluster tools: one **H**, two **J**, and one **B**.

The next clustering option to explore is the integration of three process steps, 21, 22, and 23. Eight stand-alone machines will be replaced by three **F**-type machines. Throughput will be 1.0139 lots/hr and cycle time will drop to 73.88 hours. The improvement in throughput over the non-clustered case is 0.0377 more lots an hour, or 82.3 more lots a quarter. The extra 2057.5 wafers will generate $11.3mm more in profits over the quarter ($7000 wafers with production cost of $1500/wafer).

In addition to the improved throughput with this clustering option, the fab's cycle time also saw a marked improvement; the cutback from 80.72 hours to 73.88 hours is an 8.5% reduction, meaning the fab would see 8.5% more turns a year, as well as the associated yield improvements. (See Chapter 5 for more discussion on the yield-cycle time relationship.)

In combination with the **F**-type clustering, parallel clustering of step 24 in a **B**-type machine actually has something of an adverse effect on throughput, even though it does reduce the total number of machines and causes a slight reduction in cycle time.

There is one more 3-step clustering possibility to explore: putting steps 22, 23, and 24 into either a **G**- or an **O**-type machine. If a **G**-type configuration is used, then the eight stand-alone machines are replaced by three cluster tools. Throughput will be 1.0040 lots/hr and cycle time will be 75.81 hours. If an **O**-type configuration is used, the eight machine will be replaced by two cluster tools, and performance will be somewhat superior to the **G**-type tools (1.0079 lots/hr throughput and 74.65 hours cycle time). Neither of these clusterings is quite as impressive as the **F/24** case, however.

As might be expected, the big guns come out and the best performance is achieved by clustering all four steps (21, 22, 23, and 24) in one machine. Only three **K**-type tools are needed to replace all ten of the conventional stand-alone machines. Yet these three machines by far achieve the best throughput and cycle time of all the clustering options (1.0179 lots/hr and 72.3 hours, respectively). The improved throughput over the non-clustered case generates an extra 2275 wafers over the quarter for an additional profit of $12.5mm. The improvement in cycle time is 10.4%.

It is interesting to note at this point that CMOS-1 is a small, fictitious toy fab with only one metal layer. The mind boggles to consider clustering a state-of-the-art fab with four or five metal layers and 8-15 machines per workstation.

8.6.2 Capital Cost and Reliability

With such tremendous fab performance incentives to cluster, it would seem to be no question about doing so. But the extensive amount of processing that has gone into wafers by the time they reach the metals/interconnect area actively encourages rabid conservatism. Clustering is by no stretch of the imagination a conservative approach.

Since reliability is the number one criticism leveled against integrated equipment, fab management must be made to feel confident that these tools will not adversely affect production, that the loss of a chamber on a tool will not bring down the entire tool, or that the loss of a tool will not bring down a substantial portion of the line.

Perhaps the best way to alleviate the reliability issue is to learn from other industries. With the electric utilities, for example, reliability is not optional; for decades it has been standard practice to *spare* (or provide redundancy for) any piece of equipment that is not 100% reliable. So why not do the same with cluster tools?

The standard response is that the tools are just too expensive. But are they really? Table 8.20 provides some representative cost estimates for conventional equipment versus the cost of processing chambers for a generic radial cluster tool.

TABLE 8.20: EQUIPMENT COST ESTIMATES

	Conventional		Radial Chamber	
	Step	Cost ($mm)	I.D.	Cost ($mm)
Ti Metal Dep	21	2.0	P1	0.40
W-CVD	22	1.5	P2	0.30
W-Etch	23	1.0	P3	0.25
Al Metal Dep	24	2.0	P4	0.35

Assuming the bare radial platform, without chambers, is $1.25mm, then we can use these estimates to cost the cluster tools which give the superior manufacturing performance for CMOS-1. These costs may then be used to compare the cluster-tool workstations with the conventional stand-alone workstations.

As we saw in Table 8.19, clustering achieves the same or superior performance than conventional equipment, but requires substantially fewer machines to get it. Thus, regardless of whether two, three, or all four steps are clustered, the savings in capital expenditures is so great that spare chambers can be provided for redundancy and money will still be saved. In fact, entire cluster tools can be spared with money left over.

For example, consider the case where one **H**-type machine is used to cluster step 21, two **J**-type machines are used to cluster steps 22 and 23, and step 24 is left unclustered. The cost of the **H**-type machine is $2.45mm ($1.25mm + (3 x $0.40mm)) compared to $6.0mm for three stand-alone titanium-metal sputters. The cost of the each of the **J**-type machines is $2.1mm ($1.25mm + (2 x $0.30mm) + $0.25mm), or $4.2mm for both. These two tools replace three tungsten CVD machines ($4.5mm) and two tungsten etchers ($2.0mm) that have a total price tag of $6.5mm. Thus, the total savings for clustering these three process steps is $5.85mm (($6.0mm - 2.45mm) + ($6.5mm - 4.2mm)). This $5.85mm could be used to buy an extra **H**-type machine and an extra **J**-type machine as spares and still have $1.3mm left over from the capital budget for the stand-alone equipment.

Now, consider the more radical clustering where all four metals steps are clustered in one machine. This is the case that gives the best manufacturing performance, but it is also the case likely to be the most unreliable. If reliability were not an issue, then only three **K**-type PIPEs could replace ten conventional machines. Each **K**-type tool carries a price tag of $2.55mm ($1.25mm + 0.40mm + 0.30mm + 0.25mm + 0.35mm), for a total workstation cost of $7.65mm. Since the ten conventional machines would cost $16.5mm, the capital saved by clustering would be $8.85mm, enough to buy three spare cluster tools and still have $1.2mm left over.

Table 8.21 provides cost comparisons for a number of clustering options, ranging from spared chambers (**H-1**) to multiple spared machines.

TABLE 8.21: CAPITAL COST OF CLUSTERING

Metals Workstation	# of Tools per Wkst	ΔProfits ($mm)	% Cycle Time Improvement	Cost / Wkst ($mm)	Total Metals Cost ($mm)	Cost Savings ($mm)
21/22/23/24	3 / 3 / 2 / 2	--	Base	6.0/4.5/2.0/4.0	16.50	Base
H / J / 24	1 / 2 / 2	5.4	4.7	2.45/4.2/4.0	10.65	5.85
H / J / 24	2 / 3 / 2	5.4	4.7	4.9/6.3/4.0	15.20	1.30
H-1 / J / 24	1 / 2 / 2	5.4	4.7	2.85/4.2/4.0	11.05	5.45
F / 24	3 / 2	10.3	8.5	6.6/4.0	10.60	5.90
F / 24	4 / 2	10.3	8.5	8.8/4.0	12.80	3.70
F / B	3 / 1	6.5	8.9	6.6/1.95	8.55	7.95
21 / O	3 / 2	8.6	7.3	6.0/4.9	10.90	5.60
21 / O	3 / 3	8.6	7.3	6.0/7.35	13.35	3.15
H / O	1 / 2	9.2	7.7	2.45/4.9	7.35	9.15
K	3	11.4	10.4	7.65	7.65	8.85
K	4	11.4	10.4	10.2	10.20	6.30
K	5	11.4	10.4	12.75	12.75	3.75
K	6	11.4	10.4	15.3	15.30	1.20

Profit/wafer = $5000
Performace measures are quarterly.

Considering the substantial capital savings that may result from clustering, added to the not-insignificant improvements in fab throughput and cycle time, perhaps it is time for the industry to evaluate cluster tools from a manufacturing perspective, as well as a processing perspective.

Other Manufacturing Reductions

As a final note before moving on, it should be pointed out that clustering also has ramifications on the physical size of the fab itself, the size of the clean-room environment, the decrease in the number of operators and/or technicians, the reduction in inventory, and the impact of all these factors on yield.

In the radical clustering case, the number of machines was reduced from ten to three, more than a two-thirds reduction. Added to this decrease in machines is the fact that cluster tools, in general, have a much smaller footprint than conventional equipment. Thus, physical size and the clean-room requirements should be reduced by an even greater degree. Instead of fabs getting chronically larger as flows lengthen, clustering may help to cap facility requirements, at least for a time.

Similarly, less equipment means less operators and, possibly, less technicians. Granted, the personnel dealing with the cluster tools will have to be better trained, but the shear magnitude of the decrease in manpower says that fab operating costs should be reduced.

Finally, every second, third, fourth, etc. process step that goes into a cluster tool can no longer carry work-in-process (WIP). If WIP is not appropriate for the degree of clustering, the fab will become 'clogged' with inventory, and the favorable performance gained from integration will be lost. However, if the WIP level is aligned with the fab's design and operation, then each clustering effort will eliminate WIP and move the fab closer to a just-in-time operation.

Fewer manufacturing steps, fewer opportunities for wafers to be handled, smaller fabs, less WIP, more-highly trained personnel, all should translate directly into improved yield.

8.6.3 Inappropriate Clustering

Before leaving the topic of clustering and manufacturing performance, it should be pointed out that there are certain instances where clustering can actually degrade manufacturing performance. For instance, clustering an unbalanced recipe where the first step is short and successive steps longer can be tricky, and, if not done cautiously, can lead to lower throughputs and/or cycle times. Putting this type of recipe in a buffered PIPE can be especially difficult.

Another situation where clustering may be both difficult, and detrimental to manufacturing performance, is when one of the steps to be clustered is a hub, a multi-use workstation. In such a situation, one or more of the uses may be amenable to clustering, while others may require stand-alone gear. It then becomes necessary to properly size not only the cluster tool workstation, but also the conventional-tool workstations as well. Clustering mixed-use hubs may be a temptation that is best resisted.

Similarly, whereas using the same tools for multiple metal layers may be perfectly logical, using different recipes in the same tool might be precarious. Chamber usage may be sufficiently different when switching recipes to make reliability and maintenance more of a problem, not to mention the possibility of cross contamination.

One final use of PIPEs that deserves some cautionary remarks is being seen in some of the more innovatively-aggressive wafer fabs: tools used for non-consecutive process steps. Because the cluster tool platform is expensive compared to the various process modules that may be hung on it, and because the variety of modules is becoming extensive, cross-use clustering is tempting from both a 'conservation of capital' and a 'conservation of space' perspective.

The problem in having tools used for non-consecutive steps is that more expensive (and possibly more sensitive) later lots may end up spending undue waits while machines are tied up with less-urgent earlier lots. Therefore, unless the fab has good 'look-ahead' tracking, and a set of rational and consistent loading rules, capital conserved by cross-use clustering may be bought at the cost of degraded fab performance. And, as has been repeatedly shown, it takes only a **tiny** degradation in performance to wipe out large capital savings or investments.

8.7 Summary

Just as faster cycle time means improved yield, which in turn means higher revenues, so too does the other fabrication-performance criteria, throughput, impact profits. Even slight improvements in throughput seriously boost revenues. Little wonder then that the cutting-edge wafer fabs, those making the largest profits and producing the most advanced chips, spare little or no expense when it comes to having sufficient equipment, and making sure that said equipment is well maintained.

These 'spare-no-expense' fab houses are more than just an equipment vendor's dream-come-true. Their existence points towards a possible future where manufacturing performance, though it will always take second place to process performance, may eventually get the attention and respect that it warrants. In such a future, we may find that the fabs are not only modularized, but also highly-clustered, and that the process clusters are co-developed along with the IC-technology development. But, in order to achieve compatibility of hardware, materials, software, and processing, such co-development will necessitate a closer marriage between chip maker and equipment maker.

References

[8.1] Bergendahl, A. S., D.V. Horak, P.E. Bakeman, and D.J. Miller, "Analysis and Minimization of the Drawbacks for Process Clustering of a 16Mb DRAM Program", **Semiconductor International**, September, 1990, pp. 94-102.

[8.2] Atherton, L. F., **THOR User's Manual**, In-Motion Technology, Los Altos, CA, 1989-1993.

[8.3] Cook, Rick, "A Squeeze Play in Semiconductors", **Managing Automation**, November, 1989, pp. 48-51.

[8.4] Newboe, Betty, "Cluster Tools: A Process Solution?", **Semiconductor International**, July, 1990, pp. 82-88.

[8.5] Bader, Martin E., Robert P. Hall, and George Strasser, "Integrated Processing Equipment", **Solid State Technology**, May, 1990, pp. 149-154.

[8.6] Atherton, R. W., F. T. Turner, L. F. Atherton, and M. A. Pool, "Performance Analysis of Multi-Process Semiconductor Manufacturing Equipment," **Proceedings of Advanced Semiconductor Manufacturing Conference**, September, 1990.

[8.7] Dynamics of Cluster Tools and Modular Equipment, course notes, In-Motion Technology, Mountain View, California, October, 1990.

[8.8] Pool, M. A. and L.F. Atherton, "Cluster Tools Need Performance Analysis", **Semiconductor International**, September, 1990, pp. 98-99.

Exercises

1. Fig. 8.26 shows the OPS graphs for two of eight possible routings for the four-step recipe, 24/30/30/35 sec. The two graphs are for the extreme cases of no waits and waits after every step. Draw the OPS graphs for the six intermediate wait cases.

2. Discuss why interconnect process sequences are useful applications for cluster tools.

3. The following 3-step recipe is being considered for clustering:

EXERCISE 8.3: BATCH vs SINGLE-WAFER PROCESSING

	Process Step	Process Times	
		Single (sec)	Batch (hrs)
1	PVD TiN	60	1.0
2	CVD	120	2.0
3	PVD Al	90	1.5

List the feasible groupings of steps.

4. For the 3-step recipe above, assume up to four cluster-tool chambers are available. List possible groupings.

5. Match the possible groupings in Exercise 4 to the hardware options in Figure 8.34.

6. Select the probable 'best' clustering/hardware options from Exercise 5.

7. For the options selected in Exercise 6, and based on the cluster-tool and conventional-equipment costs in Table 8.20, determine the capital expenditure for a one-on-one exchange, i.e., one cluster-tool chamber replacing one stand-alone conventional tool. (Don't forget to add in the cost of the platform.)

8. Assume each sequential step added to a cluster tool results in a 8% throughput improvement. How many fewer cluster tools are needed in Exercise 7 for each option?

Projects

1. Develop performance analysis models for the designs discussed in section 8.3

Much Ado About Nothing
-Shakespeare

Chapter 9
Performance Economics

9.0 Introduction

Libraries have been written on the topic of economics, from general treatises on the art and science of the subject, to specific business applications to justify or explain costs. The economics presented in this chapter have no such noble pretensions nor grand schemes behind them. They are offered merely as tools to aid in performance analysis of wafer fabs and, hopefully, help us make good decisions on how to achieve better performance.

When one thinks about the economics of wafer fabrication, one normally thinks in terms of profits. Unfortunately, 'profits' are subjective and overtly dependent on cost-allocation methods. Even the principal object of the profit motive varies from fab to fab. For most fabs, wafers are the principal object of the cost analyses, even though the wafers may contain die with a whole spectrum of performance characteristics, from 'zero' to 'better-than-expected'. Determining the cost of production of wafers and setting their value to the customer are more a matter of skill than good accounting.

For fabs that perform their own sort, the principal object is still the product, but in their case the product is the individual working die. Setting product value is much easier in this case; but determining the cost of production is still a challenge.

Production costs have included in them direct cost, such as direct material and labor, depreciation costs for building and equipment, and indirect costs, such as indirect labor, engineering expenses, maintenance, and miscellaneous services. Of these components, only the direct costs are easily assignable and handled adequately in most wafer fabs.

The reason the other costs are handled less than adequately goes back to the subject that was covered in some detail in Chapter 6: fabs have multiple processes that share fab resources. The consumption of resources, and the associated costs, must therefore be allocated across the various process flows. Fair cost allocation can only be made if it is clearly understood how the flows are using the fab. Conversely, once it **is** clearly understood, assigning consumption costs to the multiple flows is straightforward.

9.1 Cost Allocation by Resource Consumption

Of the various contributors to production costs, shown in Table 9.1, only the first two, materials and direct labor, are considered direct costs, directly assignable to specific products. All others are regarded as indirect costs and are allocated to products as overhead costs. There are a plethora of cost-allocation methods being used by the industry [9.1] to make this assignment.

TABLE 9.1: CONTRIBUTORS TO PRODUCTION COSTS

	BiPolar (1988) $/wafer	CMOS (1991) $/wafer	CMOS (1995) $/wafer
Materials	56	85	150
Direct Labor	38	48	150
Maintenance	28	35	105
Depreciation/Buildng	50	300	900
Misc. Services	22	28	60
Engineering Support	14	50	150
Indirect	8	10	25
Total Wafer Cost	216	556	1,540 *

Base Case = 20,000 wafers/mo
* MOR variety wafers; $7000 sale price.

The most prevalent cost-allocation methods used by the industry include:

- direct-labor ratio
- cost per 'move'
- activity-based accounting
- cost-center accounting

Of the various methods currently in use, the one most similar to apportionment by resource consumption is cost-center accounting. The allocation method suggested in this chapter is a close facsimile, except that we stand by our well-known work centers or workstations instead of cost centers.

For accounting purposes, as we earlier did for performance-analysis purposes, the wafer fab is divided into manufacturing activities, where each activity uses like equipment, e.g., diffusion, photolithography, etch, implant, dielectric, etc. Each workstation has its own equipment (for depreciation), its own personnel, its own maintenance, and its own services. Costs which were formerly considered as overhead are assigned to work centers, and then allocated to process flows based on the flows' ratioed use of the work center resources. The *ratioed use* is determined by the actual, by-process throughput of the workstation.

Invariably, there will still be some costs that can not be assigned to a specific workstation. What becomes of them?

Our recommendation, and it is only a recommendation, is that these costs be distributed among the various process flows according to the overall percentage of fab capacity being dedicated to each flow. This may sound like a complicated determination, but in actuality it is simple, logical, and for the most part accurate and fair.

Since the fab's capacity is set by the bottleneck or constraining workstation, we need only find it, find out how much each flow is using it, and allocate the unaccounted-for costs accordingly.

Table 9.2 outlines the procedure for allocating costs by the resource-consumption method.

TABLE 9.2: COST ALLOCATION BY RESOURCE CONSUMPTION

- Group equipment, personnel, etc. at workstations.
- Assign expenses to workstation, where possible.
- Allocate costs to flows by % of workstation throughput.
- Costs which can not be assigned to workstation should be allocated among flows by bottleneck-usage ratio (assuming all flows have same bottleneck).
- If different bottlenecks, use mixing rules to determine pseudo-bottleneck ratios.

When all assignable and non-assignable costs have been gathered, and a reasonable allocation method decided upon, then distributing the costs across saleable product is easy and defendable. Assuming, of course, that good tracking data exists, and workstation throughputs, by process, are known.

As we have throughout this book, we will use one of our toy fabs, CMOS-2, to illustrate our proposed approach to fab economics. Again, we emphasize that our purpose is to help analyze and improve fab performance. If the economics seem simplistic, it is for a very good reason. The real complexity of wafer fabrication is in analyzing performance of the factory, not the assignment of dollar values to that performance. If we do the former correctly, the latter **should** be simple.

9.2 Contributors to Wafer Cost

There are basically three different types of expenses that are lumped into the production cost of a wafer. There are the direct costs (labor, materials, and maintenance) that are clearly assignable to a work center or a specific process step at a work center. These costs generally account for about 25% of the total manufacturing costs.

Next, there are the capital investments for equipment, buildings, land, installation, and sometimes startup that may or may not be clearly assignable to a given work center. About 60% of wafer cost is due to this monkey.

Finally, there are the indirect costs, such as engineering support, engineering lots, miscellaneous services, indirect labor, etc. These costs are usually not assignable to specific work centers or process flows. These expenses make up about 15% of a wafer's cost.

Based on tracking data from CMOS-2, and using the resource-consumption allocation method, we will assign fair and reasonable wafer costs to the two production flows presently using the fab. First, however, we must set the scene.

CMOS-2 has been on-line about a year and is currently running two production flows, Flow-1 and Flow-2. Flow-1 is the 2-metal flow shown in Table 6.1, where the two metal steps are both performed at the constraining workstation (Wkst 8).

Flow-2 is the 3-metal flow shown in Table 6.2. Two of its three metals steps are performed at the constraining workstation (Wkst 8). Further, both of the Wkst-8 metal steps are longer than the analogous ones in Flow-1. In addition, Flow-2 has an entirely new, extra metal step performed at a new, dedicated workstation. Because of the extra metal step, this second flow has two other added steps not used by Flow-1. Flow-2 is clearly the more complex of the two flows, and consumes more of the fab's resources.

9.2.1 Direct Costs Allocation

Table 9.3 shows the direct costs that have been collected by CMOS-2 over its year of fab operation. These costs are labeled 'Labor & Matl.' and 'Maintenance', and have been directly assigned to the work centers where they were tracked. Not too surprisingly, there is a rough correlation with the number of machines in each work area. Machines require: operators, technicians, and engineers to operate and oversee them; gases and other materials to perform the required processing; and maintenance to operate properly. These are the easy costs to track and assign.

TABLE 9.3: DIRECTLY-ASSIGNABLE COSTS, CMOS-2

Two Production Flow Facility

	Workstation	# of Mach.	Cost/Mach. ($mm)	Total Equip ($mm)	Labor & Matl. ($mm/yr)	Maintenance ($mm/yr)
1	MISC.	2	0.75	1.50	0.48	0.2
2	DIFFUSION 1	5	0.75	3.75	1.20	0.5
3	MASK/ETCH 1 *	7	5.00	35.00	5.04	2.1
4	IMPLANT	3	3.00	9.00	0.72	0.3
5	DIFFUSION 2	3	0.75	2.25	0.72	0.3
6	MASK/ETCH 2 *	6	5.00	30.00	4.32	1.8
7	DIFFUSION 3	4	0.75	3.00	0.96	0.4
8	METAL DEP 1	3	2.50	7.50	0.72	0.3
9	MASK/ETCH 3 *	4	5.00	20.00	2.88	1.2
10	DIEL-FAB	5	2.00	10.00	1.20	0.5
11	METAL DEP 2	2	2.50	5.00	0.48	0.2
	Totals	44		127.00	18.72	7.8

*Actually 3 machines: photo, stepper, etcher.

Over the last quarter of CMOS-2 operation (2016 hours), the plant produced 1641 Flow-1 lots, or 41,025 wafers. Over the same 2016 hours, it also produced 510 Flow-2 lots, or 12,750 wafers. The assignment of the direct costs to these wafers may be done in one of two ways. First, we could make it by a crude straight percentage of overall production.

Flow-1 --> 1641 lots = 76.29% of total lots produced
Flow-2 --> 510 lots = 23.71% of total lots produced

Thus, taking these percentages of the total direct costs ($18.72 + 7.8 = $26.52mm) and dividing by the annual wafer throughput, TP_i, for each flow, i, gives the direct cost per wafer, dc_i:

$$dc_1 = 0.7629 * \$26.52mm / TP_1 = \$20.23mm / TP_1$$
$$dc_2 = 0.2371 * \$26.52mm / TP_2 = \$6.29mm / TP_2$$

The annual throughput for each flow, TP_i, may be taken from tracking data for the entire year or, more reasonably, annualized from some shorter tracking period. For our case, we will use the 2016 hours of tracking data.

$$TP_1 = (1641 \text{ lots}/2016 \text{ hrs}) * 25 \text{ wafers/lot} * 8760 \text{ hrs/yr}$$
$$= 178{,}263 \text{ wafers/yr}$$

$$TP_2 = (510 \text{ lots}/2016 \text{ hrs}) * 25 \text{ wafers/lot} * 8760 \text{ hrs/yr}$$
$$= 55{,}402 \text{ wafers/yr}$$

Therefore,

$$dc_1 = \$20.23mm / 178{,}263 = \$113.48/\text{wafer}$$
$$dc_2 = \$6.29mm / 55{,}402 = \$113.54/\text{wafer}$$

Not much difference in wafer cost by this method. So instead, we will try the second, more sophisticated method for cost assignment: a resource-consumption breakdown by workstation.

Table 9.4 gives the breakdowns **by workstation** for: the total direct cost (i.e., 'Labor & Matl' plus 'Maintenance'), each flow's workstation throughput, wt_{ij}, each flow's ratio of workstation consumption, r_{ij}, and the allocation of direct cost for each flow, dc_{ij}.

The ratio of resource use, r_{ij}, is simply the ratio of each flow's wafers moving through that resource. For example, the throughput of Wkst 9 is 356,418 Flow-1 wafers plus 165,336 Flow-2 wafers, for a total of 521,754 wafers. Thus, the ratio of Flow-1 use is 356,418 / 521,754 = 0.6831. The ratio of Flow-2 use is 165,336 / 521,754 = 0.3169 (or 1.0 - 0.6831).

The allocation of direct cost, dc_{ij} is calculated by multiplying workstation j's total direct cost, DC_j, by its ratio of use for Flow-i, and dividing by Flow-i's annualized throughput, TP_i. (Note that this is the **flow** not workstation throughput.) Thus, for

Flow-1 at Wkst 9, the total direct cost DC_9 is \$4.08mm, the ratioed use is 0.6831, and the annualized Flow-1 throughput is 178,263 wafers. Therefore, the direct cost assignment, dc_{19}, is:

$$dc_{19} = \$4.08mm * 0.6831 / 178,263 = \$15.63/wafer$$

Similarly for Flow-2:

$$dc_{29} = \$4.08mm * 0.3169 / 55,402 = \$23.38/wafer$$

Voila! There is the wafer cost differences we were expecting. And, as seen from all the direct cost allocations in Table 9.4, the heavier use of fab resources by Flow-2 is reflected in the numbers for Workstations #9 through #11.

TABLE 9.4: DIRECT COST ASSIGNMENTS TO WORKSTATIONS, CMOS-2

Wkst	Total Direct Costs ($mm)	Flow-1			Flow-2		
		Thrupt (wafers)	Ratio of Use	dc_1 ($/wafer)	Thrupt (wafers	Ratio of Use	dc_2 ($/wafer)
1	0.68	178,372	0.7648	2.92	54,859	0.2352	2.89
2	1.70	714,466	0.7645	7.29	220,086	0.2355	7.23
3	7.14	896,314	0.7641	30.60	276,683	0.2359	30.40
4	1.02	535,985	0.7647	4.38	164,901	0.2353	4.33
5	1.02	356,744	0.7651	4.38	109,500	0.2349	4.32
6	6.12	713,597	0.7645	26.25	219,760	0.2355	26.01
7	1.36	536,854	0.7640	5.83	165,879	0.2360	5.79
8	1.02	356,418	0.7641	4.37	110,043	0.2359	4.34
9	4.08	356,418	0.6831	15.63	165,336	0.3169	23.38
10	1.70	178,915	0.6187	5.90	110,260	0.3813	11.70
11	0.68	0	0.0000	0.00	55,076	1.0000	12.35
Totals				107.01			132.74

Perhaps the only surprise from this exercise is why Workstation #8 does not also have a higher cost for Flow-2. After all, Flow-2's process times at the work center are longer than Flow-1's. So, what gives?

The reason is that Workstation #8 is the constraining workstation (see Chapters 5 and 6) around which the fab's lot-start rates were initially determined. This workstation is setting the capacity of the entire fab. Every work center upstream in the various process flows is being gated by Workstation #8. Thus, Wkst 8 is not only setting its own usage ratio, but the usage ratio of every other equipment set that is used exclusively upstream of it. This constraining workstation already has the heavier use factored in, as do Workstations #1 through #7.

We can not emphasize too strongly the importance of knowing the fab's bottlenecks or constraining workstations, for both economic **and** performance analyses.

9.2.2 Depreciation Allocation

The second category of expenses generally lumped into production costs is depreciation of investments. Investments in a wafer fab include: equipment, buildings, land, installation, and, in some instances, startup and shakedown. As discussed earlier, some of these costs are directly assignable to work centers; most, however, are not.

Assignable Depreciation Costs

Since the equipment is, in essence, the work centers, the depreciation of this investment expense is directly assignable. Likewise, the portion of the Class-1 building that houses the particular equipment is directly assignable. Thus, to write these expenditures off against specific-process use of the work centers, we need only determine the recovery period and the method of depreciation.

Although some fairly creative and interesting games may be played with capital recovery, for our discussion, we will use a method that follows tax guidelines.

Capital Recovery Period = 5 years
Method of Depreciation = Straight-line

In general, the depreciation on a capital item would amount to the cost of the investment divided by the throughput, divided again by the recovery period. This simplistic formula needs some modification, however, to be useable for allocating depreciation across process flows and to work centers. The capital expenditure (both equipment and building) must be broken down by workstations, and multiplied by the process-usage ratios (Table 9.4) in order to apportion the investment among the flows.

The total equipment costs, given in Table 9.5, is simply the cost of each machine-type multiplied by the number of machines at the workstation. Determining the Class-1 building expenditure takes a bit more work. Inasmuch as the different machine types are different in size, workstation floor-space requirements will reflect both quantity and size of machine-type.

TABLE 9.5: ASSIGNABLE DEPRECIATION COSTS

Wkst	Capital Costs, IWj ($mm)			Flow-1 Dep.		Flow-2 Dep.	
	Equip	Bldg	Total	Ratio	dp_1 ($/w)	Ratio	dp_2 ($/w)
1	1.50	0.37	1.87	0.7648	1.60	0.2352	1.59
2	3.75	2.78	6.53	0.7645	5.60	0.2355	5.55
3	35.00	10.38	45.38	0.7641	38.90	0.2359	38.65
4	9.00	1.67	10.67	0.7647	9.15	0.2353	9.06
5	2.25	1.67	3.92	0.7651	3.36	0.2349	3.32
6	30.00	8.90	38.90	0.7645	33.37	0.2355	33.07
7	3.00	2.22	5.22	0.7640	4.47	0.2360	4.48
8	7.50	1.39	8.89	0.7641	7.62	0.2359	7.57
9	20.00	5.93	25.93	0.6831	19.87	0.3169	24.72
10	10.00	1.85	11.85	0.6187	8.23	0.3813	16.31
11	5.00	0.94	5.94	0.0000	0.00	1.0000	21.44
			165.10		132.17		165.76

As a case in point, Workstation #2 which has five diffusion furnaces requires $2.78mm-worth of Class-1 real estate, whereas Workstation #10, also having five machines (dielectric), needs only $1.85mm-worth of floor space. The difference is the larger size of the furnaces. Experienced layout specialist will know the footprint of most commercial semiconductor equipment down to the inch.

Once we have determined the capital investment by workstation, IW_j, it is possible to find the depreciation per wafer, DP_{ij}:

$$DP_{ij} = IW_j * r_{ij} / TP_i$$

where r_{ij} is the flow's ratioed use of the workstation, and TP_i is the flow's annualized throughput. The annual depreciation per wafer, dp_{ij}, is simply the total depreciation, DP_{ij}, divided by the recovery period, n:

$$dp_{ij} = DP_{ij} / n$$

Thus, Table 9.5 gives the workstation-assignable depreciation numbers, as well as the flows' annual per-wafer totals. Again, the more complex process, Flow-2, has higher per wafer depreciation, $165.76, primarily because of its more extensive use of Workstations #9-11. The assignable costs for Workstations #1-7 are basically the same for either flow, since both flows use these workstations the same.

Non-Assignable Depreciation Costs

As was seen with the direct costs earlier, Workstation #8, the constraining workstation, sets the capacity of the fab and determines what happens upstream of it. Built into the fab's capacity allocation (set by Wkst 8) is Flow-2's longer process times at the constraining workstation. Thus, Workstations #1-7 have been 'burdened' in their throughputs and ratioed usages by what Workstation #8 permits. Any deviation from the set constraint occurs downstream, and shows up as 'exceptions' in the usage ratios.

Because the bottleneck or constraining workstation governs fab performance, and because localized deviations are taken into account explicitly as exceptions, it therefore makes considerable sense to allocate the non-assignable depreciation costs by the same ratioed usages that govern the constraining workstation, and thus the fab. As was seen with Workstations #1-8, this means that, regardless of flow, the per-wafer burden will be about the same.

To illustrate, we see from Table 9.6 that grouped into the non-assignable investment costs are: the building costs not directly assignable to workstations, the land costs, installation of tank farms and other shared facilities, integration of subsystems and the entire system, and, for this particular fab, startup and shakedown costs. For CMOS-2, these costs total $84.9mm.

TABLE 9.6: NON-ASSIGNABLE COSTS, CMOS-2

	Capital ($mm)	Operating ($mm/yr)
Buildings, Land, & Installations	84.9	
Enginering Support		12.02
Misc. Services		5.01
Indirect		1.67
FAB TOTAL	84.9	18.70

We go back to the more simplistic definition to assign the depreciation, ND_i, to the two flows: ratioed investment divided by throughput, divided by recovery period:

$$ND_i = r_{i,\,btlnk} * IW_{tot} / TP_i / n$$

Thus, for Flow-1:

$$ND_1 = 0.7641*(\$84{,}900{,}000) / 178{,}263 \text{ wafs/yr} / 5 \text{ yrs} = \$72.78/\text{wafer}$$

Similarly for Flow-2:

$$ND_2 = 0.2359*(\$84,900,000)/\ 55,402 \text{ wafs/yr} / 5 \text{ yrs}$$
$$= \$72.30/\text{wafer}$$

As said earlier, the non-assignable burden is about the same, regardless of flow. But one thing that was not mentioned earlier is why the non-assignable investment costs for CMOS-2 is so high. The reason is a recent trend being observed among world-class manufacturers: putting up extra building space.

CMOS-2 will have a life cycle of approximately twelve years, and over that life cycle more equipment will be added as processes evolve and are eventually displaced by new, more-complex processes. By the time CMOS-2 is retired, building requirements could easily double. Unfortunately, the longest lead-time item for added capacity is Class-1 clean-room space. Unless planned for and built in advance, this building space is rarely ever added. Instead, equipment is added, hodgepodge, and shoe-horned in wherever it will fit. The spaghetti fab from Chapter 4 becomes more of a pizza fab *with everything on it*; it will be congested, have no clearly-demarcated work centers, and be a tracking nightmare.

The world-class manufacturers (Table 6.15) are providing the anticipated building requirements at the outset, even if it means wafers that *look* more expensive. But the companies that fall into this 'world-class' category count their per-wafer profits in the thousands of dollars, not in tens or even hundreds of dollars.

9.2.3 Indirect Costs Allocation

The final category of expenses which generally ends up included in wafer-production costs is the catchall, indirect costs. These costs include: engineering support and development costs, miscellaneous services (especially clean-room related), and indirect labor costs.

Unlike depreciation which is usually an up-front, one-time investment, the costs in this category are ongoing, too numerous for description, and the very devil to track. To exacerbate the problem of tracking massive amounts of data, most fabs have no logical methodology for allocating the costs among the products once they are collected.

As we did with the non-assignable depreciation costs, we are going to allocate CMOS-2's indirect costs, id_i, based on the constraining workstation's usage ratios:

$$id_i = r_{i,\,btlnk} * ID_{tot} / TP_i$$

where ID_{tot} is the total annualized indirect costs. For CMOS-2, the total indirect cost is $18.7mm annually (Table 9.6). Allocating this cost among the flows gives:

$$\begin{aligned} id_1 &= 0.7641 * \$18{,}700{,}000 / 178{,}263 \text{ wafs/yr} \\ &= \$80.15/\text{wafer} \end{aligned}$$

and

$$\begin{aligned} id_2 &= 0.2359 * \$18{,}700{,}000 / 55{,}402 \text{ wafs/yr} \\ &= \$79.62/\text{wafer} \end{aligned}$$

Alternatively, we can break these costs down by category as shown in Table 9.7. (The same formula holds, except that we substitute the individual costs for the $18.7mm total.)

TABLE 9.7: NON-ASSIGNABLE INDIRECT COSTS

	Flow-1 ($/waf)	Flow-2 ($/waf)
Engg Support	51.52	51.18
Misc. Services	21.47	21.33
Indirect	7.16	7.11
TOTAL	80.15	79.62

New-Process Development

Before leaving the topic of indirect costs, we should comment briefly on process development costs (engineering lots, etc.). In theory, many feel that the cost of developing a process should be borne by that process once it is in production. While the theory has a perceived fairness about it, it is nonetheless impractical and difficult, if not impossible, to implement. New-process development efforts and engineering lots are a fact of life in wafer fabrication, and becoming more so with every passing year.

Any semiconductor company wishing to survive must have a sizeable R&D effort on-going at all times. And the effort must be paid for, not at some far off time in the future, but as it is happening. For, in the future, the company may be out of business, the product may be obsolete, the market may have dried up, etc.

Since R&D must be paid for in real time, it falls to the current profit makers to cover the freight, even if the extra burden hastens the demise of one of those profit makers. But since R&D efforts, and hence the costs of them, are always around, if an older product is showing up less favorably, then perhaps its demise is warranted in order to make room for higher profit-margin products.

R&D costs should in no way discourage new process development. After all, new processes are the future of the company. The sensible, least painful methodology for handling R&D includes:

- recognizing that there will always be an on-going effort,
- make an accurate assessment of the size of the effort,
- try not to let R&D degrade production performance,
- accurately gauge the cost of R&D, and
- logically allocate R&D costs to production flows.

9.2.4 Total Cost per Wafer

We have determined all of the assorted wafer costs, assigning them to work centers where possible, and allocating them to process flows according to workstation (or bottleneck) usage ratios. To find the total cost per wafer, by product, WC_{i}, we merely sum the costs in Table 9.8.

TABLE 9.8: TOTAL WAFER COSTS

	Flow-1 ($/waf)	Flow-2 ($/waf)
Direct Costs	107.01	132.74
Depreciation:		
Assignable	132.17	165.76
Non-Assignable	72.78	72.30
Indirect Costs	80.15	79.62
TOTAL	392.11	450.42

Wafers from Flow-1 cost $392.11 and wafers from Flow-2 cost $450.42. Although wafer cost from real fabs now range between $900 and $5000 per wafer, it is interesting to note that even though CMOS-2 is a toy fab, these are still hefty wafer costs. The numbers are close to the actual costs of wafers from about the 1990-time frame (Table 9.1). Also, the breakdown is roughly as expected:

28% direct (labor, matl, maint)
53% depreciation (equip, bldgs, land)
19% indirect (engg, misc, etc.)

Further, of the total costs, about 60% could be assigned directly to work centers. The other 40% are equitably distributed between the two flows based on fab performance (i.e., how the flows are using the fab as determined by the constraining workstation).

9.3 Wafer Costing and the Evolving Factory

The wafer costs in Table 9.8 are reflective of how the two flows are using the fab during the tracking interval. Over the life cycle of a plant, however, the product slate will be evolving, older product being displaced by newer ones, and eventually being phased out altogether. During the evolution, the engineering load may change dramatically, which again may impact the cost of production wafers. Finally, five years into the plant's life, the bulk of the depreciation will be completely written off, and wafer costs should see a major change.

Thus, the bottom line on wafer costs is that they, like just about everything else involved in wafer fabrication, are dynamic variables. The open question then is how sensitive these costs are to factory dynamics.

9.3.1 Evolving Product Slate

The wafer costs in Tables 9.4-9.8 are based on a product slate in CMOS-2 of approximately 75% Flow-1 and 25% Flow-2. Over the next eighteen months of operation, production of the less-complex Flow-1 wafers will gradually be displaced by the more-advanced Flow-2 wafers. We, therefore, need to know how wafer costs will change with this evolution.

TABLE 9.9: WORKSTATION USAGE RATIOS

SLATE (F1/F2):	75% / 25%		50% / 50%		25% / 75%	
Wkst #	Flow-1	Flow-2	Flow-1	Flow-2	Flow-1	Flow-2
1	0.7648	0.2352	0.5120	0.4872	0.2543	0.7457
2	0.7645	0.2355	0.5125	0.4875	0.2570	0.7430
3	0.7641	0.2359	0.5123	0.4877	0.2543	0.7457
4	0.7647	0.2353	0.5126	0.4874	0.2545	0.7455
5	0.7651	0.2349	0.5125	0.4875	0.2544	0.7456
6	0.7645	0.2355	0.5126	0.4874	0.2545	0.7455
7	0.7640	0.2360	0.5117	0.4883	0.2542	0.7458
8	0.7641	0.2359	0.5123	0.4877	0.2542	0.7458
9	0.6831	0.3169	0.4115	0.5885	0.1853	0.8147
10	0.6187	0.3813	0.3452	0.6548	0.1463	0.8537
11	0.0000	1.0000	0.0000	1.0000	0.0000	1.0000

Table 9.9 provides the workstation usage ratios as the lot-start rates for Flow-1/Flow-2 change from 75%/25% to 50%/50%, and then later to 25%/75%. As was done in Section 9.2 for the 75%/25% case, these usage ratios are used to allocate direct costs and flow-specific depreciation to the respective products. Annualized flow throughputs (wafers/yr) for the new product slates are:

	75%/25%	50%/50%	25%/75%
TP_1:	178,263	115,257	56,162
TP_2:	55,402	110,260	164,793
	233,665	225,517	220,955

Tables 9.10 and 9.11 track the sensitivities of these assignable costs to changes in product slate.

TABLE 9.10: SENSITIVITY OF DIRECT COSTS

To Changing Product Slate

SLATE (F1/F2):	75% / 25%		50% / 50%		25% / 75%	
Wkst #	Flow-1	Flow-2	Flow-1	Flow-2	Flow-1	Flow-2
1	2.92	2.89	3.03	3.00	3.08	3.08
2	7.29	7.23	7.56	7.52	7.78	7.66
3	30.60	30.40	31.74	31.58	32.33	32.31
4	4.38	4.33	4.54	4.51	4.62	4.61
5	4.38	4.32	4.54	4.51	4.62	4.62
6	26.25	26.01	27.22	27.05	27.73	27.67
7	5.83	5.79	6.04	6.02	6.16	6.15
8	4.37	4.34	4.53	4.51	4.62	4.62
9	15.63	23.38	14.57	21.78	13.46	20.17
10	5.90	11.70	5.09	10.10	4.43	8.81
11	0.00	12.35	0.00	6.16	0.00	4.13
TOTALS	107.01	132.74	108.86	126.74	108.83	123.83

All values in $/wafer.

As seen for both direct costs and depreciation, the less-complex Flow-1 is not very sensitive to changes in product slate.

As it is gradually phased out of production, there is a slight increase in per-wafer costs simply because there are fewer being produced over which to distribute any flow-specific expenses.

The resource-intensive Flow-2, however, shows more sensitivity. As production of Flow-2 increases, there is a quantitative drop in costs, again because there are more wafers over which to distribute flow-specific expenses. And, with these wafers, there are considerably more flow-specific expenses than with Flow-1.

TABLE 9.11: SENSITIVITY OF DEPRECIATION
To Changing Product Slate

SLATE (F1/F2):	75% / 25%		50% / 50%		25% / 75%	
Wkst #	Flow-1	Flow-2	Flow-1	Flow-2	Flow-1	Flow-2
1	1.60	1.59	1.66	1.65	1.69	1.69
2	5.60	5.55	5.81	5.77	5.98	5.89
3	38.90	38.65	40.34	40.14	41.10	41.07
4	9.15	9.06	9.49	9.43	9.67	9.65
5	3.36	3.32	3.49	3.47	3.55	3.55
6	33.37	33.07	34.60	34.39	35.26	35.20
7	4.47	4.48	4.64	4.62	4.73	4.72
8	7.62	7.57	7.90	7.86	8.05	8.05
9	19.87	24.72	18.52	27.68	17.11	25.64
10	8.23	16.31	7.10	14.07	6.17	12.28
11	0.00	21.44	0.00	10.77	0.00	7.21
TOTALS	132.17	165.76	133.55	159.85	133.31	154.95

All values in $/wafer.

So, what are the total per-wafer costs as the product slate makes a radical change over the eighteen months of operation? Using the standard formulas for non-assignable depreciation,

$$ND_i = r_{i,\,btlnk} * IW_{tot} / TP_i / n$$

and indirect costs,

$$id_i = r_{i,\,btlnk} * ID_{tot} / TP_i$$

and using the appropriate TP_i for the particular product slate, the rather surprising answers are provided in Table 9.12.

TABLE 9.12: TOTAL WAFER COSTS vs PRODUCT SLATE

Slate (F1/F2):	75%/25%		50%/50%		25%/75%	
	Flow-1 ($/waf)	Flow-2 ($/waf)	Flow-1 ($/waf)	Flow-2 ($/waf)	Flow-1 ($/waf)	Flow-2 ($/waf)
Direct Costs	107.01	132.74	108.86	126.74	108.83	123.83
Depreciation:						
Assignable	132.17	165.76	133.55	159.85	133.31	154.95
Non-Assignable	72.78	72.30	75.47	75.11	76.85	76.85
Indirect Costs	80.15	79.62	83.12	82.71	84.64	84.63
TOTAL	392.11	450.42	401.00	444.41	403.63	440.26

Reducing the production of Flow-1 wafers from 75% of the fab's capacity to 25% causes a per-wafer increase in cost of $11.52, or 2.94%.

Increasing the production of the more-complex Flow-2 wafers from 25% of the fab's capacity to 75% causes a per-wafer decrease in cost of $10.16, or 2.26%.

It is worth noting at this point that, as Flow-2 consumes more of the fab's capacity, fewer wafers overall are produced from both flows (220,955 for 25%/75% vs 233,665 for 75%/25%). These fewer wafers must bear the burden of non-assignable costs. Thus, the gains Flow-2 made, by having more of that particular type of wafer to spread the assignable costs over, are wiped out by the lower overall throughput and resulting higher non-assignable burden.

A basic observation to be drawn ,rom these results is that there is certainly a variation in wafer cost depending on flow complexity and consumption of fab's resources. However, the per-wafer cost is not overtly sensitive to shifts in product slate among the various production flows.

Where shifts in product slate may have a significan impact is in the production of fewer wafers and lots of a particular type.

But if more fab resources are consumed producing fewer, but more advanced products, those products will generally have a higher sale price associated with them.

Therefore, as long as production stays within the fab's dynamic capacity, quantitative variation between the various flows should not require accounting micro-management. If, however, fab capacity is exceeded, or if the new product slate causes a shift in the bottleneck or constraining workstation, then all bets are off. Performance analyses must be re-done, and the accounting numbers must be revised to reflect the 'new' factory.

9.3.2 Engineering Lots

In contrast to product slate, engineering or development lots are almost always an accounting headache. Not only are the quantity and frequency of these lots variable, but the method of covering their costs is subjective and varies from fab to fab.

We saw in Chapter 6 how these lots, if not planned for, can severely degrade performance. But even when planned, there is a cost associated with producing them; and the cost can be quite high depending on flow complexity and frequency.

As we did in Chapter 6, we will investigate the costing impacts of running various levels of an engineering flow in CMOS-2. When no engineering lots are present in the line, the production processes, Flow-1 and Flow-2, are each consuming about 50% of the fab's resources. Over the next year of operation, CMOS-2 will, at various times, be running engineering lots at a frequency of: 1 lot/week, 1 lot/day, or 1 lot/shift.

The chip under development, i.e., the engineering flow, is an advanced 4-metal microprocessor. The process recipe and workstation assignments are given in Table 6.23. As is the case with the two production flow, the major variation with this flow is in the back end of the process, steps #21-30, and its demands on Workstations #9-11.

Even with the heaviest engineering demand (1 lot/shift), the fab's constraining workstation remains unchanged (Wkst 8).

Table 9.13 presents the workstation usages ratios at the various engineering loads. Because the fab's constraining workstation does not change, the usage ratios of Workstations #1-7 are basically the same as that of Wkst 8.

TABLE 9.13: IMPACT OF ENGINEERING LOTS ON WORKSTATION USAGE RATIOS

	---------- Usage Ratios Resulting from Engineering Rates of ----------			
	none	1 lot/week	1 lot/day	1 lot/shift
Wkst	Flow-1/Flow-2/Engg	Flow-1/Flow-2/Engg	Flow-1/Flow-2/Engg	Flow-1/Flow-2/Engg
1-8	0.5123/0.4877/0.0000	0.5093/0.4849/0.0058	0.4919/0.4685/0.0396	0.4510/0.4296/0.1194
9	0.4115/0.5885/0.0000	0.4078/0.5829/0.0093	0.3871/0.5506/0.0623	0.3404/0.4801/0.1795
10	0.3452/0.6548/0.0000	0.3417/0.6468/0.0115	0.3217/0.6022/0.0762	0.2801/0.5089/0.2110
11	0.0000/1.0000/0.0000	0.0000/0.9767/0.0233	0.0000/0.8537/0.1463	0.0000/0.6343/0.3657

The rest of CMOS-2's performance data under the various engineering loads are presented in Table 9.14. The annualized throughput of each flow, including engineering, is given, as well as the fab total for each scenario. As may be seen, there is a continual drop in overall throughput with increased rates of the more-complex development flow.

TABLE 9.14: IMPACT OF ENGINEERING LOTS ON ANNUALIZED PROCESS THROUGHPUT

	Thrupts (wafers/yr) Resulting from Engg Rates of:			
Flow:	none	1 lot/week	1 lot/day	1 lot/shift
Flow-1 TP	115,257	114,388	109,174	96,573
Flow-2 TP	110,260	109,717	103,851	92,228
Engg TP	0	1,304	8,908	26,506
TOTAL	225,517	225,409	221,933	215,307

The data in Tables 9.13 and 9.14 may be used with the 'Total Direct Cost' information, DC_j, in Table 9.4 and the 'Capital Cost' information, IW_j, in Table 9.5 to determine the costs which may be directly assignable to each wafer as a result of the flow's ratioed usage of the workstations. The direct cost totals by flow (dc_1, dc_2, and dc_{engg}) and the depreciation totals by flow (dp_1, dp_2, and dp_{engg}) are shown in Table 9.15. As may be seen, there is a slight but steady per-wafer increase in the direct costs for the production flows with increasing engineering lots. But overall, the depreciation for the production wafers seems to be unaffected.

TABLE 9.15: IMPACT OF ENGINEERING LOTS ON ASSIGNABLE COSTS (CAPITAL & DEPRECIATION)

	Direct Costs ($/waf)			Depreciation ($/waf)		
Engg Rate:	dc-1	dc-2	dc-engg	dp-1	dp-2	dp-engg
none	108.86	126.74	0.00	133.55	159.85	0.00
1 lot/week	108.97	126.21	145.48	133.69	160.22	189.40
1 lot/day	109.85	127.47	144.21	133.32	160.38	185.42
1 lot/shift	112.98	128.97	142.61	133.88	159.78	178.98

Proviso: The engineering rates would have only slight and/or negligible impact on production costs <u>if</u> there were some other way to pay for their expensive use of fab resources. But, in the final accounting, the saleable production wafers must foot their own bill, <u>and</u> the cost of engineering wafers as well. Thus, each engineering wafer's $140+ of direct cost and $178+ of depreciation must be distributed, and lumped in with the cost of the production wafers.

As mentioned earlier, the distribution of development costs is very subjective and fab-dependent. Some fabs may load the costs on the oldest production flow, some on the newest, others on the flow that is most similar, and yet others may distribute it equally across all wafers.

In the interest of fairness, CMOS-2 uses this last approach, and 'taxes' the wafers from all flows equally.

To determine the per-wafer engineering tax, we find the total cost for all engineering wafers, and distribute it over all production wafers. Thus, for the tax on direct cost at a particular engineering rate,

$$dc_{tax} = (dc_{engg} * TP_{engg}) / (TP_1 + TP_2)$$

For example, at the engineering rate of 1 lot/shift,

$$dc_{engg} = \$142.61$$
$$TP_{engg} = 26{,}506 \text{ wafers}$$
$$TP_1 = 96{,}573 \text{ wafers}$$
$$TP_2 = 92{,}228 \text{ wafers}$$

Thus,

$$dc_{tax} = (142.61 * 26{,}506) / (96{,}573 + 92{,}228)$$
$$= \$20.02$$

The engineering tax on the assignable depreciation costs are derived in much the same way:

$$dp_{tax} = (dp_{engg} * TP_{engg}) / (TP_1 + TP_2)$$

For the same engineering rate of 1 lot/shift,

$$dp_{engg} = \$178.98$$
$$TP_{engg} = 26{,}506 \text{ wafers}$$
$$TP_1 = 96{,}573 \text{ wafers}$$
$$TP_2 = 92{,}228 \text{ wafers}$$

Thus,

$$dp_{tax} = (178.98 * 26{,}506) / (96{,}573 + 92{,}228)$$
$$= \$25.13$$

Table 9.16 gives the engineering-burdened wafer costs for the two production flows. Mirroring results seen previously in Chapter 6, at the rate of 1 lot a week, the production costs are barely affected (< 1% increase). At 1 lot a day, the engineering effort is starting to be felt (5.6% increase); and at 1 lot per shift, the effort is putting a substantial burden (17-20%) on the cost of making saleable product.

TABLE 9.16: WAFER COSTS WITH ENGINEERING 'TAX'

	--------- Wafer Costs ($/waf) Resulting from Engg Rates of ---------							
	none		1 lot/week		1 lot/day		1 lot/shift	
	Flow-1	Flow-2	Flow-1	Flow-2	Flow-1	Flow-2	Flow-1	Flow-2
Direct Cost	108.86	126.74	108.97	126.21	109.85	127.47	112.98	128.97
Engg Tax	0.00	0.00	0.84	0.84	6.03	6.03	20.02	20.02
Depreciation:								
Assignable	133.55	159.85	133.69	160.22	133.32	160.38	133.88	159.78
Engg Tax	0.00	0.00	1.10	1.10	7.75	7.75	25.13	25.13
Non-Assign	75.47	75.11	76.05	75.48	79.66	79.76	90.06	89.81
Indirect Cost	83.12	82.71	83.75	83.14	87.75	87.84	99.20	98.91
TOTAL	401.00	444.41	404.40	446.99	424.36	469.23	481.27	522.62

The non-assignable costs are derived using the standard formulas for non-assignable depreciation,

$$ND_i = r_{i,\,btlnk} * IW_{tot} / TP_i / n$$

and indirect costs,

$$id_i = r_{i,\,btlnk} * ID_{tot} / TP_i$$

The only modification comes from the fact that these costs also must be borne solely by the production flows. Therefore, the ratioed usage of the bottleneck or constraining workstation should be normalized on the production flows. For instance, from Table 9.13, at the 1 lot/day engineering rate, the usage ratios for Workstation #8, the bottleneck, are:

Flow-1: 0.4919
Flow-2: 0.4685
Engg: 0.0396

But since Flow-1 and Flow-2 will be picking up the entire tab on non-assignable costs, we normalize their usage ratios:

$$r_{1,\,btlnk} = 0.4919 \;/\; (0.4919 + 0.4685) = 0.5122$$
$$r_{2,\,btlnk} = 0.4685 \;/\; (0.4919 + 0.4685) = 0.4878$$

Thus, for Flow-1:

$$\begin{aligned} ND_1 &= r_{1,\,btlnk} * IW_{tot} \;/\; TP_1 \;/\; n \\ &= 0.5122 * 84{,}900{,}000 \;/\; 109{,}174 \;/\; 5 \\ &= \$79.66 \end{aligned}$$

Ditto, for the indirect costs,

$$\begin{aligned} id_1 &= r_{1,\,btlnk} * ID_{tot} \;/\; TP_1 \\ &= 0.5122 * 18{,}700{,}000 \;/\; 109{,}174 \\ &= \$87.75 \end{aligned}$$

Not too surprising, since fewer production wafers are sharing the non-assignable burden, the per-wafer costs increase accordingly.

Perhaps from an accounting perspective, these results make engineering or development lots look like an extravagance the fab should forego. But nothing could be further from the truth. As was shown extensively in Chapter 6, the cost of producing a wafer, even with a 20-40% engineering burden, pales in comparison to the revenues generated from the sale of a state-of-the-art, high-yielding 6 or 8" wafer. Sale prices now range from $5,000-15,000 for wafers with a fully-burdened fabrication cost of $900-3000. And it is the engineering or development efforts in most fabs that keep this gap widening.

9.3.3 Adding a Process Flow

Well, CMOS-2 has just reached the mature age of three. Flow-1 and Flow-2 are now old technologies, and the new kid on the block is Flow-3, formerly the 4-metal engineering flow from above. An entirely new production flow has been added to the fab.

In our haste to bring Flow-3 on-line, no new equipment was added to the original equipment set. Thus, the total assignable direct costs and depreciation cost remain unchanged. The workstation usage ratios and annualized flow throughputs, however, have changed radically. These are given in Table 9.17.

TABLE 9.17: ADDITION OF A NEW PRODUCTION FLOW

Wkst	Total DCj ($mm)	IWj ($mm)	Workstation Usage Ratios		
			Flow-1	Flow-2	Flow-3
1	0.68	1.87	0.1554	0.1554	0.6892
2	1.70	6.53	0.1575	0.1575	0.6851
3	7.14	45.38	0.1597	0.1594	0.6809
4	1.02	10.67	0.1573	0.1573	0.6853
5	1.02	3.92	0.1563	0.1563	0.6874
6	6.12	38.90	0.1575	0.1575	0.6849
7	1.36	5.22	0.1605	0.1604	0.6790
8	1.02	8.89	0.1600	0.1600	0.6800
9	4.08	25.93	0.0918	0.1376	0.7706
10	1.70	11.85	0.0639	0.1281	0.8079
11	0.68	5.94	0.0000	0.1067	0.8933
Annualized TPi (wafers):			29,439	29,330	120,906

Based on these performance values, we see that direct costs and depreciation (Table 9.18) are quite high, not only on wafers from the new flow, but from the old flows as well. Production of the new, more-complex wafers has brought overall fab throughput down to 179,675 wafers a year, a significant degradation in fab performance (see Table 9.14).

TABLE 9.18: FAB COST WITH NEW PRODUCTION FLOW

Wkst	Directly-Assignable Costs ($/wafer)					
	Flow-1		Flow-2		Flow-3	
	dc-1	dp-1	dc-2	dp-2	dc-3	dp-3
1	3.59	1.97	3.60	1.98	3.88	2.13
2	9.10	6.99	9.13	7.01	9.63	7.40
3	38.73	49.24	38.80	49.33	40.21	51.11
4	5.45	11.40	5.47	11.44	5.78	12.10
5	5.42	4.16	5.44	4.18	5.80	4.46
6	32.74	41.26	32.86	41.78	34.67	44.07
7	7.41	5.69	7.44	5.71	7.64	5.86
8	5.54	9.66	5.56	9.70	5.74	10.00
9	12.72	16.17	19.14	24.33	26.00	33.05
10	3.69	5.14	7.42	10.35	11.36	15.84
11	0.00	0.00	2.47	4.32	5.02	8.78
Total	122.39	152.04	137.33	170.13	155.73	194.80

Furthermore, this substantial drop in fab throughput means that there are significantly fewer wafers to absorb the non-assignable costs. Finally, we know from the performance analyses performed in Chapter 6, that the addition of this new production flow has caused a true bottleneck shift. Wkst 8 has been displaced by Wkst 11 as the constraining workstation. Which leaves us exactly where?

At this point, we have three options available, none of which, at first glance, look too appealing:

- allocate non-assignable costs based on non-exception workstation usage (not bottleneck),
- re-analyze fab and revise accounting packages based on new and/or split bottlenecks, or
- add necessary equipment to restore adequate factory performance.

The first option is the simplest and, on the surface, appears the least costly. The non-exception workstations are Wksts 1-7, i.e., those used identically across all flows. Based on these workstations, averaged usage ratios may be determined:

$$r_{1,avg} = 0.1577$$
$$r_{2,avg} = 0.1577$$
$$r_{3,avg} = 0.6846$$

With these usage ratios, the non-assignable costs may be allocated across the flows. The total wafer costs using this method are presented in Table 9.19.

TABLE 9.19: TOTAL WAFER COSTS FOR 3 FLOWS

	Flow-1 ($/waf)	Flow-2 ($/waf)	Flow-3 ($/waf)
Direct Costs	122.39	137.33	155.73
Depreciation:			
Assignable	152.04	170.13	194.80
Non-Assignable	90.96	91.30	96.14
Indirect Costs	100.17	100.55	105.88
TOTAL	465.56	499.31	552.55

However, we know that since Flow-1 is not using the new bottleneck, it is probably carrying more than its fair share of the non-assignable costs. By the same token, Flow-2 and especially Flow-3 are probably carrying less than their shares. Thus, we may want to consider the rather labor-intensive second option.

The second option, re-analyzing the fab and revising accounting packages, will most certainly have to be done if no new equipment is forthcoming. Unfortunately, analyzing fab performance for a split bottleneck is tricky, and will require a first-rate analyst to be done correctly. (A *split bottleneck* exists when some flows do not use the primary bottleneck or constraining workstation.)

Before excessive time and effort are spent pursuing cost allocations for a bottlenecked or poorly performing fab, however, the option of buying more equipment should be evaluated first. After all, we have added a new, more-complex flow. It is unrealistic to expect to get something for nothing, or in this case, to get good or adequate performance with no capital investment in the equipment set.

From the performance analyses in Chapter 6, we know that one additional machine at Workstation #10 and one at Workstation #11 will debottleneck the fab and restore good performance. Table 9.20 presents the new workstation usages ratios and flow throughputs that are achieved when this equipment is added. The table also reflects the added direct costs and depreciation that result from the acquisitions.

TABLE 9.20: ADDED PRODUCTION AND EQUIPMENT

Wkst	Total DCj ($mm)	IWj ($mm)	Workstation Usage Ratios		
			Flow-1	Flow-2	Flow-3
1	0.68	1.87	0.1250	0.1250	0.7500
2	1.70	6.53	0.1272	0.1272	0.7456
3	7.14	45.38	0.1292	0.1293	0.7415
4	1.02	10.67	0.1268	0.1268	0.7464
5	1.02	3.92	0.1262	0.1262	0.7475
6	6.12	38.90	0.1271	0.1271	0.7458
7	1.36	5.22	0.1300	0.1301	0.7399
8	1.02	8.89	0.1301	0.1301	0.7397
9	4.08	25.93	0.0732	0.1095	0.8173
10	**2.04**	**14.22**	0.0507	0.1007	0.8486
11	**1.02**	**8.91**	0.0000	0.0823	0.9177
Annualized TPi (wafers):			28,353	28,461	156,211

As a result of the much-improved performance, the directly assignable direct costs and depreciation decline dramatically (Table 9.21) for all flows.

In fact, the total wafer costs for Flow-1 and Flow-2 are lower than when the plant was making those products alone (Table 9.22 vs 50%/50% case in Table 9.12). Perhaps even more remarkable, the Flow-3 wafers have a lower production cost than the heavily-engineering-burdened, less-complex Flow-1 and

Flow-2 wafers (Table 9.16). However, these Flow-3 wafers are worth a whole lot more money. And we are making 30% more of them (156,211 vs 120,906) with the addition of the two machine.

TABLE 9.21: FAB COST WITH NEW EQUIPMENT

	Directly-Assignable Costs ($/wafer)					
	Flow-1		Flow-2		Flow-3	
Wkst	dc-1	dp-1	dc-2	dp-2	dc-3	dp-3
1	3.00	1.65	2.99	1.64	3.26	1.80
2	7.63	5.86	7.60	5.84	8.11	6.23
3	32.54	41.36	32.44	41.23	33.89	43.08
4	4.56	9.54	4.56	9.51	4.87	10.20
5	4.54	3.49	4.54	3.48	4.88	3.75
6	27.43	34.88	27.33	34.74	29.22	37.14
7	6.24	4.79	6.22	4.77	6.44	4.94
8	4.68	8.16	4.68	8.13	4.83	8.42
9	10.53	13.39	15.70	19.95	21.35	27.13
10	3.63	5.09	7.22	10.06	11.08	15.45
11	0.00	0.00	2.95	5.15	5.99	10.47
Total	104.78	128.21	116.23	144.50	133.92	168.61

TABLE 9.22: WAFER COSTS WITH NEW EQUIPMENT

	Flow-1 ($/waf)	Flow-2 ($/waf)	Flow-3 ($/waf)
Direct Costs	104.78	116.23	133.92
Depreciation:			
Assignable	128.21	144.50	168.61
Non-Assignable	77.42	77.62	80.40
Indirect Costs	85.81	85.48	88.55
TOTAL	396.22	423.83	471.48

Thus, by debottlenecking the fab, we markedly lower the per-wafer fabrication costs for all production flows, we improve overall factory performance, and we have lots more high-priced goodies to sell.

Further, we now know that, of the three options listed above, by far the wisest **and** least costly way to deal with a bottlenecked factory is to buy more equipment.

9.3.4 Rework

Rework in photolithography is treated in some detail in Chapter 3, particularly in the problem set. We will mention it here merely as a means to illustrate how work-center costing directly evaluates a fab's specific operating policies.

As a result of rework, additional process steps are added to a process flow. These extra steps involve photoresist stripping and repeated photolithography operations. The costs of these steps must be addressed.

By tracking each flow's use of a given work center, all process steps, even the extra, sometimes unexpected ones, are explicitly accounted for. If the extra steps require more equipment, which in turn requires more operators, more building space, etc., then the cost of performing the steps is reflected in the flow's usage ratio of the work center(s) involved. The usage ratio will insure that the cost of the extra steps is not overlooked, and is indeed correctly assigned to the appropriate wafers.

Similarly, the costs of all such operating policies are determined definitively as a function of the policy's consumption of fab resources.

9.4 Cost per Good Die

Thus far, our simplistic economic analyses have been based on wafer costs. But once the wafers are made, one of two things will happen to them based on who did the fabrication. The wafer will either be sent to a Sort operation for testing, slicing, and preparation for assembly; or they will be shipped intact to the buyer. The former is the standard procedure for semiconductor companies with their own fabs; the second describes a foundry operation that primarily services fabless semiconductor companies.

For companies with their own fabs, control of the wafer, and the wafer parts, remains with the manufacturer until the final product, the packaged chip, is ready for the customer. For foundries, the customer is the fabless semiconductor company, and the product is the entire wafer, not the individual chips.

For either type of semiconductor company, the value of the wafer lies in the number of good chips (or die) that it contains. In the final accounting, the functioning chips are the ultimate saleable product and thus the entity which must bear the expense of production.

The number of good die is known explicitly by companies that fab their own wafers. Fabless houses rely on an expected-yield value that comes along with their wafers from the foundry.

9.4.1 Die Costing by Fab Houses

The die cost calculation for the semiconductor companies owning their own fab is straightforward since the number of good die per wafer, gd_i, is explicitly known. Using the total wafer costs, WC_i, we obtain the cost per good die, c_i by:

$$c_i = WC_i / gd_i$$

Testing determines the number of good die, gd_i; and we have spent the bulk of this chapter learning how to determine the total wafer costs, WC_i. Summarizing briefly:

- We assign direct cost, DC_j, and capital costs, IW_j, to work centers j whenever possible.

- Through good tracking practices, we know how each flow i is using the work center j. Thus, we calculate a flow usage ratio, r_{ij}, for each work center.

- We assign the sums of the direct cost, dc_{ij}, and the assignable depreciation, dp_{ij}, to the flow-specific wafers.

- Through performance analysis and/or operating data, we know the fab's bottleneck or constraining workstation. We determine the flow usage ratio, $r_{i,\,btlnk}$, of the bottleneck.

- We use the bottleneck's usage ratio to allocate normally non-assignable costs (some depreciation, ND_i, and all other indirect costs, id_i) to the flows.

- We sum all assignable and non-assignable costs to determine WC_i.

9.4.2 Die Costing by Foundries

Foundries typically sell unsorted wafers, not individual die. The value of the wafers is based on the expected number of good die. But since the wafers are unsorted, this number is a guesstimate that assumes some standard yield. If the actual yield of delivered wafers falls below the standard, then price is either adjusted downward or additional wafers are provided by the foundry.

9.5 Summary

A fab producing state-of-the-art microprocessors has more money-making potential than a key to the mint and a license to print cash. On a per ounce basis, these advanced chips are often more valuable than gold.

The keys to fab profitability are: the correct choice of product(s) and good operating policies. The former is generally the result of astute market awareness, rapid response, and outstanding development efforts. Good operating policies are the result of a thorough understanding of factory dynamics and comprehensive performance analyses.

Despite this promising picture, there is one big cloud hanging over the wafer fabrication industry: the multi-billion dollar price tag of a new plant. But given the immense opportunities for profits, investment funds will be found, and plants will be built. The high *ante* to get into the game, however, should preclude over-expansion of capacity. It is extremely unlikely that there will be a repeat of the rollercoaster ride taken by the industry in the 1980's. (Even that ride was a result of an artificial situation, a subsidized capacity gluts.)

Of course, there is one other factor making capacity over-expansion unlikely: the market is not only growing, but appears to be virtually unbounded. When one considers that the per-person consumption of microprocessors has gone from zero in the mid 1970's to somewhere in the hundreds by the mid 1990's, the slope of this astonishing growth curve may be exponential. At a few dollars a chip, little wonder the public expects, and is willing to pay for, that extra bit of 'smarts' in everything from toys to automobiles. With demand expected to double by the turn of the century, the industry is much more likely to have a bad case of the *shorts*.

Which brings us to the subject of much of this book: getting the best performance we can from the capacity we have. As this and earlier chapters have shown, good operating policies and fab dynamics are dominated by factory capacity, **not** costing considerations. If fabs are designed, operated, and analyzed correctly, then costing concerns are indeed...

....much ado about nothing.

Further Reading and Notes

1. A comprehensive treatment of cost accounting practices in actual wafer fabs is provided by Niles Hatch (Chapter 11) in [9.1].

References

[9.1] Leachman, R. C., (ed.), **The Competitive Semiconductor Manufacturing Survey: Second Report on Results of the Main Phase**, Report CSM-08, Engineering Systems Research Center, UC Berkeley, CA, September, 1994.

[9.2] Busey, Lynn E., **The Economic Analysis of Industrial Projects**, Prentice-Hall, Englewood Cliffs, NJ, 1978.

[9.3] Park, C. S., and G. P. Sharp-Bette, **Advanced Engineering Economics**, Wiley, New York, 1990.

Exercises

1. Using the equipment cost data provided in Tables 3.3 and 5.5, determine the total equipment cost for the PiZMOS-1 factory in Exercise Table 6.0. Assume the assignable building costs are:

EXERCISE 9.1: ASSIGNABLE COSTS, PIZMOS-1

Workstation		Bldg Cost/ Mach ($mm)	Direct Costs/ Mach ($mm)
1	SPUTTER	0.50	0.3
2	PATTERN	0.75	0.5
3	DEVELOP	0.50	0.3
4	ETCH	0.50	0.3
5	PVD	0.75	0.3
6	Strip (opt)	0.25	0.2

determine the total assignable depreciation cost by workstation.

2. Determine the total assignable costs in $/wafer for FLOW-1 in PiZMOS-1. Assume that the factory is not bottlenecked; FLOW-1 is the only process in the fab; and its lot-start rate is 0.5 lots/hr.

3. Referring to problem 6.5, add one machine at Wkst 4 and recalculate workstation-assignable costs. (For building costs and direct costs pro-rate according to increase in machine total.)

4. If PiZMOS-1 is now running lot-start rates of 0.25/0.25/0.05 lots/hr of FLOW-1/FLOW-2/FLOW-3, respectively, and FLOW-3 is an engineering flow, determine the FLOW-1 and FLOW-2 assignable costs in $/wafer. (Again, assume the fab is not bottlenecked and can produce as much as it starts.)

5. If PiZMOS-1 is now running three full production flows with lot-start rates of 0.10/0.10/0.30 lots/hr for FLOW-1/FLOW-2/FLOW-3, respectively, recalculate the assignable costs in $/wafer for all three flows.

6. If the typical production costs in PiZMOS-1 hold true to form and break downs as:

25% direct (labor, matl, maint)
60% depreciation (equip, bldgs, land)
15% indirect (engg, misc, etc.)

determine the total wafer costs for the scenarios in Exercises 2, 4, and 5.

Index